T0211538

Workbook for Organic Synthesis:
Strategy and Control

Workbook for Organic Synthesis: Strategy and Control

Paul Wyatt

Reader and Director of Undergraduate Studies,
School of Chemistry, University of Bristol, UK

and

Stuart Warren

Chemical Laboratory, Cambridge University

John Wiley & Sons, Ltd

Copyright © 2008   John Wiley & Sons Ltd, The Atrium, Southern Gate, Chichester,
West Sussex PO19 8SQ, England

Telephone (+44) 1243 779777

Email (for orders and customer service enquiries): cs-books@wiley.co.uk
Visit our Home Page on www.wileyeurope.com or www.wiley.com

All Rights Reserved. No part of this publication may be reproduced, stored in a retrieval system or transmitted in
any form or by any means, electronic, mechanical, photocopying, recording, scanning or otherwise, except under the
terms of the Copyright, Designs and Patents Act 1988 or under the terms of a licence issued by the Copyright
Licensing Agency Ltd, 90 Tottenham Court Road, London W1T 4LP, UK, without the permission in writing of the
Publisher. Requests to the Publisher should be addressed to the Permissions Department, John Wiley & Sons Ltd,
The Atrium, Southern Gate, Chichester, West Sussex PO19 8SQ, England, or emailed to permreq@wiley.co.uk, or
faxed to (+44) 1243 770620.

Designations used by companies to distinguish their products are often claimed as trademarks. All brand names and
product names used in this book are trade names, service marks, trademarks or registered trademarks of their
respective owners. The Publisher is not associated with any product or vendor mentioned in this book.

This publication is designed to provide accurate and authoritative information in regard to the subject matter
covered. It is sold on the understanding that the Publisher is not engaged in rendering professional services. If
professional advice or other expert assistance is required, the services of a competent professional should be sought.

The Publisher and the Author make no representations or warranties with respect to the accuracy or
completeness of the contents of this work and specifically disclaim all warranties, including without limitation any
implied warranties of fitness for a particular purpose. The advice and strategies contained herein may not be suitable
for every situation. In view of ongoing research, equipment modifications, changes in governmental regulations,
and the constant flow of information relating to the use of experimental reagents, equipment, and devices, the reader
is urged to review and evaluate the information provided in the package insert or instructions for each chemical,
piece of equipment, reagent, or device for, among other things, any changes in the instructions or indication
of usage and for added warnings and precautions. The fact that an organization or Website is referred to in this
work as a citation and/or a potential source of further information does not mean that the author or the publisher
endorses the information the organization or Website may provide or recommendations it may make. Further,
readers should be aware that Internet Websites listed in this work may have changed or disappeared between
when this work was written and when it is read. No warranty may be created or extended by any promotional
statements for this work. Neither the Publisher nor the Author shall be liable for any damages arising herefrom.

Other Wiley Editorial Offices

John Wiley & Sons Inc., 111 River Street, Hoboken, NJ 07030, USA

Jossey-Bass, 989 Market Street, San Francisco, CA 94103-1741, USA

Wiley-VCH Verlag GmbH, Boschstr. 12, D-69469 Weinheim, Germany

John Wiley & Sons Australia Ltd, 42 McDougall Street, Milton, Queensland 4064, Australia

John Wiley & Sons (Asia) Pte Ltd, 2 Clementi Loop #02-01, Jin Xing Distripark, Singapore 129809

John Wiley & Sons Canada Ltd, 6045 Freemont Blvd, Mississauga, Ontario, L5R 4J3, Canada

Wiley also publishes its books in a variety of electronic formats. Some content that appears
in print may not be available in electronic books.

Library of Congress Cataloging-in-Publication Data:

Wyatt, Paul
  Workbook for organic synthesis : strategy and control / Paul Wyatt and
Stuart Warren
    p. cm.
  Includes bibliographical references and index.
  ISBN 978-0-470-75883-0 (cloth) – ISBN 978-0-471-92964-2 (pbk. : alk. paper)
  1. Organic compounds–Synthesis. I. Wyatt, Paul. II. Title.
  QD262.W287 2007
  547′.2 – dc22
                                                                2008009594

British Library Cataloguing in Publication Data

A catalogue record for this book is available from the British Library

ISBN 978-0-470-75883-0

Typeset in 10.5/13pt Times by Laserwords Private Limited, Chennai, India.

Contents

Preface

Our text (Paul Wyatt and Stuart Warren, *Organic Synthesis: Strategy and Control*, John Wiley & Sons, Ltd, Chichester, 2007) is a work of instruction and there is a real danger that someone might think 'all this is really quite simple, I understand it completely' after reading a chapter. But, much like learning to swim, organic synthesis cannot be understood by reading a text-book. We need to test our understanding by trying problems and the main function of this workbook is to provide a graded series of problems, some derived directly from the textbook and some from different examples. Though there will be material in every chapter of this workbook derived from the corresponding chapter in the textbook, there is also much more as the planning for organic syntheses does not come conveniently packaged into labelled pieces, called chapters in a textbook, and we need to be able to draw on a wide range of ideas when planning the synthesis of a new compound.

As with the textbook, we are indebted to many people for the material and the presentation. Some of the problems come from our own lecture courses or those of our colleagues in the form of lecture problems and many were devised by us for our courses in industry and have been tried and tested in that more severe environment. We acknowledge with gratitude the staff and students of Bristol and Cambridge chemistry departments and the many people we met at Organon (Oss and Newhouse), AstraZeneca, (Alderley Park, Avlon works, Mölndal and Macclesfield), DSM (Geleen), Lilly (Windelsham), Novartis (Basel), and Solvay (Weesp) and on the SCI (Young Chemists Panel) courses. For many of the chemists on those courses it was the problem sessions they enjoyed the most and we hope you too will enjoy this workbook.

References

References to the original literature appear at the end of each chapter.

General References

Full details of important books referred to by abbreviated titles in the chapters to avoid repetition.

Clayden, *Lithium*: J. Clayden, *Organolithiums: Selectivity for Synthesis*, Pergamon, 2002.

Clayden, *Organic Chemistry*: J. Clayden, N. Greeves, S. Warren and P. Wothers, *Organic Chemistry*, Oxford University Press, Oxford, 2000.

Collins, *Chirality in Industry*: eds. A. N. Collins, G. N. Sheldrake and J. Crosby, *Chirality in Industry: The Commercial Manufacture and Applications of Optically Active Compounds*, Vol I, 1992, Vol II, 1997, Wiley, Chichester.

Comp. Org. Synth.: eds. Ian Fleming and B. M. Trost, *Comprehensive Organic Synthesis*, Pergamon, Oxford, 1991, six volumes.

Corey, *Logic*: E. J. Corey and X.-M. Cheng, *The Logic of Chemical Synthesis*, Wiley, New York, 1989.

Designing Syntheses: S. Warren, *Designing Organic Syntheses*, Wiley, Chichester, 1978

Disconnection Textbook: S. Warren, *Organic Synthesis, The Disconnection Approach,* Wiley, Chichester, 1982.

Disconnection Workbook: S. Warren, *Workbook for Organic Synthesis, The Disconnection Approach*, Wiley, Chichester, 1982.

Drauz and Waldmann: K. Drauz and H. Waldmann, *Enzyme Catalysis in Organic Synthesis*, VCH, Weinheim, Two Volumes, 1995, ISBN 3-527-28479-6.

Eliel: E. L. Eliel and S. H. Wilen, *Stereochemistry of Organic Compounds*, Wiley, New York, 1994.

Fieser, *Reagents:* L. Fieser and M. Fieser, *Reagents for Organic Synthesis*, Wiley, New York, 20 volumes, 1967–2000, later volumes by T.-L. Ho.

Fleming, *Orbitals*: Ian Fleming, *Frontier Orbitals and Organic Chemical Reactions,* Wiley, London, 1976.

Fleming, *Syntheses*: Ian Fleming, *Selected Organic Syntheses*, Wiley, London, 1973.

Houben-Weyl; *Methoden der Organischen Chemie*, ed. E. Müller, and *Methods of Organic Chemistry*, ed. H.-G. Padeken, Thieme, Stuttgart, many volumes 1909–2004.

House: H. O. House, *Modern Synthetic Reactions*, Benjamin, Menlo Park, Second Edition, 1972.

Joule and Mills: J. A. Joule and K. Mills, *Heterocyclic Chemistry*, 4th Edition, Blackwell, Oxford, 2000.

Morrison: *Asymmetric Synthesis*, ed. J. D. Morrison, Academic Press, Orlando, 5 Volumes, 1983-5.

Nicolaou and Sorensen: K. C. Nicolaou and E. Sorensen, *Classics in Total Synthesis: Targets, Strategies, Methods.* VCH, Weinheim, 1996. Second volume now published.

Ojima, *Asymmetric*: I. Ojima (ed.) *Catalytic Asymmetric Synthesis*, Second Edition, Wiley, New York, 2000.

Saunders, *Top Drugs*: J. Saunders, *Top Drugs: Top Synthetic Routes*, Oxford University Press, Oxford, 2000.

Vogel: B. S. Furniss, A. J. Hannaford, P. W. G. Smith, and A. R. Tatchell, *Vogel's Textbook of Practical Organic Chemistry*, Fifth Edition, Longman, Harlow, 1989.

We do not put in many references to the main text for which this is the workbook. However, this workbook will make little sense unless you have a copy of the main text to hand.

A
Introduction: Selectivity

1

Planning Organic Syntheses: Tactics, Strategy and Control

Introduction: The Purpose of this Workbook

You have made a good start in your investigation of more advanced organic synthesis by, we suppose, reading the textbook *Organic Synthesis: Strategy and Control*. A lot can be learnt from reading but, to gain a real understanding of a subject, more involvement is needed. The workbook gives you more examples of the chemistry discussed in the main text to expand your experience. More importantly, it also offers the opportunity to put your understanding to the test by providing sets of problems with worked answers. Each chapter contains further details of, and recent developments in, the chemistry discussed in the chapter. There is little to add to this short chapter, however.

Other Literature on Organic Synthesis

The list of general references gives you many valuable resources. We would like to draw your attention here to other ways of looking at organic synthesis. Many organic chemists think that the synthesis of natural products is the highest goal for chemistry. A masterly and entertaining account[1] of the synthesis of the 'CP molecules' – naturally occurring cholesterol-lowering fungal metabolites – uses the analogy of Theseus hunting the minotaur target molecule through the labyrinth (organic synthesis).

Sharpless[2] has put forward the challenging and interesting idea that organic synthesis, particularly the discovery of new drugs, should focus not on natural products but on molecules that are easy to make. He uses an estimate by Guida that there are about 10^{62} potential 'reasonable' drug molecules and there are not enough atoms in the universe to make even one molecule of each. Random searching is doomed. Sharpless proposed a new type of chemistry–'click' chemistry–that uses only kinetically controlled and very favourable reactions of alkenes so that the amount of material from each step increases rather than the typical arithmetic decrease in so many natural product syntheses. This idea has many adherents.

Another challenging article[3] on the success or otherwise of organic synthesis questions the philosophy behind much of the synthetic work of the 1990s–the title 'Dead Ends and Detours en Route to Total Syntheses' reveals the line taken by the authors.

Workbook for Organic Synthesis: Strategy and Control   Paul Wyatt and Stuart Warren
© 2008 John Wiley & Sons, Ltd

The Synthesis of Fostriecin

This interesting molecule has continued to attract attention and you can read about other syntheses.[4]

Examples of Problems

In our minds we have categorised problems as 'simple', 'tricky' or 'taxing'. This is partly to provide problems of progressive difficulty and partly to set your mind at rest when you find a problem too difficult. Here are some examples to give you the idea:

Simple: Identify which atom in the final intermediate **19** becomes which atom in flexibilene TM **13**. Suggest mechanisms for the reactions giving **15** and **16**.

Tricky: Which reactions control the stereochemistry of each double bond in flexibilene TM **13**?

Taxing: Suggest how compound **26** might be combined with compound **25** in the synthesis of fostriecin. What problem(s) of selectivity do you foresee?

Examples of Answers

These will usually give full details and literature references or at least enough, as here, to put you on the right track.

Simple: The best technique is to number (arbitrarily) the atoms is either the starting material or the product. This technique is also helpful in solving complicated mechanistic problems. Here the linear starting material is most easily numbered. Inspection of flexibilene makes atom 3 easy to identify. The counting round to the nearer alkene (or the nearer methyl group) reveals which way round you should number flexibilene. The new alkene is evidently between C-1 and C-16.

Tricky: Two alkenes (C-11/C-12 and C-8/C-9) are already present in the starting material **18**. One (C-4/C-5) first appears in **19** but is really made by the hydrozirconation reaction giving **17**. As you will see in chapter 16, such reactions occur with retention of configuration. The final alkene is formed stereoselectively in the McMurry cyclisation of **19**. Notice that the 15-membered ring can accommodate *four* E-alkenes – and prefers to do so. This is a bit surprising as one E-alkene becomes possible only in an eight-membered ring.

Taxing: You might have made several different suggestions here and it is important for you to realise at this early stage that *there is no 'correct' answer to a synthesis question*. After all, there are many syntheses of fostriecin and only this one couples these fragments. The most obvious thing is to make the lithium derivative of the dithian **26** and combine it with the epoxide or the ketone in **25**. The selectivity problem is which reaction is preferred. Chavez and Jacobsen[5] did make the lithium derivative **W1** but then combined it with the bigger fragment **W2** without the ketone to give **W3**. In doing this they showed that the ketone in **25** is actually more electrophilic than the epoxide. It is not always possible to be certain which of two functional groups is the more reactive but this can be determined experimentally and the strategy altered accordingly.

Compound Numbers in the Workbook

Since much of the workbook refers directly to the main text *Organic Synthesis: Strategy and Control*, all plain compound numbers in the workbook refer to the same compound in the same chapter of the main book. Compounds in the workbook but not in the main text are given numbers with a **W** prefix. So compound **34** is the same as in the main text but compound **W34** is a workbook compound. Workbook numbers start afresh with **W1** in each chapter.

References

1. K. C. Nicolaou and P. S. Baran, *Angew. Chem., Int. Ed.*, 2002, **41**, 2678.
2. H. C. Kolb, M. G. Finn and K. B. Sharpless, *Angew. Chem., Int. Ed.*, 2001, **40**, 2005.

3. M. A. Sierra and M. C. de la Toore, *Angew. Chem., Int. Ed.*, 2000, **39**, 1539.
4. J. Cossy, F. Pradaux and S. BouzBouz, *Org. Lett.*, 2001, **3**, 2233; Y. K. Reddy and J. R. Falck, *Org. Lett.*, 2002, **4**, 969; Y. G. Wang and Y. Kobayashi, *Org. Lett.*, 2002, **4**, 4615; T. Esumi, N. Okamoto and S. Hatakeyama, *Chem. Commun.*, 2002, 3042; K. Maki, R. Motoki, K. Fujii, M. Kanai, T. Kobayashi, S. Tamura and M. Shibasaki, *J. Am. Chem. Soc.*, 2005, **127**, 17111.
5. D. E. Chavez and E. N. Jacobsen, *Angew. Chem., Int. Ed.*, 2001, **40**, 3667.

2

Chemoselectivity

Introduction

Examples of selectivity of all three kinds are given in *The Disconnection Approach*: Chemoselectivity in chapter 5, Regioselectivity in chapter 14, and Stereoselectivity in chapters 12 and 38.

Problems and Further Examples Relating Directly to the Text

You will find it helpful to have chapter 2 from the textbook open as you look at this first section.
Problem 2.1. Suggest ways to convert the ketoacid **2** selectively into the hydroxyacids **1** and **3**.

Answer 2.1: The obvious ways are reduction with $NaBH_4$ to give **3** and protection of the ketone, say as an acetal, before reduction with $LiAlH_4$ and deprotection to give **1**. There are many other ways.

Problem 2.2: The textbook reveals that combining enamine **21** with acrolein gives initially adduct **24**. On distillation, a 75% yield of the bicyclic amino-ketone **23** can be isolated. Only if **24** is isolated without distillation and hydrolysed, can the simple conjugate addition product **22** be isolated. Give mechanisms for the formation of **24** and **23** and explain why distillation favours the formation of **23** but hydrolysis favours the formation of **22**.

Answer 2.2: Adduct **24** has one molecule each of enamine and acrolein with an extra molecule of pyrrolidine.[1] Presumably the first formed intermediate **W2** from conjugate addition **W1** gives the aldehyde **W3** while the iminium salt is intact and, after addition of pyrrolidine, cyclises **W4** to **24**. Distillation drives out water and pyrrolidine and favours the alternative cyclisation **W6**. On hydrolysis **24** reverts to **W4** and the iminium salt is hydrolysed to **22**.

Problem 2.3: Corey's longifolene synthesis starts from the bicyclic enone **57**. How would you make **57**?

Answer 2.3: This cyclohexenone **57** is clearly a Robinson annelation product but you could have got there by simple aldol **57a** and 1,5-diCO **W6** disconnections if you did not notice the Robinson. The diketone **W7** is a good starting material as it will form a stable enol(ate) and will be good at conjugate additions.[2] Treatment with acid or base gives **57**.

Problem 2.4: Still on Corey's longifolene synthesis,[2] why does the Wittig reaction on **58** give a mixture of *E* and *Z*-**59**? Why does this not matter? Draw a mechanism for the rearrangement of **60** and check that you understand why it gives this product **61**.

Answer 2.4: The ylid is unstablised and would give mostly *Z*-alkenes with aldehydes. The difference between the two sides of the ketone **58** is so slight that there is little selectivity. Though the diol **W8** and tosylate **60** will be mixtures of *syn* and *anti* diastereoisomers, it does not matter because the stereochemistry disappears in the rearrangement to **61**.

The rearrangement could take place in one step **60a** with preferential migration of the vinyl group rather than the saturated side chain. However, participation by the vinyl group **W9** shows more clearly why unsaturated groups migrate well. A reasonably stable intermediate **W10** is better than a transition state.

Reactions Involving Enolisation; Discussion and Further Explanation

Traditional solutions to these problems are found in *The Disconnection Approach*, particularly chapter 13, page 106; chapter 19, page 144; chapter 25, page 209; and chapter 28, page 229.

Reactions Involving Lithium Enolates

Reactions involving lithium enolates initially form the lithium alkoxide, e.g. **78**. The reaction mixture is then worked up by addition of acid and water, separation of the amine *i*-Pr$_2$NH, used to make LDA in the first step, and the lithium (as its salt with the counterion of the added acid) in the aqueous layer, and isolation of the product **79** from the organic layer. Work-ups are not usually specifically described in papers and books, so beware!

Aldol Reactions with Silyl Enol Ethers

Here again the aldol product, e.g. **107** is usually shown without any work-up. In fact the silyl group is transferred during the aldol reaction and the product under the reaction conditions is the silyl ether **W11**. Work-ups are not usually described in papers and books.

The Lewis acid catalyst, here $TiCl_4$, first links the two oxygen atoms together and the aldol reaction **W12** ensues giving initially the titanium alkoxide **W13** of the product and a molecule of Me_3SiCl. These combine to form the first stable product **W14**. Simplified structures are used to make this easier to see. Work-ups are not usually described in papers and books.

Synthesis of Mannicone

We said 'It is particularly impressive that the optically active aldehyde **119** has its stereogenic centre at the enol position and yet optically active mannicone is formed by this route without racemisation.' The stereogenic centre (another name for chiral centre, see chapter 20 in the text) in **36** is marked with a circle. The enol of **119** is planar and achiral and evidently not formed at all during the synthesis. The base (Et_3N) and the Lewis acid ($TiCl_4$) are mild enough.[3]

Mannicone: Synthesis

Examples of Chemoselective Reactions in Synthesis

Problems and examples from this section of chapter 2.

Chemoselective Reduction in the Synthesis of Lipstatin

The first step in the lipstatin synthesis was the chemoselective reduction of one methyl ester of dimethyl malate **127** to give the diol **128**. Sodium borohydride does not normally reduce esters so the clue is the presence of borane as its dimethyl sulfide complex.[4]

127; (S)-(–)-malic acid dimethyl ester → 128; 83% yield

The borane forms a chelate with the carbonyl group of the ester and the OH group. The five-membered ring **W16** is preferred to the six-membered ring **W15** we should get with the other ester. In the complex **W16** the carbonyl group is more electrophilic and is easily reduced by $NaBH_4$.

Rubrynolide Synthesis

Problem 2.5: In the textbook we said about the first step (**136** to **138**) that 'The lithium enolate was too basic and the aluminium enolate was used instead.' What might too basic an enolate do in this reaction that is unacceptable? Suggest a synthesis of the epoxide **137**.

Answer 2.5: If the problem is basicity, the lithium enolate **W17** must remove a proton from somewhere. An E2 elimination on the epoxide **W18** is a possibility or it might even remove a proton from the epoxide itself to give the anion **W20** that can do many reactions. You might reasonably also have suggested that the lithium enolate **W17** might react with the ester **136** in a Claisen condensation, though lithium enolates of esters are usually. well behaved. In the paper[5] they suggest that the lithium enolate might deprotonate the acetylene in **137**.

There are many possible syntheses of the epoxide **137**, the only chemoselectivity problem is whether epoxidation of the alkene might affect the alkyne at the other end of the molecule. They actually did it this way: dec-9-en-1-ol **W21** was tosylated and treated with lithium acetylide (as its ethylene diamine complex). The resulting enyne **W22** was epoxidised with *m*CPBA without any chemoselectivity problems (i.e. no reaction at the alkyne).

Hirsutine Synthesis

Problem 2.6: Complete the mechanism for the conversion of **150** into **151**.

Answer 2.6: This must be an E1cB elimination: the alkoxide in **150** is basic enough to form the very stable enolate **W22** of the Meldrum's acid part of the molecule.[6]

Problem 2.7: In the synthesis, **161** was not isolated but reacted immediately in a 'tandem' process (chapter 36) with the enol ether **W23** to give first **W24** and then the product **W25** in one step as a single diastereoisomer. What is happening? (PMB = *para*-methoxybenzyl.)

Answer 2.7: The s-*cis* enone system of **161** acts as a heterodiene in a Diels-Alder reaction[6] **W26** to give **W27**. This is an unstable ketene acetal and hydrolyses to something like **W28** which promptly loses acetone and CO_2 to give **W24**.

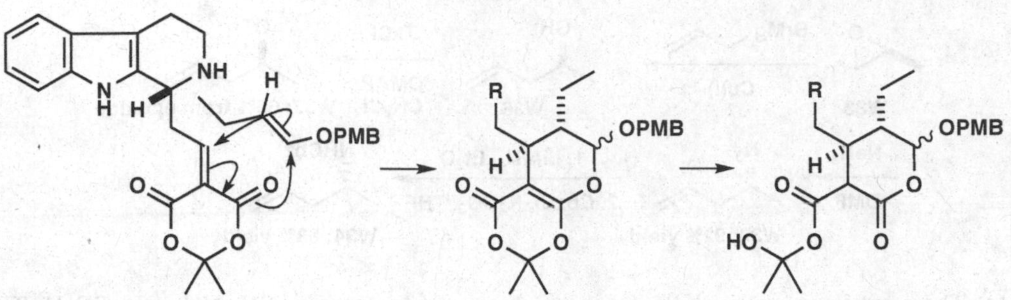

W26 W27 W28

Basic methanol opens the lactone **W24** to reveal a hemiacetal **W29**, the PMB group is removed by hydrogenation which also catalyses the intramolecular reductive amination of the aldehyde **W30** onto the ring amine to give **W25**.

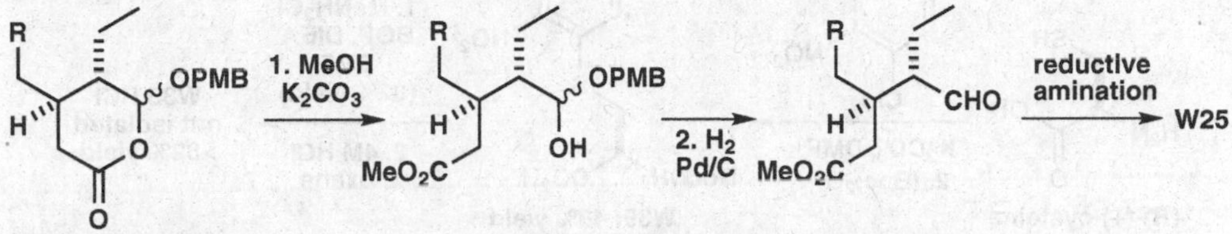

W24 W29 W30

Further Problems and Examples on Chemoselectivity

Problem 2.8: Propose reagent(s) for these transformations, identifying any problems of chemoselectivity and explaining how they are overcome by your reagent(s).

(a)

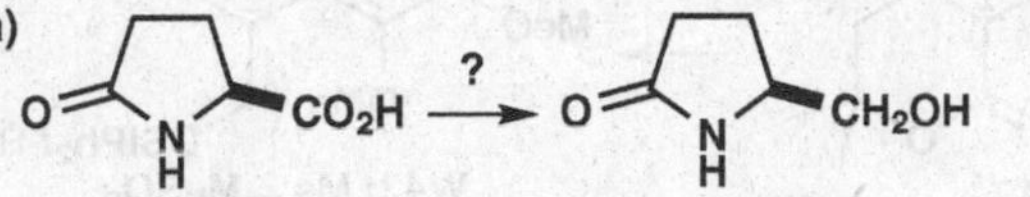

W31 W32

(b)

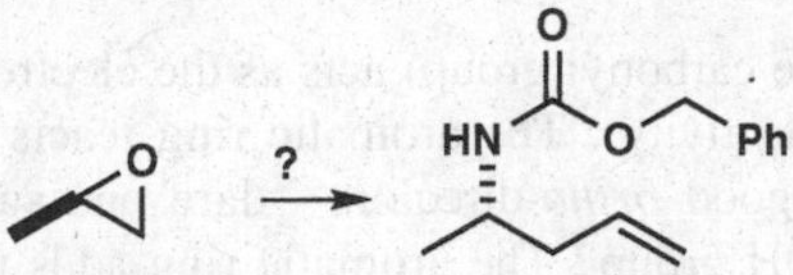

W33 W34

(c)

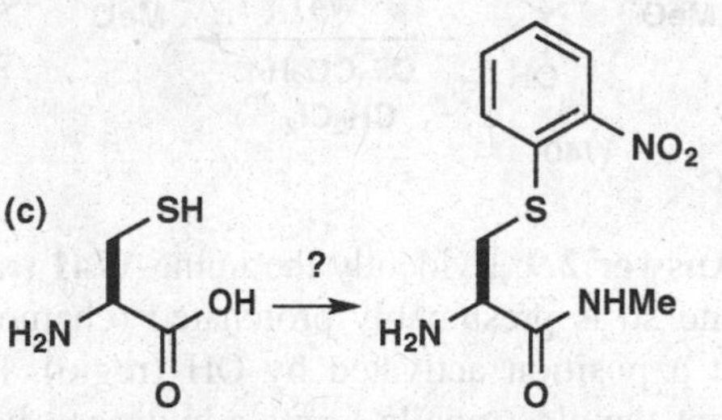

(*R*)-(+)-cysteine W35

Answer 2.8: (a) Typically amides and carboxylic esters are reduced by the same reagents (borane, DIBAL) because they are similarly unreactive towards nucleophilic reducing agents. It is difficult to change the amide but the carboxylic acid can easily be made more electrophilic by conversion to an ester or acid chloride. The published solution[7] is: 1. $SOCl_2$, MeOH (85% yield of acid chloride) and 2. $NaBH_4$, i-PrOH (86% yield of target molecule).

(b) Two groups must be introduced: amino and allyl at two electrophilic centres. The chiral centre must be inverted. There are many solutions: the published one[8] is to add allyl Grignard and then use azide to invert at the other centre.

(c) There are three nucleophilic groups in the starting material – SH, NH$_2$ and CO$_2$H. The ortho-nitrophenyl group must be introduced by nucleophilic aromatic substitution and SH is best at that. Now we must protect the NH$_2$ group while the amide is made otherwise polymerisation will occur. The published solution[9] is to use a Boc group.

Problem 2.9: Propose a mechanism for the first step, commenting on any selectivity, and suggest how the second step could be carried out with full control.

Answer 2.9: Evidently the imine **W41** (rather then the carbonyl group) acts as the electrophile and so is presumably protonated (chemo- or regioselectivity). The aromatic ring reacts **W44** at a position activated by OH (regio). Phenols are good *ortho*-directors – dare one suggest that the electrophile forms a hydrogen bond to the OH group? The aromatic ring adds to the opposite side of the electrophile to the phenyl group.[10]

There is lots to do in the second part: the lactone opened and reduced, and one OH out of three must be silylated while the phenol must be mesylated. This is not too difficult if the mesylation is done first as, after reduction, one OH is primary and one is tertiary **W47**. Only the primary can react with such a large electrophile. Here are the published details, no intermediates were isolated.

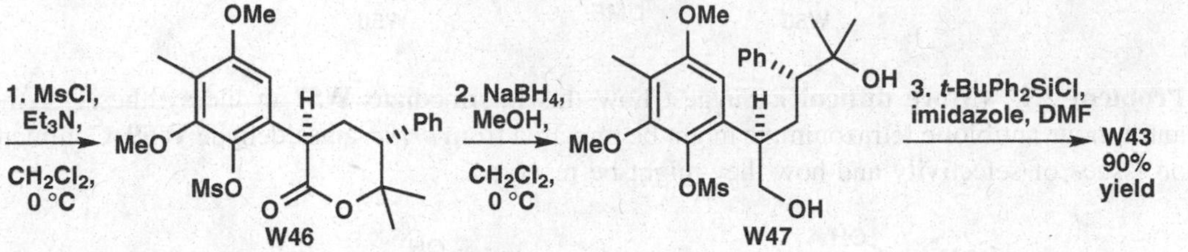

Problem 2.10: Propose a mechanism for the following reactions, identifying and explaining any control.

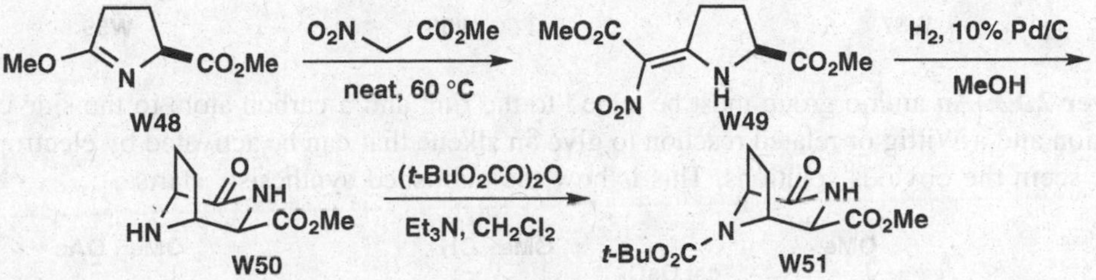

Answer 2.10: *Step 1*: The nitroester is largely in the enol form (one nitro is worth two carbonyl groups) and can select between the methyl ester and imino ester (chemoselectivity) of **W48**. Both are conjugated to the same extent but the C=N is weaker than the C=O. *Step 2*: Hydrogenation reduces the alkene and the nitro group but not the ester groups (chemoselectivity). This is thermodynamic control: C=C and N=O are weaker than C=O. There are now two amino groups and two esters **W52**. Cyclisation could give three-, four- or six-membered rings but the six-membered is preferred thermodynamically and probably kinetically. *Step 3*: The amine is more nucleophilic than the conjugated amide.[11]

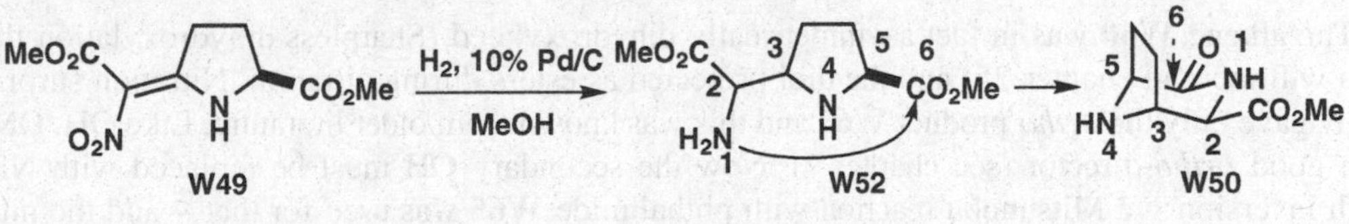

Problem 2.11: Suggest how arabinose could be selectively protected thus:

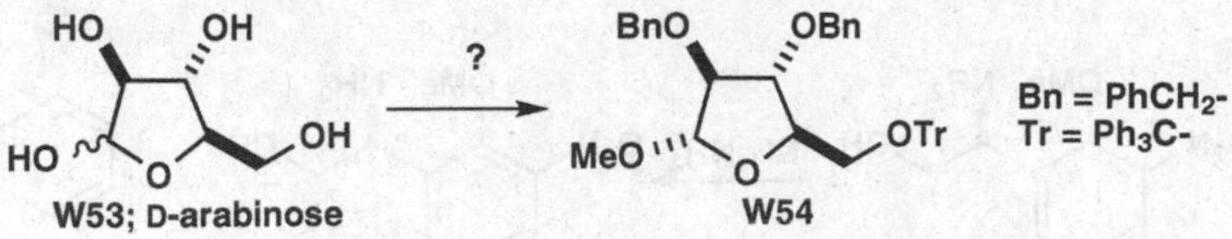

Answer 2.11: The hemiacetal (anomeric) position must be dealt with first as this forms the acetal with methanol under thermodynamic control. The preferred conformation **W56** has two groups equatorial and the OMe axial – this is preferred because of the anomeric effect. Then the primary alcohol can be alkylated with trityl chloride (chemoselectivity from steric hindrance) by an S_N1 mechanism **W56** and finally the two remaining secondary alcohols benzylated.[12]

W53 $\xrightarrow[\text{HCl}]{\text{MeOH}}$ **W55** $\xrightarrow[\substack{\text{pyridine}\\\text{DMAP}\\\text{DMF}}]{\text{TrCl}}$ **W56** $\xrightarrow[\text{2. BnBr}]{\text{1. NaH}}$ **W54**

Problem 2.12 (More difficult): Suggest how this intermediate **W57** in the synthesis of the antitumour antibiotic tetrazonimine might be prepared from *ortho*-anisaldehyde **W59**. Comment on issues of selectivity and how they might be resolved.

W57 $\Longrightarrow$ **W58** $\overset{?}{\Longrightarrow}$ **W59**

Answer 2.12: An amino group must be added to the ring and a carbon atom to the side chain. Nitration and a Wittig or related reaction to give an alkene that can be activated by electrophilic attack seem the obvious solutions. This is how the published synthesis[13] starts:

W59 $\xrightarrow{Ph_3P=CH_2}$ **W60; 82% yield** $\xrightarrow[\substack{\text{chiral}\\\text{ligand}}]{\substack{\text{cat OsO}_4\\K_3Fe(CN)_6}}$ **W61; 93% yield** $\xrightarrow[\text{pyridine}]{Ac_2O}$ **W62; 94% yield**

$\xrightarrow[\substack{(CF_3CO)_2O\\CHCl_3}]{KNO_3}$ **W63; 83% yield** $\xrightarrow[\substack{H_2O\\MeOH}]{K_2CO_3}$ **W64; 93% yield**

The alkene **W60** was in fact asymmetrically dihydroxylated (Sharpless dihydroxylation that you will meet in chapter 25) and the diol protected as esters during nitration. Nitration surprisingly gave only the *ortho* product **W63** and this was known from older literature. Like OH, OMe is a good *ortho*-director (see chapter 7). Now the secondary OH must be replaced with NH_2 with inversion – a Mitsunobu reaction with phthalimide **W65** was used for that – and the nitro group reduced. Forming the side chain NH_2 group before the ring NH_2 group allows selective reactions before reduction if required. The nitro group was reduced with H_2 gas and Pd/C.

W64 $\xrightarrow[\substack{\text{W65}\\R_2NH\\\\Ph_3P\\DEAD}]{}$ **W66; 89% yield** $\xrightarrow{NH_2NH_2}$ **W67; 100% yield**   **W65; phthalimide (R$_2$NH)**

References

1. A. Z. Britten and J. O'Sullivan, *Tetrahedron*, 1973, **29**, 1331.

2. E. J. Corey, M. Ohno, R. B. Mitra and P. A. Venkatacherry, *J. Am. Chem. Soc.,* 1964, **88**, 478.
3. T. Mukaiyama, *Chem. Lett.*, 1976, 279.
4. A. Pommier, J.-M. Pons and P. J. Kocienski, *J. Org. Chem.*, 1995, **60**, 7334.
5. S. K. Taylor, J. A. Hopkins, K. A. Spangenberg, D. W. McMillen and J. B. Grutzner, *J. Org. Chem.*, 1991, **56**, 5951.
6. L. F. Tietze and Y. Zhou, *Angew. Chem., Int. Ed.*, 1999, **38**, 2045; L. F. Tietze, Y. Zhou and E. Töpken, *Eur. J. Org. Chem.*, 2000, 2247.
7. D. de J. Oliveira and F. Coelho, *Tetrahedron Lett.*, 2001, **42**, 6793.
8. S. Randl and S. Blechert, *J. Org. Chem.*, 2003, **68**, 8879.
9. C.-B. Xue, M. E. Voss, D. J. Nelson, J. J.-W. Duan, R. J. Cherney, I. C. Jacobson, X. He, J. Roderick, L. Chen, R. L. Corbett, L. Wang, D. T. Meyer, K. Kennedy, W. F. DeGrado, K. D. Hardman, C. A. Teleha, B. D. Jaffee, R. -Q. Liu, R. A. Copeland, M. B. Covington, D. D. Christ, J. M. Trzaskos, R. C. Newton, R. L. Magolda, R. R. Wexler and C. P. Decicco, *J. Med. Chem.*, 2001, **44**, 2636.
10. K. Mori, K. Rikimaru, T. Kan and T. Fukuyama, *Org. Lett.*, 2004, **6** 3095.
11. S. Pichlmair, K. Mereiter and U. Jordis, *Tetrahedron Lett.*, 2004, **45**, 1481.
12. A. N. Cuzzupe, R. Di Florio and M. A. Rizzacasa, *J. Org. Chem.*, 2002, **67**, 4392.
13. P. Wipf and C. R. Hopkins, *J. Org. Chem.*, 2001, **66**, 3133.

3

Regioselectivity: Controlled Aldol Reactions

Problems and Further Examples Relating Directly to the Text

You will find it helpful to have chapter 3 from the textbook open as you look at this first section.

A More Detailed Study of the Baeyer-Villiger Rearrangement

The mechanism for the Baeyer-Villiger rearrangement of ketones[1] involves the pre-equilibrium formation of a hemiacetal-like adduct **W2** between the ketone and the peroxyacid followed by rate-determining migration **W3** of one of the C–CO bonds from carbon to oxygen. In the migration step, electrons from the C–C σ-bonding orbital (the HOMO) interact in a bonding fashion with the empty σ*-orbital of the weak O–O bond to form a new C–O σ-bond.

Some people prefer to think of the rate-determining step as a concerted migration and proton transfer **W6**. In either interpretation, the transition state, e.g. **W7**, has a partial negative charge on the departing carboxyl portion and a partial positive charge on the carbon skeleton. The more highly substituted the migrating group, the better the carbon skeleton can stabilise the positive charge and the faster the migration. Hence more substituted carbon atoms and those with nearby sources of electrons such as lone pairs on heteroatoms or π-bonds, migrate better.

Workbook for Organic Synthesis: Strategy and Control   Paul Wyatt and Stuart Warren
© 2008 John Wiley & Sons, Ltd

Examples of Regioselective Baeyer-Villiger Reactions

An important consequence of this understanding is that the migration occurs with retention as it is the populated σ-orbital that reacts and not the unpopulated σ*-orbital as in the S_N2 reaction. Here are four examples[1] showing both the regioselectivity (which group migrates) and that the migrating group retains its configuration. Compounds **W8** and **W12** show[2] that the order of migration is *t*-alkyl > *s*-alkyl > *n*-alkyl. Compound **W10** shows[3] that aryl migrates better than alkyl and **W14** shows[4] that electron-rich aryl migrates better than phenyl. Compound **W12** shows the retention and also that even such a reactive group as an epoxide survives the rearrangement.

Chemoselectivity in Baeyer-Villiger Reactions

Epoxidation competes with the Baeyer-Villiger reaction. The reagents are the same peroxyacids and many compounds have both ketone and alkene functionality. It might seem at first sight that there should be good selectivity between the two reactions as the Baeyer-Villiger uses the peroxyacid as a nucleophile while epoxidation uses it as an electrophile. Think again. A simplified version of the mechanism shows that the essence of epoxidation is the electrophilic attack of the O–O bond of the peracid on the C=C bond of the alkene **W16**.

Just as in the Baeyer-Villiger reaction, a better mechanism **W19** shows concerted proton transfer and we have drawn the transition state **W20** for comparison with **7**. The two transition states are remarkably similar having partial positive charges on the organic part and partial negative charges on the departing carboxylate ion. It is rarely possible to direct which reaction is to occur by choice of reagent. Pertrifluoroacetic acid is the best reagent both for epoxidation and the Baeyer-Villiger reaction since the breaking of the O−O bond is involved in the rate-determining step in both reactions.

The delicate balance between the two is shown by the reactions of unsaturated ketones **W21** and **W23**. In both cases the functional groups are well apart and are not conjugated. Enone **W21** has a trisubstituted alkene and an unstrained ketone. Epoxidation wins. Bicyclic enone **W23** has a slightly strained but only disubstituted alkene and a very strained ketone. Baeyer-Villiger wins. The formation of the epoxy-lactone **25** with an excess of peracid shows that there is nothing wrong with the epoxidation, it is just slower.

These are the questions you need to ask when you are considering the reaction of an enone with a peracid:

The alkene: How many substituents does it have? The more substituted the more reactive. Is it strained? The more strained the more reactive.

The ketone: Is it strained? The more strained the more reactive.

Both: Are they conjugated with each other? If so, the carbonyl will dominate and the Baeyer-Villiger reaction usually predominates. Epoxidation needs H_2O_2 and NaOH.

Chemoselectivity Examples: Epoxidation by Peroxy Acids vs. Baeyer-Villiger

Sometimes choice of reagent is effective. Treatment of the ene-dione **W27** with *t*-BuO_2H in base gives just the Baeyer-Villiger product **W28** from the strained cyclobutanone with the usual regio- and stereoselectivity. An excess of buffered *m*-CPBA gives all possible reactions: two Baeyer-Villigers and an epoxidation. The stereochemistry of the product **W26** shows retention at both migrating groups but the stereochemistry of the epoxide is surprising.[5]

W26   **W27**   **W28**

Problem 3.1: Identify and explain the selectivities shown in these examples.

W29   **W30; 75% yield**   **W31**   **W32; 100% yield**

Answer 3.1: The strained ketone **W29** undergoes the Baeyer-Villiger oxidation easily but the less substituted group migrates presumably because of the electron-withdrawing chlorine atom. No epoxidation occurs on the disubstituted and only slightly strained alkene.[6] Enone **W31** has a tetrasubstituted alkene and is easily epoxidised while the slightly strained ketone has a tertiary migrating group so both reactions occur and the *t*-alkyl group migrates.[7]

Viagra Synthesis

In our discussion of the synthesis of viagra in chapter 3 of the text, we gave some details of the Claisen ester condensation between pentan-2-one **15** and diethyl oxalate $(CO_2Et)_2$. We said 'condensation of **15** with diethyl oxalate in base without any control gives **14** because the true product of the reaction is the stable enolate of **14**. The enolate **16** at the methyl group is preferred and the stable enolate of **14** is also preferred to the alternative because it is less substituted.' Now we can explore this further.[8] **Problem 3.2**: What is the product of the alternative condensation and why is its enolate less stable than that of **14**?

15   **16**   **14; R = Et**   **enolate of 14; R = Et**

Answer 3.2: We need to draw the alternative enolate **W33** and condense it with oxalate in the same way to give the isomeric diketoester **W34**. This too can form a stable enolate but it has an extra destabilising alkyl group on the middle carbon and this is enough to drive the equilibrium towards **14**.

15   **W33**   **W34**   **enolate of W34**

Lipoic Acid Synthesis

In the textbook we gave a synthesis of lipoic acid **67** without going into much detail. It uses an enamine **72** to control an alkylation and a Baeyer-Villiger reaction showing the regioselectivity we have already discussed.[9]

Lipoic Acid: Synthesis

74; 88% yield

1. LiAlH$_4$ (96%)
2. Ac$_2$O (91%)
3. TsOH (90%)

75

MeCO$_3$H

76; 65% yield

1. H$_2$N—C(S)—NH$_2$
2. KOH

W35; 80% yield

FeCl$_3$

67; 80% yield

Problem 3.3: What is the purpose of the formation of **74**? Draw the compounds formed at each stage in the conversion of **74** into **75** and in the conversion of **75** into **76**.

Answer 3.3: It is easier to answer the second question first: The intermediates are the alcohol **W36** and the acetate **75**. The last step hydrolyses the cyclic acetal which had to be there to stop the reduction of the ketone with LiAlH$_4$ in the formation of **W36**.

74 → 1. LiAlH$_4$ → W36; 96% yield → 2. Ac$_2$O → 75; 91% yield → 3. TsOH → 76; 90% yield

Thiourea reacts through sulfur (just one reaction shown) in S$_N$2 reactions (regioselectivity) because sulfur has higher energy lone pair electrons than nitrogen and the S$_N$2 reaction is orbital and not charge controlled. The intermediate *bis*-thiouronium salt **W37** is stable until treated with alkali – just one mechanism is shown. This is like a carbonyl substitution *via* a tetrahedral intermediate with the least basic group leaving (pK_a RSH < H$_2$O < NH$_2$).

75 → 1. H$_2$N—C(S)—NH$_2$ → *bis*-thiouroniumsalt W37 → 2. KOH → W35

Epoxide Opening by Enol(ate)s

Two reactions in the synthesis of the spirocyclic lactone described in the textbook are interesting: the formation of the epoxide and its transformation into the lactose **100a**. **Problem 3.4**: Draw a mechanism for the formation of epoxides, e.g. **W39**, from ketones with sulfur ylids **W38** using cyclopentanone. Can you suggest another method for this conversion?

Answer 3.4: Nucleophilic attack by the ylid **W38** on the ketone is followed by cyclisation **W40** forming a strong C–O bond at the expense of a weak S–O bond. The simplest alternative is a Wittig reaction followed by epoxidation of the alkene **W41**.

Problem 3.5: What determines the stereochemistry of the new spirocyclic centre in the lactone **100a**?

Answer 3.5: The dilithium enolate **96** attacks the less substituted end of the epoxide **102** without changing the stereochemistry and the cyclisation of the product **W42** on work-up also occurs without affecting the chiral centre. It is the formation of the epoxide **102** that decides the stereochemistry. The ylid **W38** must approach the bottom face of the ketone **101**, opposite the pseudo-axial methyl group.[10]

102 → W42 → W43 → 100a

Conjugate (Michael) Addition

Problem 3.6: Why is the *cis* isomer *syn*-**59** favoured in this conjugate addition?

57 → 58 → *syn*-59

Answer 3.6: You have a choice: you could suggest that the methyl group in **103** will prefer to be axial to get out of the way of the pyrrolidine ring and that the conjugate addition prefers an axial approach or you could say that the second centre equilibrates and the diequatorial product is preferred thermodynamically.[11]

A Final Example

Problem 3.7: Suggest reagents for the conversion of the phenol **120** into the ketone **121**.

120 → ? → 121

Answer 3.7: The simplest route is total reduction by hydrogenation under pressure with a nickel catalyst followed by oxidation of the alcohol with, say, Cr(VI). Alternatively, Birch reduction might give the ketone directly. Ayres and colleagues[12] actually reduced the aromatic ring completely with a rhodium on alumina catalyst and then oxidised the alcohol with CAN (ceric ammonium nitrate) to give **121**. You were not expected to predict the precise reagents.

120 →[H₂, Rh/Al₂O₃, EtOH, 100% yield] →[Ce(IV) (CAN), NaBrO₃, MeCN/H₂O, 74% yield] 121

Further Problems and Examples on Regioselectivity: Enolates and Conjugate Addition

Problem 3.8: The first two steps in a recent synthesis of the immunosuppressant FR901483 are shown here. What aspects of regioselectivity can you identify? How is control exerted? DBU is a strong amidine base.

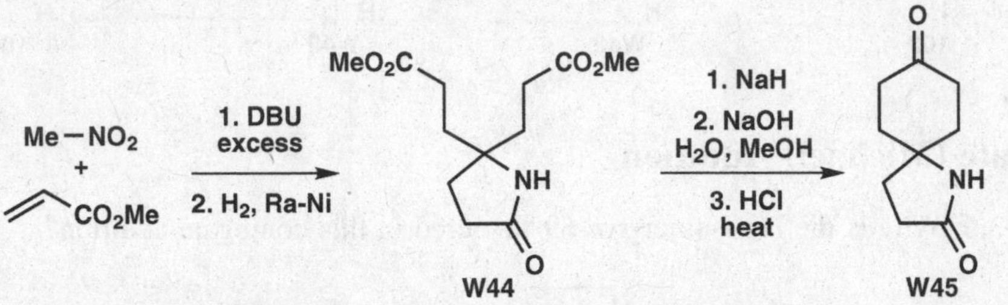

Answer 3.8: No less than three molecules of methyl acrylate must add to the easily made carbanion ('enolate') of nitromethane in a conjugate fashion **W46**. The immediate product of the addition, the enolate anion **W47**, is basic enough to deprotonate itself next to the nitro group **W48** so that the second conjugate addition can occur, and so on.[13] The initial product **W50** is reduced to the amine **W51** that cyclises to the lactam **W44**.

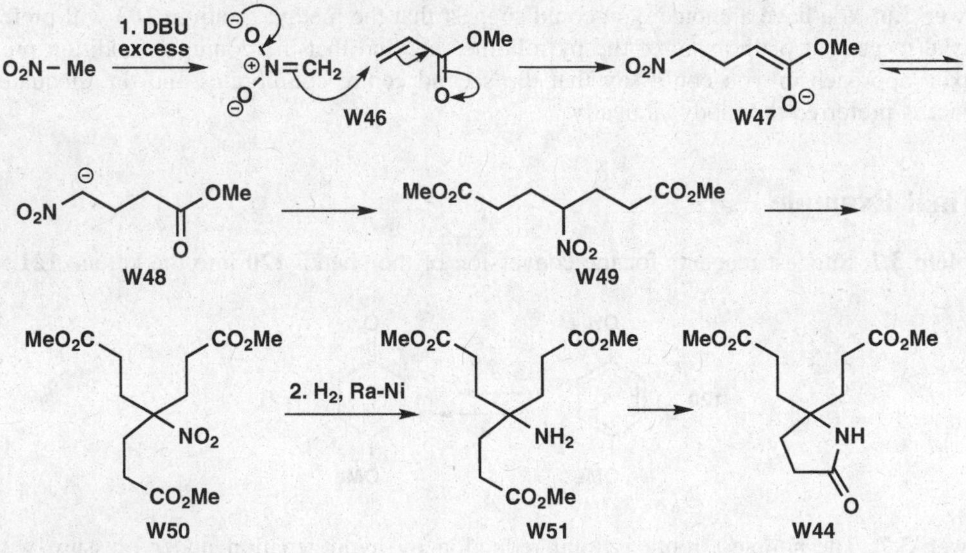

Sodium hydride is a strong enough base to make the enolate of one of the remaining esters that cyclises **W52** onto the other in a Claisen condensation to give the ketoester **W53**, formed as its enolate anion. Hydrolysis to the free acid **W54** and decarboxylation gives **W45**.

Problem 3.9: Suggest mechanisms for these reactions, explaining any regioselectivity.

Answer 3.9: The enolate of malonate can be formed even with as weak a base as an amine.[14] Formaldehyde cannot enolise but is very electrophilic. Presumably the unstable conjugated alkene **W57** accepts another enolate **W58** in a conjugate addition. Acid hydrolyses all four esters in **W59** but only two decarboxylate, e.g. **W60**, as an enol product **W61** is needed for the reaction to go.

Problem 3.10: If the same reaction is applied to ethyl acetoacetate **W62**, a quite different product **W63 is** formed. Give mechanisms for the reactions, structures for the intermediates, and explanations for the regioselectivity.

Answer 3.10: The sequence starts in the same way with the formation of the Michael acceptor **W64** and the addition of another enolate to give **W65**. Now the difference[15] is that the ketones can do an aldol reaction of the enolate formed at the methyl group of one onto the carbonyl of the other in an intramolecular reaction to give **W66**. Hydrolysis and decarboxylation is as before except that the second (γ-) ester group cannot use the cyclic mechanism **W60**.

Problem 3.11: Here is a third reaction, very similar to the last one, but with the addition of ammonia. Yet another product **W67** is formed in good yield. How is the regioselectivity controlled this time?

Answer 3.11: This is an important reaction (the Hantzsch pyridine synthesis) used to make a series of drugs to reduce blood pressure so quite a lot is known about it.[16] The difference is of course the presence of ammonia that forms an enamine **W68** with the acetoacetate and it is this enamine that may do the conjugate addition on **W64**. Cyclisation by nucleophilic nitrogen is preferred to cyclisation via the enolate, present in only low concentration. This is a summary of the mechanism.

Further Example

Some of these reactions involving enolate formation and conjugate addition have caused genuine problems in research. The NaOEt catalysed reaction between octan-2-one **W71** and ethyl acrylate was first reported to give the enone **W72** (54% yield) and then to give the cyclic dione **W70** (27% yield). A modern reinvestigation[17] showed that the product was the dione **W70** with substantial amounts of conjugate addition of ethoxide to ethyl acrylate as by-product. Good yields (88%) of **W70** were obtained using t-BuO$^-$ as the base.

Further exploration with other ketones showed that good yields of cyclic diones were generally formed with t-BuO$^-$ in THF. However, ketone **W73** gave an uncyclised by-product **W75**.

W73   W74; 71% yield   W75; 29% yield

Problem 3.12: Suggest a detailed mechanism for the reactions of **W71** and **W73** with acrylate esters. You might consider some important questions.

1. Is **89** a likely intermediate in the formation of **87**?
2. Are ketoesters like **W75** intermediates in the formation of **W70** and **W74**?
3. If so, why does **W75** itself not cyclise?

Answer 3.12: There are two steps needed: conjugate addition and Claisen condensation. If Claisen condensation occurred first, **W72** would be an intermediate and might be isolated as a by-product but if conjugate addition occurred first, ketoesters such as **W75** would be intermediates. Isolation of the ketoester **W75** suggests that conjugate addition occurs first. This was confirmed by the attempted cyclisation of **W72** (prepared by a different route) under the same conditions: no **W70** was formed.[17] Why does **W75** not cyclise to **W77**? Under the conditions of the reaction, **W75** exists as its stable conjugated enolate **W76**. Cyclisation of **W75** would give **W77**, which is unable to form such an enolate as it has no hydrogen atoms between the two ketones.

W74   W76   W77

The Choice Between Two Ends of an Alkene

When the choice is between two such similar positions as the two ends of an alkene, it can be more difficult to decide which product will be formed. If the alkene is unsymmetrical, the electrophile goes for the less substituted end.[18] The non-conjugated diene **W79** was made by Birch reduction[19] and alkylation of **W78**. Treatment with iodine gives one product – the iodolactone **W80** with the electrophile on the less substituted end of the alkene. Note the chemoselectivity too: the iodine attacks the more highly substituted, and hence more nucleophilic, alkene.[20]

W78   W79   W80

That this is largely because we have an intramolecular reaction would be obvious in the iodolactonisation of the symmetrical compound **W82**. Now the two alkenes are equally substituted as are the ends of each alkene. So the choice of iodine on one end and carboxylate on the other is simply because the molecule prefers to form a five- **W83** rather than a four-membered lactone. The electrophile (I_2) is external but the nucleophile (carboxylate anion) is tethered to the rest of the molecule.

The Choice Between Two Ends of an Epoxide

The last example involved three-membered cyclic iodonium ion intermediates and the next two concern epoxides, also reacting with nucleophiles. You will be familiar with the usual regioselectivity: nucleophiles attack the less hindered end of an epoxide.[21] Suppose the epoxide has equal numbers of substituents at both ends? Again it is possible to use a tether to direct the nucleophile to one end or the other. Epoxides **W85** of allylic alcohols **W84** are particularly important (chapter 25) and usually nucleophilic attack is preferred at C-3 to give **W86**.

This selectivity can be reversed by delivering the nucleophile from the coordination sphere of a metal (B and Ti are best) for which the alcohol is a ligand.[22] Nucleophiles include azide and cyanide ions and thiolates (RS^-). All are directed to C-2. Perhaps the most striking results are with the even more nearly symmetrical ether **W87** which gives a 92:8 ratio of C-2 **W88** to C-3 products in 96% yield. The probable mechanism is intramolecular delivery of azide **W89**.

Diols as Intermediates in Regioselective Functionalisation of Conjugated Alkenes

The situation with epoxides **W91** of alkenes conjugated to a carbonyl group **W90** is even more clear-cut and is reversed. Displacement by S_N2 is preferred at C-2, activated by the carbonyl group. The diols **W92**, made by osmylation of **W90** can be tosylated selectively at C-2 **W93** or converted into cyclic sulfates **W94** that also react at C-2.

W90   W91   W92   W93   W94

Reaction at C-3 has been achieved by the Mitsunobu reaction[23] using 'DEAD' **W95** and Ph_3P in non-basic solution. The azide **W96** is formed with inversion at C-3 and the tosyl derivative **W97** made by Mitsunobu reaction with pyridinium tosylate gives the other diastereoisomer **W98** by conventional S_N2 displacement with sodium azide.[24] You should notice that the regio-selectivity problem with epoxide **108** has been solved by using chemoselectivity with the diol **W92**.

Bromination of Conjugated Alkenes

Problem 3.13: Bromination of the closely related conjugated alkenes **W99** and **W102** gives different products **W101** and **W104** after reaction with base. Identify aspects of regioselectivity in these two reactions. Why are the results so different?

Answer 3.13: Bromination of the ester gives the *anti*-dibromide **W100** that forms an enolate in alcoholic base which eliminates bromide by the E1cB mechanism **W105**. The stereochemistry of **W100** is lost in enolate formation and the elimination gives the more stable Z isomer (Br and Ph *cis* but the two large groups *trans*) which can eliminate again by the E2 mechanism **W106** to give the triple bond.[25] Adding water to the strongly alkaline solution hydrolyses the ester **W107** to give the anion **W108** of the acid **W101** released on acidification.

Compound **W103** is also the *anti*-dibromide but, by contrast, base treatment of the acid gives the carboxylate anion **W109** that eliminates CO_2 and bromide to give **W104**. The regioselectivity is the choice of proton removed by base: the carboxylic acid proton is preferred to any C-H but, if there is no CO_2H, enolate formation occurs.

Enolate Regioselectivity

One simple form of regioselectivity we have not discussed is reactions of enol(ate)s at carbon or oxygen. You will be familiar with the usual reactions at carbon: with halogens, alkyl halides, aldehydes, ketones and esters as well as with the usual reactions at oxygen: protons (kinetically), acid chlorides and silyl chlorides and we assume you know the reasons for this selectivity.[26] Sometimes this regioselectivity is a real problem. When Dow Agrosciences set out to develop a large scale synthesis for their insecticide **W110** the first disconnections were easy. Ureas are usually made by addition of an amine, here **W112**, to an isocyanate, here **W111**. The heterocyclic amine **W112** can be made from the enone **W113** and hydrazine **W114**. One nitrogen adds in conjugate fashion and one direct – it does not matter which.

Enones such as **W113a** with exo-methylene groups, are usually made by a Mannich reaction from the corresponding saturated ketone **W115** and the obvious way to make this is by reaction of the enolate of the methyl ketone **W116** with a pyridine electrophile **W117** where X is a leaving group such as a halide. There is no problem with **W116** as fluorine is *o,p*-directing so a Friedel-Crafts reaction will do the job. We have not discussed the regioselectivity of electrophilic aromatic substitution as we assume you are familiar with that.[27]

What sort of electrophile is **W117**? In particular, will it react with the enolate of **W116** at C or O? The easiest way to find out is to do the reaction.[28] With the familiar combination of NaH in DMSO to make the enolate and fluoride as the leaving group, the enol ether **W118** was the major product by 64:36. This is obviously unacceptable on a large scale.

W117; X = F   W118; 64%   W115; 36%

Two simple changes solved this regioselectivity problem. Using NaH in THF instead of DMSO and the chloride **W115; X = Cl** instead of the fluoride changed the ratio to >95:<5 in favour of **W113**. However, their problems were not over. They found that the Mannich reaction with Me_2NH and CH_2O worked well but that the pyridine encouraged elimination of Me_2NH to give **W111**. A good thing you might say. The conjugate addition of the enolate **W115** to **W111** was faster than the Mannich reaction. This is a chemoselectivity problem: the product reacts faster than the starting material. They solved this problem by taking advantage of a reaction they wanted and going in one step from **W113** to **W108**. This was the final method:

W119

Formaldehyde was replaced by its aminal with Me_2NH, the Mannich base **W119** was immediately reacted with hydrazine in the same solvent without isolation and the isocyanate **W111** added to **W112** again without isolation to give crystalline **W110** in nearly 80% yield in one pot from **W115**. The solution to selectivity problems is as vital in production as in the laboratory.

References

1. G. R. Crow, *Organic Reactions*, 1993, **43**, 251.
2. C. Houge, A. M. Frisque-Hesbain, A. Mockel, L. Ghosez, J. P. Declercq, G. Germain and M. Van Meerssche, *J. Am. Chem. Soc.*, 1982, **104**, 2920; G. Grethe, J. Sereno, T. H. Williams and M. R. Uskokovic, *J. Org. Chem.*, 1983, **48**, 5315.
3. P. E. Sonnet and J. E. Oliver, *J. Het. Chem.*, 1974, **11**, 263.
4. L. Horner and D. W. Baston, *Liebig's Annalen*, 1973, 910.
5. P. A. Grieco, T. Oguri and S. Gilman, *J. Am. Chem. Soc.*, 1980, **102**, 5886.
6. S. M. Ali and S. M. Roberts, *J. Chem. Soc., Perkin Trans. 1*, 1976, 1934.

7. L. Skattebøl and Y. Stentstrom, *Acta Chem. Scand. B*, 1985, **39**, 291.
8. D. J. Dale, P. J. Dunn, C. Golightly, M. L. Hughes, P. C. Levett, A. K. Pearce, P. M. Searle, G. Ward and A. S. Wood., *Org. Proc. Res. Dev.*, 2000, **4**, 17.
9. A. Segre, R. Viterbo and G. Parisi, *J. Am. Chem. Soc.*, 1957, **79**, 3503.
10. P. L. Creger, *J. Am. Chem. Soc.*, 1967, **89**, 2500; *J. Org. Chem.*, 1972, **37**, 1907.
11. N. F. Firrell and P. W. Hickmott, *J. Chem. Soc., Chem. Commun.*, 1969, 544; J. W. Huffmann, C. D. Rowe and F. J. Matthews, *J. Org. Chem.*, 1982, **47**, 1438.
12. F. D. Ayres, S. I. Khan, O. L. Chapman and S. N. Kaganove, *Tetrahedron Lett.*, 1994, **35**, 7151.
13. T. Kan, T. Fujimoto, S. Ieda, Y. Asoh, H. Kitaoka and T. Fukuyama, *Org. Lett.*, 2004, **6**, 2729.
14. Vogel, p. 684.
15. Vogel, p. 1098.
16. Vogel, p. 1168.
17. T. Ishikawa, R. Kadoya, M. Arai, H. Takahashi, Y. Kaisi, T. Mizuta, K. Yoshikai and S. Saito, *J. Org. Chem.*, 2001, **66**, 8000.
18. Clayden, *Organic Chemistry*, chapter 20.
19. Clayden *Organic Chemistry*, p. 628.
20. D. A. Ellis, D. J. Hart and L. Zhao, *Tetrahedron Lett.*, 2000, **41**, 9357.
21. *Disconnection Approach*, chapter 6.
22. M. Sazaki, K. Tanino, A. Hirai and M. Miyashita, *Org. Lett.*, 2003, **5**, 1789.
23. Clayden, *Organic Chemistry*, p. 431.
24. S. Y. Ko, *J. Org. Chem.*, 2002, **67**, 2689.
25. Vogel, pp. 510–512.
26. Clayden, *Organic Chemistry*, chapter 21.
27. Clayden, *Organic Chemistry*, chapter 22.
28. J. M. Renga, K. L. McLaren and M. J. Ricks, *Org. Process. Res. Dev.*, 2003, **7**, 267.

4

Stereoselectivity: Stereoselective Aldol Reactions

Problems and Further Examples Relating Directly to the Text

You will find it helpful to have chapter 4 from the textbook open as you look at this first section.

Enolate Geometry and Aldol Diastereoselectivity

Why is the requirement that the t-Bu group must occupy an axial position in the Zimmermann-Traxler transition state **14** from the Z-enolate not so bad as appears at first sight? It is because there is no axial substituent on the trigonal (blob) C atom. There appears to be no axial substituent on Li either. But there is – a THF ligand.

Zirconium Enolates

In the textbook we promised an explanation of the remarkable fact that *all* zirconium enolates give *syn* aldols regardless of geometry. These were the examples given in the textbook and Evans's explanation follows.[1]

Workbook for Organic Synthesis: Strategy and Control   Paul Wyatt and Stuart Warren
© 2008 John Wiley & Sons, Ltd

Possible Explanation of the Stereoselectivity of Zirconium Enolates

The crowded zirconium complex Cp_2ZrCl_2 **W1**, having 16 electrons on Zr and a Cl–Zr–Cl bond angle of only 97° adds the lithium enolate to give a mixture of *E* and *Z* enolates **W2** with 18 electrons round the Zr atom. One chlorine is now exchanged for the aldehyde to form a new 18 electron complex **W3** with an O–Zr–O of about 71°. The enolate and the aldehyde are thus forced into very close proximity and are thought to react through transition states such as **W4** and **W5**, both giving *syn* aldols.

Enone Stereochemistry

The nopinone **49** story is a little more interesting than we revealed in the textbook. Direct aldol condensation of nopinone **49** with a large excess of acetaldehyde gives a more or less 4:1 mixture of *E*- and *Z*-enone **52** but treatment of this mixture with acid in the same solvent gives *E*-**52** alone.[2]

Problem 4.1: Why is a large excess of acetaldehyde used? Why is the *E*- more stable than the *Z*-enone **52**? Suggest how the mixture becomes one compound in acid solution. What kind of control is this?

Answer 4.1: Acetaldehyde is very prone to self condensation and the excess allows for this.

There appears little difference between *E*- and *Z*-**52** but in *Z*-**52** the methyl group eclipses the carbonyl oxygen in the planar enone. Protonation of the enone **15** reduces the double bond character in the ene part and allows rotation. If you suggested that conjugate addition could do the same thing, you may be right.

Another Nootkatone Synthesis

Problem 4.2: A different synthesis of (+)-nootkatone from (−)-β-pinene illustrates another aspect of enolate stereochemistry.[3] Reaction of nopinone **49** with diethyl carbonate gives one diastereoisomer of the ketoester **W8**. Methylation using only K_2CO_3 as base gives one diastereoisomer of the ketoester **W9**. Explain both these selectivities. We have helpfully used conformational drawings of the compounds.

Answer 4.2: The CMe_2 bridge across the top of the molecule makes the bottom face the favoured site for a substituent both thermodynamically (formation of **W8** by reversible protonation of the stable enolate **W10**) and kinetically, in the irreversible formation of **W9** via the same enolate **W11**.

Recent Developments in Stereoselective Aldol Reactions

In the textbook we said: 'the cyclic ester **56** must of course form an '*E*' enolate and when the boron enolate *E*-**57** reacts with aldehydes the *anti*-aldol products **58** are formed with good stereoselectivity ranging from 4:1 to >20:1. The predominate isomer is that expected from the Zimmerman-Traxler transition state'. **Problem 4.3**: Draw the Zimmerman-Traxler transition state and show it leads to the *anti*-aldol.

Answer 4.3: While **56** can have a chair-like conformation **56a**, the enolate **57** has both ring oxygen atoms conjugated with the enolate double bond the whole molecule is close to being planar. The Zimmerman-Traxler transition state presumably looks something like **W12** with the aldehyde R choosing to go equatorial and the aldehyde adding to the face of the enolate opposite the nearer Ph group.[4]

A particularly interesting case reported by Denmark and Pham[5] reveals several types of selectivity. The methyl ester **W13** of natural (*S*)-lactic acid forms a Weinreb amide **W14** with Me$_3$Al as catalyst. These amides are designed to react only once with Grignard reagents and organolithiums thus solving a classical chemoselectivity problem (chapter 2). Here the starting material **W16** for the aldol reaction is prepared by reaction of ethyl Grignard with **W14** and protection of the OH group by silylation. **Problem 4.4**: Explain the selectivity of this formation of **W15** in detail. Why was it unnecessary to protect the OH group?

Answer 4.4: Presumably the first molecule of EtMgBr forms the magnesium alkoxide **W17** but the second adds to the carbonyl group giving adduct **W18** stabilised by chelation. Aqueous acidic work-up gives the new ketone. The silyl enol ether of **W16** gives excellent stereoselective aldols and you are referred to Denmark and Pham's paper[5] for the details.

Juvabione Synthesis

Problem 4.5: Draw the mechanisms for the conversion of **69** into **71**, the rearrangement of **71** to **72**, and the isomerisation of **72** to **73**. What questions of stereochemistry arise during this sequence?

69; mixture of diasts

71; 80% yield

72; 80% yield

73; 87% yield; 78:22 diasts

74; 75% yield; juvabione

Answer 4.5: Oxalyl chloride $(COCl)_2$ converts the free acid **69** into the acid chloride and we hope you can draw the mechanism for that. If you cannot, a clue is that the other products are CO and CO_2. The acid chloride reacts with diazomethane CH_2N_2 acting as a nucleophile through carbon **W19** to give an unstable diazonium salt that is deprotonated by a second molecule of diazomethane to give the diazoketone **71**. Silver catalyses the loss of nitrogen gas from **71** to give the α-ketocarbene that rearranges (Wolff rearrangement, see *Clayden* page 1072) to give the ketene **W22**. Addition of water to this very electrophilic species gives the acid **72**. The only step that affects stereochemistry is the rearrangement **W21** and this goes with retention at the migrating group as usual.

W19

W20

71; diazoketone

W21; α-keto-carbene

W22; a ketene

These compounds are all mixtures of diastereoisomers at the CO_2Me centre but treatment with base (DBU) moves the alkene into conjugation with CO_2Me by deprotonation **W23** to give the extended enolate and reprotonation **W24** at the thermodynamically favoured γ-position (see chapter 11). DBU is too weak to form the enolate exclusively so thermodynamic control applies.[6]

W23

W24; extended enolate

Lintetralin

Stereochemical control may be needed in an aldol reaction at sites other than the two we have been considering so far. A synthesis[7] of the lignan lintetralin **W25** will illustrate the point. The molecule has three rings and three adjacent stereogenic centres. Disconnecting the middle, non-aromatic, ring is good strategy if we want a short synthesis, and cyclisation of **W26** by a Friedel-Crafts alkylation should be a good reaction as the alcohol in **W26** is secondary and benzylic and will easily give a good cation. By the same token there is no point in trying to control the stereochemistry of the alcohol as it will be lost when the cation is formed. If we want to use an aldol reaction we must have a carbonyl group in a 1,3 relationship with the benzylic alcohol, i.e. as the acid derivative in **W28**.

The lactone **W30** looks like a good starting material. We have discussed how to control the stereochemistry of the aldol stereogenic centres themselves during this chapter, but here we want to control one aldol centre (next to the C=O group) and one adjacent centre. As these are arranged *anti* to one another, the simple lithium enolate gives this relationship without any special provision as the new bond is formed on the opposite face of the ring to the existing substituent.

Compound **W31** is formed as a mixture of diastereoisomers at the alcohol centre, but both alcohols give the same cation in acid solution, and the cyclisation continues the pattern of *anti* substituents, now round the six-membered ring **W32**. The cyclisation may be reversible and hence controlled thermodynamically. **Problem 4.6**: What geometry of enolate is formed from **W30** and what aldol stereochemistry would you expect to predominate in **W29**? What reagents would you suggest to complete the synthesis of lintetralin **W23**?

Answer 4.6: As a lactone, **W30** must form an *E*-enolate **W33** so we should expect the *anti*-aldol with the alkyl group *anti* to the new OH group *anti*-**W31** (more easily seen by looking at the Hs). Control is evidently not very good here but the chemists would take no interest in the stereochemistry as it disappears in the next step. All that needs doing in the final stages is the reduction of the lactone to the diol and methylation of the two OH groups. You might have suggested a number of reagents for this, the chemists used LiAlH$_4$ in THF for the reduction and methylated with MeI and NaH in DMSO.

Stereoselectivity occurs in reactions other than those described in the chapter. Here are some more problems related to reactions that you certainly know.

Problems Related to the Stereochemistry of Epoxides

The epoxide **W34** of *cis*-but-2-ene rearranges[8] to the acetal **W35**. Note, the acetal **W36** can be transformed into the same epoxide **W34** by a different sequence.[9] **Problem 4.7**: Give mechanisms for these reactions and explain the stereochemistry. Why is the *cis* epoxide **W34** related to the *trans* acetal in one sequence but to the *cis* acetal in the other? NBS is a radical generator (see *Clayden* chapter 39).

Answer 4.7: In the first reaction BF$_3$ acts as a Lewis acid to help acetone open the epoxide **W38** with inversion. The product cannot cyclise as first formed **W39** but can do so after rotation **W40**. One inversion and one retention (no change) means *cis* epoxide gives *trans* acetal.

The second sequence involves radical bromination of **W36** at the benzylic centre by NBS to give a very unstable intermediate that expels bromide **W41** to give the stable cation and a bromide ion that attacks (at either centre) with inversion **W42** to give **W43** which is just **W38** drawn in a different way (make sure you agree!). NaOH hydrolyses the ester releasing the oxyanion that carries out a second S$_N$2 reaction **W44** at the same centre: two inversions – compare **W42** and **W44** – result in overall retention.

Problem 4.8: Reaction of the *bis-cis*-epoxide **W45** with Na$_2$S gives the bicyclic sulfide **W46**. Draw a mechanism for this reaction and explain the stereochemistry of the product. Is the product enantiomerically pure?

Answer 4.8: The key is that cyclohexane epoxides open in a *trans*-diaxial manner.[10] The first product must be **W48**, as all four epoxide carbons are the same, and this closes intramolecularly to give a five-membered ring rather than the four-membered alternative (regioselectivity). The stereochemistry is again inversion and the sulfide bridge requires a boat conformation for the six-membered ring. Rotation of **W46a** by 180° about a horizontal axis gives the diagram used in the question. The product cannot be enantiomerically pure as the starting material **W45** is achiral (having two planes of symmetry) and the reagent (Na$_2$S) is also achiral.[11]

Problem 4.9: Now a question that revises other forms of selectivity but ends with two stereoselective examples. Identify and explain the selectivity in each step of this synthesis.

Answer 4.9: Obviously the best answer is an annotated mechanism, something on these lines, but any reasonable explanation is fine. The formation of **W51** has several kinds of selectivity. Regioselective formation of the less substituted lithium enolate is kinetic control. Then chemoselective attack at the C=O group in a charge-controlled reaction between hard species **W55**. Under the reaction conditions, EtO⁻ forms the stable enolate **W56** – this is thermodynamic control. The reactive enolate **W57** is just another form of **W56**. Cyclisation by 5-*exo-tet* process **W57** is favoured (Baldwin's rules). *C*-Alkylation would give a three-membered ring and is reversible.

The stereoselectivity of the aldol is tricky as you are not used to cyclic double enol ethers. You should draw the titanium chelate **W58** and the enol can have only the *E* geometry, so that helps. The O in the ring and the OSiMe₃ group must be *anti* and the *t*-Bu chooses to go equatorial. Our suggested drawing of the chair transition state **W59** has bits removed for clarity. Putting in the Hs at the new chiral centres may help.

The bromo-lactonisation looks complicated but is actually simple as far as stereoselectivity is concerned. Bromination on the top face **W60**, opposite the large side-chain, gives one bromonium ion that must be attacked from underneath **W61** to give **W53**. We have cheated by drawing the intermediate in the shape of the product. The regioselectivity is mildly interesting as **W59** is an ene-diol derivative. But the Me₃SiO is much more electron-donating than the aromatic O. This is not really needed for a good answer.

Epoxidation Reactions

Now we move to the formation of epoxides. **Problem 4.10**: Define and explain the types of stereoselectivity shown in these two epoxidation reactions on closely related compounds. If R^1 and R^2 are both H, alkyl, or aryl groups the reaction is unsuccessful. Why? What is the role of the carbonyl group in ensuring successful epoxidation?

Answer 4.10: Both epoxidations are stereo*specific* in that a *cis* epoxide is formed in both cases. But the stereo*selectivity* is different. Both compounds have an acyl group – the difference is that one has an NH and one an N-Bn group.[12] This suggests that **W62** is formed simply by reaction on the less hindered side opposite the amide while **W64** is formed when *m*CPBA is H-bonded to the NH of the substituent. The carbonyl group is needed because simple amines are readily oxidised – usually to black tars.

Problem 4.11: In a third case **W65**, epoxidation of the related carbamate gives a product **W66** which is not an epoxide. Suggest a mechanism for the reaction and suggest why yet a third route is followed.

Answer 4.11: We should expect epoxidation *anti* to the amide as it has no NH group. This would give **W67**. The product is clearly formed by attack of the carbonyl group on the epoxide **W68** followed loss of the t-Bu group **W69** as a cation. The cyclisation can happen only with the *anti*-epoxide (**W62** or **W67**) as inversion is required by the intramolecular S_N2 reaction.[12] There are two differences between this case and **W62**. The benzyl group has been replaced by a p-MeO-benzyl group making the nitrogen rather more nucleophilic and the solution is buffered with $NaHCO_3$. In the cases above it was difficult to stop the cyclisation reaction and it could be initiated on **W62** with acid. In this case, even with the $NaHCO_3$ buffer, some cyclisation occurred so it was better to let it happen.

References

1. D. A. Evans and L. R. McGee, *Tetrahedron Lett.*, 1980, **21**, 3975; Y. Yamamoto and K. Maruyama, *Tetrahedron Lett.*, 1980, **21**, 4607.
2. T. Yanami, M. Mayashita and A. Yoshikoshi, *J. Org. Chem.*, 1980, **45**, 607.
3. T. Inokuchi, G. Asanuma and S. Torii, *J. Org. Chem.*, 1982, **47**, 4622.
4. M. B. Andrus, B. B. V. S. Sekhar, E. L. Meredith and N. K. Dalley, *Org. Lett.*, 2000, **2**, 3035.
5. S. E. Denmark and S. M. Pham, *Org. Lett.*, 2001, **3**, 2201.
6. N. Soldermann, J. Velker, O. Vallat, H. Stoeckli-Evans and R. Neier, *Helv. Chim. Acta*, 2000, **83**, 2266.
7. P. A. Ganeshpure and R. Stevenson, *J. Chem. Soc., Perkin Trans. 1*, 1981, 1681.
8. B. N. Blackett, J. M. Coxon, M. P. Hartshorn, A. J. Lewis, G. R. Little and G. J. Wright, *Tetrahedron*, 1970, **26**, 1311.
9. D. A. Seeley and J. McElwee, *J. Org. Chem.*, 1973, **38**, 1691.
10. Clayden, *Organic Chemistry*, 468–470.
11. T. W. Craig, G. R. Harvey and G. A. Berchtold, *J. Org. Chem.*, 1967, **32**, 3743.
12. P. O'Brien, A. C. Childs, G. J. Ensor, C. L. Hill, J. P. Kirby, M. J. Dearden, S. J. Oxenford and C. M. Rosser, *Org. Lett.*, 2003, **5**, 4955.

5

Alternative Strategies for Enone Synthesis

Problems from the Textbook

You will need chapters 1–4 of the textbook for the first question and chapter 5 for the next few questions.

The Aldol Strategy

Problem 5.1: Search chapters 2–4 for examples of this disconnection.

Answer 5.1: Here is a short list of possibilities: you may have spotted more examples.

Lithium enolates: The exomethylene lactone **80** in chapter 2 was made by an aldol reaction using the lithium enolate of the parent lactone **81** and formaldehyde.

Equilibrium methods: The dienone **93** in chapter 2 was made by an equilibrium method. No selectivity problems arise and an equilibrating aldol reaction between the enolisable ketone **94** and the unenolisable and very electrophilic aldehyde **95** catalysed by NaOEt in EtOH gives **93** in 89% yield.

Enamines: Aldol cyclisation of **100** in chapter 2 followed spontaneously from an enamine-directed conjugate addition of **99** to a simple enone.

Workbook for Organic Synthesis: Strategy and Control   Paul Wyatt and Stuart Warren
© 2008 John Wiley & Sons, Ltd

99   **100**   **101**

Enone examples from chapter 3: Chapter 3 concerns regioselectivity and there are many examples of the aldol reaction in enone synthesis. Target molecules (*E*)-**7**, **11**, **32**, **33**, the intermediates for gingerol **43** and **45**, cyclohexenones **107** and **108** from Robinson annelation were all made by some version of the aldol strategy.

Enone examples from Chapter 4: These often concern the geometry of the new alkene as compounds (*E*)- and (*Z*)-**3**, **46** and **47**, and (*E*)-**52**.

49; (+)-nopinone   **Z-52**   **E-52**

Example from Chapter 5: Corey's Antheridic Acid Synthesis

We simply stated that the final step in the synthesis was the addition of the exomethylene group in **21** by a Mannich reaction on **22**. **Problem 5.2**: Suggest how this might be carried out, explaining why no control is apparently needed.

21   **22**

Answer 5.2: The direct method would be reaction with formaldehyde and a secondary amine to give the Mannich base **W1**, methylation and elimination in base. It appears as if no control is needed as only one CH_2 group can be enolised.

W1

However, the position next to the CO_2Me group (H marked in **22**) is also allylic and might be enolised. Corey and Myers chose to make the silyl enol ether first and then to react it with a pre-formed Mannich reagent. The epoxide in the second step is there to scavenge the Me_3Si-I by-product.[1]

22
1. Me_3SiCl
   Et_3N
2. *i*-Pr_2NLi

W2

i-Pr_2NEt, MeI

$Me_2N=CH_2$

21
60% yield

There was evidence from earlier in the synthesis that enolisation of the ester could occur as the lactone **W3** prepared by a Diels-Alder reaction, isomerised to conjugated lactone **W4** before hydrolysis to **W5**. Make sure you see how the alkene moves.

W3   KOH
              EtOH

W4   KOH
              EtOH

W5

Later in the synthesis, the tertiary alcohol in **W6** is acylated and the double bond moves out of conjugation again to give **W7** in 90% yield. This isomerises to the more stable diastereoisomer **W8** also in 90% yield. **Problem 5.3**: Suggest what intermediate is involved in these two operations.

W6
$(CF_3CO)_2O$
pyridine
CH_2Cl_2

W7
cat.
DBU
THF
$-22\ °C$

W8

Answer 5.3: Some sort of extended enolate **W10** or enol must be formed by removal of the γ-proton of **W9** (R = H or $COCF_3$). Kinetic protonation at the α-carbon gives **W11**. Addition of the proton is faster on the top face as ring A **W11** is on the bottom face. By the same token **W8** is more stable as the ester also prefers to be *anti* to ring A. You will meet more chemistry of this sort in chapter 11. This strategy of making unsaturated carbonyl compounds by moving alkenes around is quite general.

W9 → W10 → W11

Another strategy we did not consider in the textbook is making enones from aromatic compounds. Inevitably this gives cyclohexenones and an interesting example occurs in another Corey synthesis[2] of antheridic acid. The aromatic triol **W12** is oxidised to the cyclohexadienone **W13** by an I(III) compound. **Problem 5.4**: Suggest a mechanism for the reaction.

W12 → W13 → W14

Answer 5.4: Nucleophilic substitution at iodine by the phenol with displacement of acetate gives the new I(III) compound **W15**. This decomposes by loss of PhI and participation of one primary alcohol **W16** to give **W13**. Chemoselectivity in this step is determined by the inability of the other primary alcohol to form a stable ring.

Problem 5.5: How might you make **W14** from **W13**?

W12 → W15 → W16 → W13 + PhI

Answer 5.5: Protection of the remaining primary alcohol with t-BuMe$_2$SiCl and a base (imidazole) is obvious but how do you reduce one of the alkenes and not the other? To avoid reducing the carbonyl group we should use catalytic hydrogenation and, if we put the large TBDMS group on first, steric hindrance favours reduction of the more distant alkene. They used one atmosphere of hydrogen in EtOAc at 23 °C with a Rh–C catalyst. This would come from experience or published examples.

Vinyl Anion Strategy

In the textbook we used the vinyl anion strategy to make the enone **61** from the Weinreb amide **60**. The enone **61** was in fact treated with a ruthenium carbene complex to make a new enone **61a** by way of a metathesis reaction (see chapter 15) on the way to the target fumagillol. This is yet another strategy to make enones and, as it happens, is also used for a cyclohexenone. Note the (chemo?)selectivity in that two of the alkenes get involved in the cyclisation and the other two are ignored. As you will see later, metathesis typically works better on less substituted alkenes.

Aliphatic Friedel-Crafts Reaction

This synthesis of enone **72** appeared in the textbook. **Problem 5.6**: What mechanism do you suggest for the elimination on **71** to give the enone **72**? The intermediate **71** is a mixture of diastereoisomers.

Answer 5.6: Forgive us for such an elementary question, but it is all too easy to draw the E2 mechanism forgetting that only the *syn* isomer **W18** with H and Cl axial can eliminate this way. The *anti* isomer can have Cl axial but now only a hydrogen on the other side is axial **W19**. The E1cB mechanism, always to be preferred if possible, forms the stable intermediate enolate **W20** with no diastereoisomers.

We said in the textbook that enone **74**, required for a synthesis of chrysanthemic acid **73**, was made by an aliphatic Friedel-Crafts reaction. We also said that 'an aldol approach would require a cross-condensation between two enolisable ketones, one of them **75** unsymmetrical, and although this could easily be done an alternative was sought.' **Problem 5.7**: Suggest now how the aldol strategy might be realised to make the enone **74** from **75**.

Answer 5.7: Some specific enol equivalent of **75** is needed with enolisation occurring only into the methyl group. An obvious solution is a lithium enolate **W21** made with LDA. If this deprotonates acetone instead of doing the aldol, it can easily be converted into a silyl enol ether **W22**. Do not forget the Lewis acid (which one is your choice) in aldols with silyl enol ethers. This route would probably give the aldol **W23** easily dehydrated to **74**. If you suggested an equilibrium aldol with an oxygen base and a large excess of acetone, you might be right.

Acylvinyl Cation Strategy

In the textbook we discussed the use of enol ethers of 1,3-diketones, such as **101** without giving details of their preparation or reaction. **Problem 5.8**: Draw mechanisms for **100** to **101** and for **102** to **103**.

Answer 5.8: Either **99** or **100** could react with *i*-PrOH but in either case the reaction is like the start of acetal formation. We will add *i*-PrOH to the carbonyl group of **100**. These acid catalysed reactions are all equilibria and the hemiacetal intermediate can lose water **W26** with assistance from the enol. That is why a 1,3-dicarbonyl relationship is needed.

Reaction with RLi or RMgBr also occurs at the carbonyl group to give **102**, protonation and loss of water **W27**, addition of water at the other end of the allylic system **W28** and hydrolysis of the hemiacetal **W29** gives **103**. Notice how similar the structures **W25** and **W29** are. Again all these steps are equilibria and all resemble acetal hydrolysis.

One example of this strategy in the textbook was the addition of the alkyl lithium **96** to the enol ether **106**. The nucleophile **96** was made from **104**. **Problem 5.9**: Why is **104** readily available? Suggest a synthesis.

Answer 5.9: Though **104** is an enone, it is not conjugated so this question is revision rather than anything to do with chapter 5. The simplest disconnection imagines alkylating an acetone enolate equivalent **W32** with 'prenyl' bromide **W31** and a popular choice is the enolate of ethyl acetoacetate **W33**. Allylic halides such as **W31** generally react at the less hindered end with nucleophiles (chapter 19) and the specific enolate avoids self-condensation of acetone. There are of course many other excellent solutions.

The synthesis simply involves forming the enolate of **W33** with ethoxide, alkylation, then hydrolysis and decarboxylation of the product.

Rearrangement on Oxidation of Allylic Alcohols

Problem 5.10: In this oxidation from the textbook, why was the PCC solution carefully buffered with sodium acetate?

Answer 5.10: The compound contains two acetals that could easily be hydrolysed if the acid catalyst were too strong.

Problem 5.11: The next steps were conversion of **119** into **W34**, hydrolysis of the acetals, and cyclisation in acid solution to another enone **W36**. How might **119** be converted into **W34**? Identify and explain the selectivity in the formation of **W36**.

W34   W35; 91% yield   W36; 99% yield

Answer 5.11: Conjugate addition of some methyl copper derivative (the chemists used Me_2CuLi and got 96% yield) gives **W34** with the right stereochemistry by axial addition.[3] In the cyclisation of **W35**, there are five different enolisable positions and three electrophilic ketones. Only cyclisation to give **W36** gives a six-membered aldol that can dehydrate. The others all have medium rings, or bridges. Draw the structures if you are not convinced.

General Questions Not Related to Chemistry in the Textbook

Problem 5.12: Suggest a mechanism for this reaction. What strategy is being used?

Answer 5.12: The LDA must remove a proton from somewhere and the only place that leads to the product is on the furan ring to give a lithium derivative that can cyclise **W39** onto the carbamate.[4] The intermediate **W40** may decompose directly to **W38** or may do so on work up in acid. The lithiation is no doubt assisted by the furan oxygen but is mainly directed across space by the carbamate (chapter 7).

Problem 5.13: A new enone synthesis was developed in 2001 by Ballini.[5] Oxidation of the furan **W41** with a peracid gives the *cis*-enedione **W42** in very high yield (see workbook chapter 15). The new enone synthesis is to treat **W42** with an aliphatic nitro-compound and base. Enones **125** are formed at room temperature in good yield. Suggest a mechanism for this reaction.

W41   Z-**W42**; 97% yield   **W43**; 'good yield'

Answer 5.13: The paper is devoid of mechanisms but the first step must be the conjugate addition **W44** of the anion of the nitroalkane to the ene-dione **W42**. The intermediate enolate **W45** is in equilibrium with other enolates, including one that can eliminate nitrite ion by the E1cB mechanism **W46** to give **W43**.

W44   **W45**   **W46**

In general the enones **W43** were immediately hydrogenated to the saturated diones **W47** and the yield measured at that point. Hence **W47**; R = n-Bu was formed from **W42** in 88% yield and **W47**; R = $(CH_2)_2CO_2Me$ was formed from **W42** in 90% yield. These saturated 1,4-diketones **W47** were treated with TsOH in Et_2O to give new furans **W48** in good yield.

W43; 'good yield'   **W47**; 60–90% yield from 124   **W48**

Problem 5.14: In 1996 Fleming[6] wanted the enone **W51** for a study of conjugate additions by silyl cuprates. He used a method invented by Tunisian chemists[7] in which a 1,3-dicarbonyl compound was combined with an active alkylating agent and then with aqueous formaldehyde and K_2CO_3 as catalyst, all in water. An example follows. Suggest mechanisms for these reactions.

W49   **W50**; 72% yield   **W51**; 78% yield

Answer 5.14: The simple alkylation to form **W50** uses a 1,3-diketone as a stable non-basic enolate but is otherwise straightforward. The second reaction is more interesting. The chemists suggest an oxetane intermediate **W53**. Indeed they draw a concerted opening of the oxetane without involving the carbonyl group, but we prefer this:

W52 → W53 → W54 → W51

A Question Looking Forward to Asymmetric Synthesis

Problem 5.15: In the synthesis of the enone **47** the textbook says that there is a danger of the aldehyde being racemised. How might his happen? Why is there much less danger of **46** being racemised in base?

Answer 5.15: The aldehyde could racemise by enolate formation if a stronger base were used.[8] The phosphonate could in principle racemise by the same process but the key proton is the third most acidic proton. Even such a weak base as i-Pr$_2$NEt can form the enolate from the ketophosphonate **W55** and the next most acidic proton is that of the amide NH. It is very unlikely that the dianion **W56** would lose a third proton.

References

1. E. J. Corey and A. G. Myers, *J. Am. Chem. Soc.*, 1985, **107**, 5574.
2. E. J. Corey and H. Kigoshi, *Tetrahedron Lett.*, 1991, **32**, 5025.
3. S. M. Abdul Rahman, H. Ohno, T. Murata, H. Yoshino, N. Satoh, K. Murakama, D. Patra, C. Iwaka, N. Maezaki and T. Tanaka, *J. Org. Chem.*, 2001, **66**, 4831.
4. B. A. Chauder, A. V. Kalinin, N. J. Taylor and V. Snieckus, *Angew. Chem., Int. Ed.*, 1999, **38**, 1435.
5. R. Ballini, G. Bosica, D. Fiorini and G. Giarlo, *Synthesis*, 2001, 2003.
6. I. Fleming and D. Lee, *Tetrahedron Lett.*, 1996, **37**, 6929.
7. T. Ben Ayed and H. Amri, *Synth. Commun.*, 1995, **25**, 3813.
8. M. K. Edmonds and A. D. Abell, *J. Org. Chem.*, 2001, **66**, 3747.

6

Choosing a Strategy: The Synthesis of Cyclopentenones

You may find it helpful to have chapter 6 of the textbook open to start with.

Nitroalkenes

In the textbook we described a synthesis of the fused cyclopentenone **12**. The key step is the reaction of the silyl enol ether **20** with the nitroalkene and the immediate product from this reaction (not isolated) is probably **W1** rather than **21**.

Problem 6.1: Suggest mechanisms for the formation and hydrolysis of **W1**. Note that nitroalkanes are not normally hydrolysed to ketones in boiling water. Do you consider the 1,4-diketone **13** a likely intermediate?

Answer 6.1: The first step must surely be the TiCl$_4$-catalysed conjugate addition **W2** of the silyl enol ether **20** to the nitroalkene **16**. The first intermediate is ideally placed for cyclisation to **W3**.

If **W1** hydrolyses to **21** it is difficult to see how the reaction continues. However, intermediate **W1** has an imine bond that could easily hydrolyse to **W4** and hence by attack of water on Si or NO to the hydrate of **13** or **13** itself. The cyclisation of **13** to give **12** is what one would expect so, yes, **13** is a likely intermediate.[1]

Synthesis of Roseophyllin

In Robertson's synthesis of roseophyllin 'events took a twist' when **43** cyclised to **41** rather than the isomer **42** needed for roseophyllin. **Problem 6.2**: How and why does cyclisation of **43** give **41**?

Answer 6.2: No direct aldol gives **41** so we must be more devious.[2] The likeliest cyclisation is of the enolate of the ketone into the aldehyde **W5**. Now formation **W6** of the conjugated enolate (extended enolate, see chapter 11) allows reprotonation **W7** to give the unconjugated enone **W8**. Removal of one of the marked protons from **W8** to give a new extended enolate and reprotonation at the γ-position gives **41**. Why? Under these equilibrating conditions the most stable enone will form: **41** has a trisubstituted alkene rather than the disubstituted alkene in **42**.

One solution was to make **42** from tartaric acid via the acetal **48** and the cyclopentenone **W9**. An excess (2.3 equivalents) of **49** was used to make **50**. **Problem 6.3**: Give mechanisms for the reactions in the conversion of **48** to optically active **W9**.

Answer 6.3: The lithium derivative is alkylated once with either iodide and the excess of **49** forms some **W12** that cyclises rapidly.

The hydrolysis of the acetal in **50** gives **W13** and the dithioacetal is easy to hydrolyse because one of the sulfur atoms has already been oxidised to a sulfoxide. This is a standard procedure often used with dithians. Protonation on oxygen is easier than on sulfur. Finally one of the hydroxyl groups must be eliminated by the E1cB mechanism.[3] Note that it does not matter which OH is lost as **W14** is C_2 symmetric. This is an early example of the chiral pool strategy (chapter 23). The rest of this synthesis appears in the textbook.

d³ Reagents

We said in the textbook: 'the dilactone **60**, generates the acylating agent and the double bond needed for the Friedel-Crafts reaction by acid-catalysed dehydration.' **Problem 6.4**: Draw mechanisms for the conversion of **60** to **61**.

Answer 6.4: You might have drawn mechanisms that varied in detail from the one below but in essence the lactone must open in acid **W15** and eliminate by E1 **W16** or E2. Then the aliphatic Friedel-Crafts reaction occurs, maybe on the acylium ion **W17** but it will be an easy one as it is intramolecular. Finally the whole thing happens again on the other side.[4]

The Synthesis of Modhephene

Problem 6.5: Explain what this statement from the chapter means: 'The product **70** is ideally functionalised for development into modhephene **75**.'

Answer 6.5: The bridge must be added and a methyl group added to the ketone with elimination. These are precisely the places where **70** has functionality. The bridge must be added first as we need conjugate addition and alkylation. This was the sequence used by Oppolzer and Bättig.[5]

The product from conjugate addition is trapped by PhSeBr **W18** and oxidative elimination (chapter 33) then gives the key enone intermediate **W19**. Pyrolysis of **W19** follows an ene reaction to give the correct diastereoisomer of **W20** needed to make **75**. The Alder ene reaction **W22** is like a Diels-Alder reaction with a C–H bond replacing one of the alkenes in the diene component. When intramolecular, as here **W23**, both the position of the new alkene and the stereochemistry of two centres can be controlled. The bridge can be delivered only in a *cis* fashion and the new methyl group has to be to the left (as drawn) since it requires a hydrogen atom from the allylic position, also to the left. The synthesis of modhephene **75** concludes with a Wittig reaction and an acid-catalysed isomerisation of the alkene into the ring.

Other Methods: Merck's MK-0966 by Suzuki Coupling

The enone **150** was made from **147**; X = OEt by the reaction shown here. **Problem 6.6**: Suggest a mechanism. *Hint*: the tertiary amine starts things off by conjugate addition.

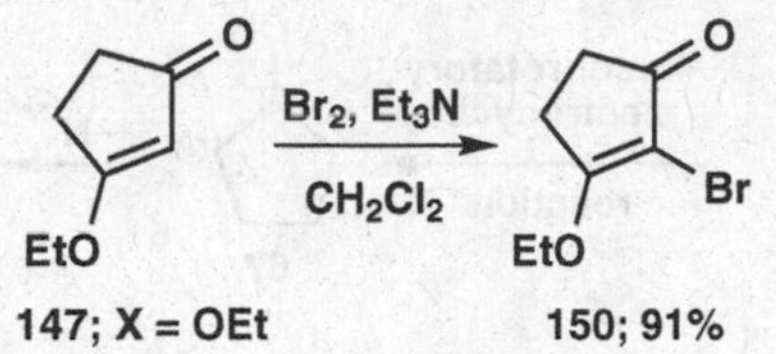

Br2, Et3N
CH2Cl2
EtO
Br
147; X = OEt
150; 91%

Answer 6.6: As hinted, start by conjugate addition of Et_3N **W24** to give an enolate that is trapped **W25** by bromine. Enolisation of the product **W26** allows elimination of Et_3N by the E1cB mechanism. This rather like a Baylis-Hillman reaction.

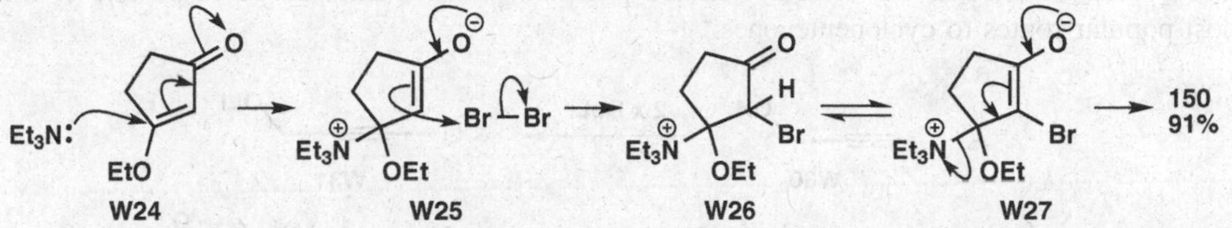

Et3N:
EtO
W24
Et3N
OEt
Br—Br
W25
H
Br
W26
W27
150
91%

Problem 6.7: Two couplings follow: Why was the sulfur atom oxidised after rather than before the first coupling? Suggest a mechanism for the details of the formation of **151**.

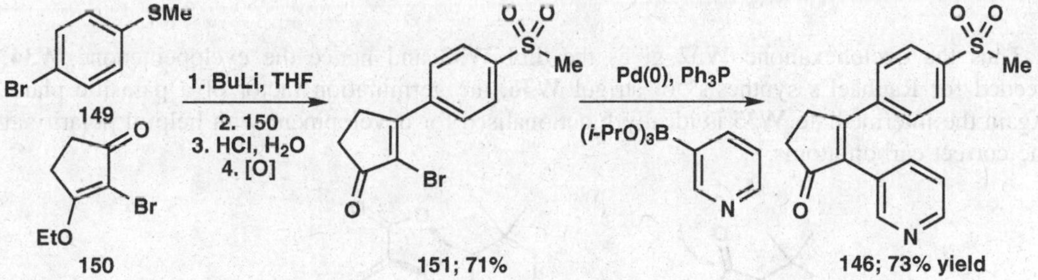

SMe
Br
149
1. BuLi, THF
2. 150
3. HCl, H2O
4. [O]
EtO
Br
150
Me
151; 71%
Pd(0), Ph3P
(i-PrO)3B
N
146; 73% yield

Answer 6.7: Sulfone is a good activator for nucleophilic aromatic substitution and there would be a danger that Br might be displaced by a nucleophile.[6] In addition, BuLi might remove a proton from SO_2Me so there are too many possible side reactions. Lithium–bromine exchange evidently occurs on the benzene ring and the Li derivative adds to the ketone in **150**. Dehydration of the product **W28** gives the enone **W29** and oxidation turns SMe into SO_2Me.

149
BuLi
SMe
Li
Br
EtO
OH
W28
HCl
H2O
W29
[O]
151

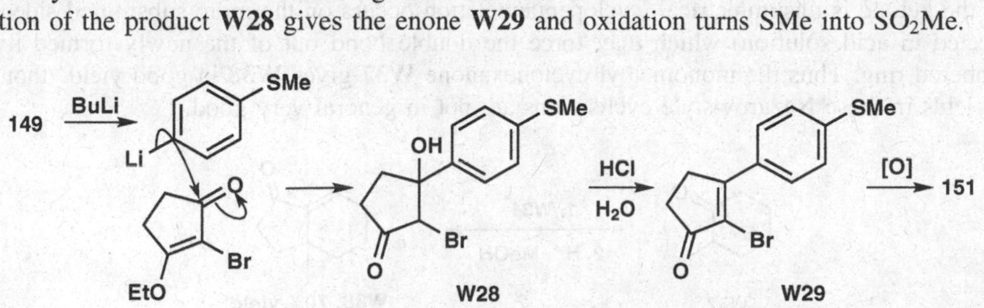

The Nazarov Reaction

The basic Nazarov reaction makes a cyclopentenone **69** from a dienone **65** by acid-catalysed cyclisation **66**.

Here is an alternative strategy that relies on the same basic idea. Instead of having two double bonds, it is possible to pack both degrees of unsaturation into a triple bond. The dilithium derivative **W31** of propargyl alcohol **W30** adds to ketones at the carbon end of the nucleophile forming alkyne-1,4-diols which cyclise in acid to cyclopentenones. Again, the double bond finishes in the more substituted position, and this has proved to be one of the most popular routes to cyclopentenones.[7]

Thus the cyclohexanone **W32** gives the diol **W33** and hence the cyclopentenone **W34**, needed for Raphael's synthesis[8] of strigol **W36**, the germination factor of a parasitic plant. Again the intermediate **W35** is ideally functionalised for development with helpful polarity at the correct carbon atoms.

If the ketone is unsymmetrical, cyclopentannelation occurs on the more substituted side (as expected in acid solution) which may force the double bond out of the newly formed five-membered ring. Thus the monomethyl cyclohexanone **W37** gives **W38** in good yield[9] though the yields in these Nazarov-style cyclisations are not in general very good.

Selectivity in the Nazarov Cyclisation

Problem 6.8: Comment on the fact that the first two examples of this silyl triflate activated Nazarov cyclisation give only one product whereas the last gives a 1:1:1 mixture of three products.

Answer 6.8: It is easy to see how the three products from **W43** are formed. Silylation at carbonyl oxygen gives an pentadienyl cation that cyclises **W47** to give the allylic cation **W48**. Loss of each of the marked protons gives one of the three products, so the mechanism drawn gives **W49**. No proton is lost from the methyl group as this would give an unstable exomethylene compound. The three products are all trisubstituted alkenes of roughly equal stability.[10]

What is different with **W39** and **W41** is clearly the CF_3 group. The intermediate cation **W51** is otherwise similar and is delocalised. However, evidently unsymmetrical **W51** has more positive charge away from the CF_3 group so that the proton lost in **W51** is as shown. It is not surprising that no exomethylene compound is formed but the selective formation of **W42** rather than an isomer corresponding to **W46** is very impressive.

The Tandem Nazarov-Dieckmann Sequence of Shindo

This is the promised exploration of the details of the synthesis[11] of cyclopentenones from ynolates **152**. These enolate-like molecules are made from dibromoesters **W63** by treatment with *t*-BuLi at low temperature and warming to $0\,°C$. Suggest a mechanism for this process.

Suggest two mechanisms for the formation of **W65** and say which you prefer. The addition of **152** to PhCHO (or other aldehydes and ketones) is described as in the paper as a '2 + 2' process. **Problem 6.9**: What do you think this means?

Answer 6.9: Exchange of Li for Br might give **W66** or might give **W67** directly. A second Br/Li exchange gives **W64**. You might prefer to write both **W66** and **W67** with covalent OLi bonds.

On warming to 0 °C, **W64** loses ethoxide **W68** or **W69** to give the ynolate **152**. You might prefer to draw this as the lithium alkynolate **152a** or as the ketene tautomer **152b**.

The two mechanisms are an aldol like addition **W70** followed by cyclisation **W71** or a one-step mechanism **W72**. This is what they mean by a 2 + 2 cycloaddition and it depends on the ketene-like properties of the ynolate.

Problem 6.10: Similarly suggest two mechanisms for the conversion of **154** into **155**. Which do you prefer? Suggest an experiment to tell them apart. Could you use this chemistry to make *cis*-jasmone **1**?

Answer 6.10: There are various possibilities including an S_N1 opening of the ring **W73** followed by an E1 loss of CO_2 or a concerted mechanism **W74**.

The difference is that any stereochemistry in **164** is lost in the cation formation while it should be retained in the concerted mechanism. The stereochemistry of both **164** and **165** is fixed (as *cis*) by the five-membered ring so a good experiment would be to see what happens in an open chain compound where free rotation is possible. This was done[12] by trapping the enolate **W65** in an aldol reaction with benzaldehyde. The product **W76** had defined stereochemistry where it mattered on the four-membered ring but was a mixture of two diastereoisomers at the alcohol. Thermal elimination preserved the stereochemistry in the resulting alkene **W77** so the concerted mechanism **W74** is strongly suggested.

Synthesising *cis*-jasmone **1** by this route simply requires making **154** with R as the jasmone side chain. This requires **W63** with the same side chain. In fact Shindo made only dihydrojasmone by this route: presumably it is too difficult to prepare the necessary dibromoester when there is an alkene in the molecule. Did you find a solution?

Problem 6.11: Suggest mechanisms for the various stages in this synthesis of prostaglandin B1 **W86** and explain any selectivity. What strategy is being used to build the cyclopentenone?

Answer 6.11: Lithiation occurs on the methyl group of the phosphonate **W81** and it is acylated twice by the diester. Then the weak base (K_2CO_3) makes the anion (enolate) at one end of **W82** which cyclises **W87** to the ketone at the other end in a Horner-Wadsworth-Emmons olefination (chapter 15). Alkylation gives **W84** and the very stable anion **W88** reacts with **W85** to give PGB1 **W86**. The cyclopentenone is built in the conversion of **W82** to **W83** by the aldol strategy with the necessary 1,4-dicarbonyl compound being bought as the diester.[13]

de Meijere's Chromium Carbonyl Chemistry

Problem 6.12: There is some discussion about the mechanism[14] of the formation of **161** but you should be able to give a mechanism for the second step.

Answer 6.12: Hydrolysis of the acetal to give **W89** is obvious but the enol ether will also be hydrolysed either to the ketone or the enol which then cyclises on an aldol reaction **W90**.

These last two examples remind us that, while there are many ingenious ways to make masked 1,4- or 1,6-dicarbonyl compounds, the aldol strategy remains one of the very best ways to make enones, especially when it is intramolecular.

References

1. M. Miyashita, T. Yanami and A. Yoshikoshi, *J. Am. Chem. Soc.*, 1976, **98**, 4679.
2. J. Robertson, R. J. D. Hatley and D. J. Watkin, *J. Chem. Soc., Perkin Trans. 1*, 2000, 3389.
3. K. Ogura, M. Yamashita and G. Tsuchihashi, *Tetrahedron Lett.*, 1976, 759; A. G. Myers, M. Hammond and Y. Wu, *Tetrahedron Lett.*, 1996, **37**, 3083.
4. P. E. Eaton and R. H. Mueller, *J. Am. Chem. Soc.*, 1972, **94**, 1014.
5. W. Oppolzer and K. Bättig, *Helv. Chim. Acta*, 1981, **64**, 2489.

6. W. C. Black, C. Brideau, C.-C. Chan, S. Charleson, N. Chauret, D. Claveau, D. Ethier, R. Gordon, G. Greig, J. Guay, G. Hughes, P. Jolicoeur, Y. Leblanc, D. Nicoll-Griffith, N. Ouimet, D. Riendeau, D. Visco, Z. Wang, L. Xu and P. Prasit, *J. Med. Chem.*, 1999, **42**, 1274.
7. M. Ramaiah, *Synthesis*, 1984, 529, table 19, 544.
8. G. A. MacAlpine, R. A. Raphael, A. Shaw, and A. W. Taylor, *J. Chem. Soc., Perkin Trans. 1*, 1976, 410.
9. T. Hiyama, *J. Am. Chem. Soc.*, 1979, **101**, 1599; *Bull. Chem. Soc. Jpn*, 1981, **54**, 2747.
10. J. Ichikawa, M. Fujiwara, T. Okauchi and T. Minami, *Synlett.*, 1998, 927.
11. M. Shindo, Y. Sato and K. Shishido, *J. Org. Chem.*, 2001, **66**, 7818.
12. M. Shindo, *Tetrahedron Lett.*, 1997, **38**, 4433.
13. M. Mikolajczyk, M. Mikina and R. Zurawinski, *Pure Appl. Chem.*, 1999, **71**, 473.
14. H. Schirmer, F. J. Funke, S. Müller, M. Noltemeyer, B. L. Flynn and A. de Meijere, *Eur. J. Org. Chem.*, 1999, 2025.

B
Making Carbon–Carbon Bonds

7

The *Ortho* Strategy for Aromatic Compounds

The first nine questions in this chapter concern directed lithiation and this chapter is longer than most in this workbook. This is because directed lithiation is likely to be the least familiar method so far introduced. **Problem 7.1**: How would you synthesise the aromatic acid **W1** using an *ortho*-lithiation strategy?

Answer 7.1: Given that we know we are going to use an *ortho*-lithiation strategy, the first thing to do is perhaps to look for *ortho*-directors. There are two *ortho*-directing groups in this molecule and they are the same, which is good. We can remove both the carboxylic acid group and the methyl group using an *ortho*-lithiation strategy. However, considering that the presence of an acid complicates matters – since it would react with the BuLi – it is perhaps sensible to remove that first. Thus we gain compound **W2**. We can now remove the methyl group to yield the methylated catechol **W3**.

In the synthesis we do not need to worry about which *ortho* site lithiates since they are equivalent. Once the methyl group has been introduced one site is blocked and there is one site only remaining for *ortho*-lithiation. The carboxylate is introduced there.

Problem 7.2: Suggest how you would convert **W4** into the aldehyde **W5** using an *ortho*-lithiation strategy.

Answer 7.2: We are going to need an *ortho*-lithiation and the lithiated aromatic compound must react with something like Me_3SiCH_2Cl. Something else has happened too – the acid has been reduced to an aldehyde.

The acid **W4** could be converted to oxazoline **W7** in the first instance *via* the acid chloride **W6**. We now have a powerful *ortho*-director (the oxazoline) and a fair director flanking the site we wish to lithiate so there should be no problem. Following the introduction of the electrophile comes the conversion of the oxazoline to the aldehyde. This can be done[1] by first reacting the oxazoline with Me_3O^+ BF_4^- to give the oxazolinium salt **W9** and then reducing it to the corresponding aldehyde with $NaBH_4$.

Problem 7.3: Suggest a synthesis for **W10**. If you need a clue, a possible starting material, with substituents in the correct positions in comparison with the product, is **W11**.

Answer 7.3: The dihydrofuran **W10** looks suspiciously like a cycloadduct of a benzyne **W12** and furan which is, of course, what it is. If we imagine the reverse cycloaddition then the two starting materials that we get are benzyne **W12** and furan itself.

The question now is how to make the benzyne **W12**. One way to make benzynes is by elimination (loss of LiBr or LiF). In this case the leaving group is a triflate (instead of bromide or fluoride) and a silyl group is removed *ortho* to the leaving group with a fluoride ion. There are two arrangements of the triflate and the silyl group that would lead to benzyne formation but we will place the TMS group as it appears in **131** to form **133** – the precursor to the benzyne.

All we have to worry about now is how we make **W13** from our starting material **W11**. The answer is to use an anionic Fries rearrangement **W14**. The phenol **W15** that results[2] from the anionic Fries rearrangement can be deprotonated and reacted with triflic anhydride to give **W13**. When **W13** is treated with TBAF in the presence of furan, elimination **W16** gives the benzyne **W12** and hence **W10** in 63% yield.

Problem 7.4: A question from chapter 7 in the textbook. In the synthesis of the quinolone **134**, only two of the four fluorine atoms can be displaced by nucleophiles and of those two, one was displaced in preference to the other. Explain.

Answer 7.4: The answer to the first part of this question is not too difficult. For the fluorine atoms to function as leaving groups we need an additional electron-withdrawing group which is a ketone in this case. In the synthesis the nucleophile attacks **W17** to give the stable anion **W18** and elimination of fluoride restores the aromatic ring **W19**.

A similar sequence **W20** and **W21** leads to *ortho* products. However, the π-electrons in the aromatic ring cannot delocalise into the carbonyl group if a nucleophile attacks either of the other carbon atoms with a fluorine atom attached such as **W22**.

It is a little more tricky to justify why one of the two fluorine sites is more reactive than the other. The simplest explanation is that steric hindrance, particularly with a large nucleophile such as the piperazine, hinders attack at the *ortho* position. **Problem 7.5**: Suggest how you would synthesise phenol **W24** starting from the carbamate **W23**. Phenol **W24** was used in the synthesis of ochratoxin B **W25**.

Answer 7.5: Just before we begin our analysis, note that there is a big clue here[3] in that we end up with a phenolic OH.

Analysis: Right, first of all we can do the fairly obvious disconnection and hydrolyse the ester to give the diacid **W26**. Before we go any further we now do three FGIs. Here is why.

Since it is easier to do *ortho*-lithiation reactions with amides we convert both of the acids to amides. We then need to introduce the side chain. We might imagine that this could be done by reacting a lithiated ring with propene oxide (disconnection **W27a**). Instead, we do a third FGI and convert the alcohol to an alkene. We can then do the next disconnection **W27b** which removes an allyl side chain. Now we are faced with the phenol **W28**.

If you did not spot it, we need to use the anionic Fries rearrangement here again. Notice also that we need to introduce a carbonyl on both sides of the hydroxyl group. The disconnection **W28a** is a reverse anionic Fries rearrangement and takes us back to **W29**. We can use to carbamate to introduce the amide using an *ortho*-lithiation.

Synthesis: The first step is to lithiate the carbamate **W23**. The lithium derivative is not allowed to warm up but reacted with diethylcarbamoyl chloride to give **W29**. Notice in fact, that if the lithiated derivative *had* been allowed to warm up and so *had* rearranged, then the electrophile would have reacted with the phenoxide to give the *same* product as when rearrangement did not occur. Thus the whole synthesis would use the anionic Fries rearrangement twice. The new amide **W29** is lithiated again and the carbamate is more potent than the amide at directing the lithiation. This time the lithiated derivative is allowed to warm up so that the anionic Fries rearrangement occurs. The phenol is alkylated with methyl iodide to give **W30**.

One more *ortho*-lithiation (and the two sites on the benzene ring are equivalent) followed by a reaction with allyl bromide leads to product **W31**. And then in one step using 6M HCl, the two amides are hydrolysed and the double bond undergoes an electrophilic addition reaction to give our product **W24** in 50% yield.

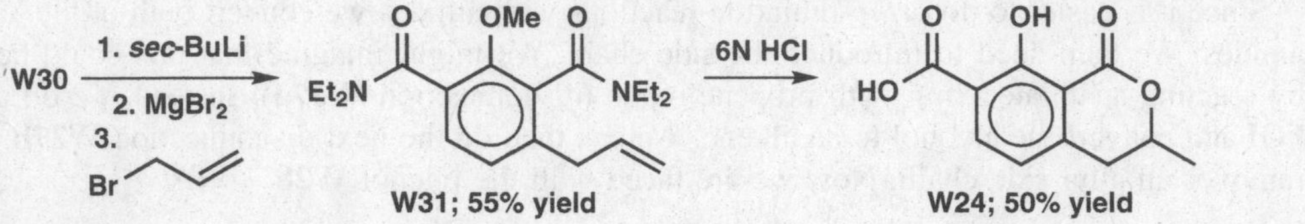

W31; 55% yield

W24; 50% yield

Problem 7.6a: When oxazoline **W32** is treated with MeI followed by CsF and acrylonitrile the adduct **W33** is obtained as a mixture of diastereoisomers. Suggest a mechanism.

Problem 7.6b: If you think you understand what is going on, predict what would happen under the same conditions with **W34**.

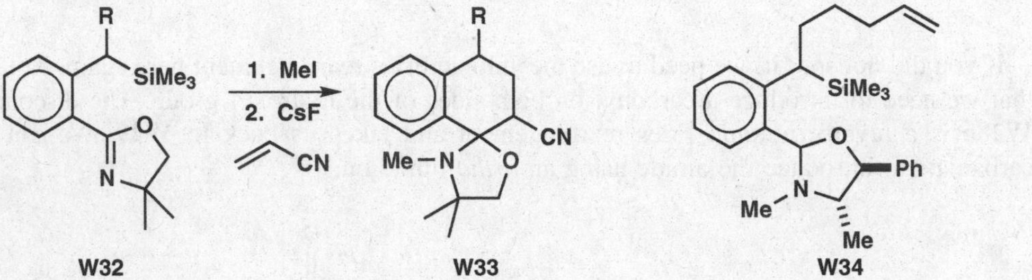

Answer 7.6a: First the nitrogen atom is methylated by MeI. The fluoride ion from CsF will then remove the silyl group to generate an anion **W35**. We could view the resulting species as a zwitterion, but it is more useful to think of it **W36** as a reactive tetraene.[4] Aromaticity is disrupted when the tetraene is formed so, when generated in the presence of acrylonitrile, it will do a Diels-Alder reaction **W37** to give **W33**.

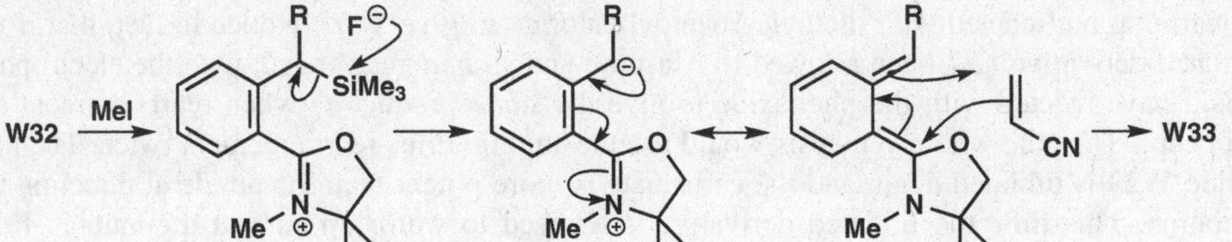

Answer 7.6b: One difference with **W34**, is that, unlike **W32**, the five-membered ring contains no unsaturation. Thus, when the methylated compound **W38** is treated with CsF, the five-membered ring is sprung open to give the reactive intermediate **W39**. An intramolecular cycloaddition then gives compound **W40** with a diastereomeric excess of 28–55%.

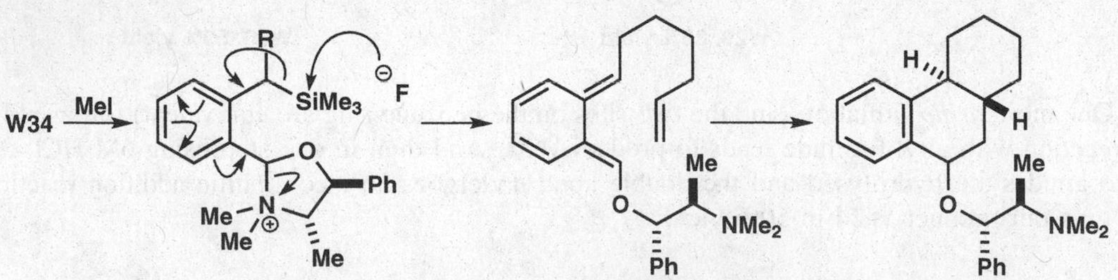

Problem 7.7: Suggest how oxazoline **W41** may be converted into lactone **W42**.

Answer 7.7: We can use an aromatic nucleophilic substitution reaction to introduce the substituted benzene ring. This nucleophilic substitution is a reaction which both *ortho*-methoxy oxazolines and *ortho*-fluoro oxazolines can do. The first reaction[4] forms the biaryl **W44** from **W41** and lithiated veratrole **W43**.

We now need to introduce the carbinol part of the final lactone and we need to get rid of most of the carbon of the oxazoline ring. The methoxy groups of **W44** are *ortho*-directors too, of course, but not as good as the oxazoline. Lithiation is followed by formylation using DMF and the aldehyde is reduced *in situ* using $NaBH_4$ to give **W45**. Finally, the oxazoline is hydrolysed and the lactone formed in 6M HCl.

Problem 7.8: Suggest a synthesis for amine **W46**.

Answer 7.8: Yes? So, you noticed that there is no *ortho* relationship between the two substituents. If this is a problem and you need a big clue, consider this: in **Problem 7.3** we saw how **W11** could be used to make **W10**. We can use similar chemistry here. Think about this before reading on.

Nucleophiles can add to benzynes **W47**. They can add to either end of the benzyne but attack which leads to a *meta* substituted product **W49** will give the more stabilised anionic intermediate **W48** – thermodynamic control.

For the compound we want to make, all we need to do is have the nucleophile around when the benzyne forms. The nucleophile used[5] was silylated *N*-methylaniline **W51**. Presumably, under the desilylating reaction conditions, it is the *N*-methylaniline anion which attacks the benzyne.

Problem 7.9: Suggest a synthesis for buflavine **W52**. Naturally your synthesis should include *ortho*-lithiation as one of the key reactions. If you find that this is simply too difficult, the first line of the answer contains a clue and the starting material of this several-step synthesis.

Answer 7.9: The clue is that the bond linking the two aromatic rings was formed by using *ortho*-lithiation and by using another key reaction which meditates the formation of biaryls. The starting material was the amide **W53**. This is a hard question and there will be several ways the synthesis could be done so it is unlikely that you came up with the precise synthesis used by Snieckus.[6]

Analysis: The first thing to realise is that neither of the methoxy groups is going to be much use. They are in an *ortho* relationship to each other but nothing else. The nitrogen atom is going to direct an *ortho*-lithiation instead. The retrosynthetic strategy first breaks up the eight-membered ring and then breaks the biaryl C–C bond. First we do FGA to form enamine **W54**. This double bond could be formed by Peterson olefination.

Next we will cleave the biaryl bond. The two important functional groups that we generate are the boronic acid **W57** and a bromide **W56** – we have in mind a palladium catalysed reaction here. The boronic acid group has to be introduced by some means and *ortho*-lithiation is one method. In order to direct the lithiation we do a FGI to convert the benzylic amine into an amide **W58**. The amide will do the job better.

We hope you notice that there may well be a problem of regioselectivity in the lithiation of **W58**. One of the *ortho* positions is also *ortho* to a methoxy group and therefore we would expect this position to be doubly activated. However, we want the lithiation to occur at the *other* site which is not *ortho* to a methoxy group.

Synthesis: This problem was overcome by using isopropyl ether **W59** as the starting material instead. The proton *ortho* to the isopropoxy group is now much more hindered. The isopropyl group will eventually have to go but we will worry about that later. Amide **W59** was lithiated using *tert*-BuLi and TMEDA and the lithiated intermediate reacted with $B(OMe)_3$. Notice that regiocontrol was achieved. The palladium-catalysed reaction is a Suzuki-Miyaura reaction and the biaryl **W61** was formed in 70% yield. The next stage, the formation of the double bond is more than just an ordinary Peterson olefination. The two silyl groups do two different jobs. The first one is picked off by fluoride to give an anion which attacks the aldehyde. The molecule is then set up to do an ordinary Peterson olefination (chapter 15) because the negative oxygen atom can attack a silyl group in the usual way to give a double bond.

The skeleton of the molecule is now complete but a few more manipulations have to be done to finish off the synthesis. Firstly the carbon–carbon double bond is reduced using hydrogen on Pd. The amide is reduced to the amine using LiAlH$_4$ and then of course there is the isopropyl group. This is selectively removed using the ether cleaving reagent, BCl$_3$ to give **W63**. Finally reaction with sodium hydride and methyl sulfate gives the target material **W52**.

The following problems do not necessarily involve *ortho*-lithiation but may require some reactions in the other part of textbook chapter 7. **Problem 7.10**: Suggest a synthesis for compound **W64**.

Answer 7.10: *Analysis 1:* Many syntheses are possible – the one we had in mind was the introduction of the nitro group at the end and the formation of the ether **W65** by addition of an alcohol to an alkene **W66** or **W67**. This reaction should be easy since a tertiary carbocation would be an intermediate.

Of the two possible alkenes **W66** or **W67**, the former is a possible product of a Claisen rearrangement **W66a** and this would solve the *ortho* selectivity problem inherent in both compounds.

Now we need to remove the allyl portion from the allyl ether **W68a** and consider its synthesis from two possible allylic halide precursors **W69** or **W70**. Either should give **W68** because the alkene has more substituents than **W66** and is *inside* the six-membered ring.

Synthesis 1: We shall leave the synthesis of **W69** or **W70** to you. The synthesis by this approach is straightforward. Notice the weak base (K_2CO_3) used in the first alkylation because of the acidity of the phenol. The stereochemistry of the acid catalysed cyclisation to give **W65** requires no control – the dihydrofuran ring is flat and much prefers a *cis* ring fusion to the flexible six-membered ring.

Analysis 2: If you preferred the alternative alkene **W67**, you cannot use the Claisen rearrangement but an *ortho*-lithiation approach looks good with the oxygen atom available as director. The electrophile could be a simple ketone **W72** as dehydration will almost certainly give the most substituted alkene.

Synthesis 2: The ketone **W72** is the one we have used extensively in discussions on regioselective enolisation (chapter 3) and we have discussed the lithiation of anisole in this chapter. Even if the dehydration does not give **W67** alone, the conditions of the acid-catalysed cyclisation are likely to equilibrate the various alkenes and the five-membered ring is the most stable product (thermodynamic control).

So far as we are aware, neither of these syntheses has been tried out in practice. You may well have suggested a better one. You may, for example, have thought of starting with the nitro group in place (*p*-nitrophenol as starting material) and thereby ensuring that the *para* position was blocked. This is all right for the Claisen approach, but probably not for the *ortho*-lithiation as you might get nucleophilic addition to the benzene ring. **Problem 7.11**: Discuss the possible products which might be obtained from a Fries rearrangement on aspirin **W73** and develop a synthesis for the Glaxo-SmithKline (GSK) anti-asthma drug salbutamol **W74**, available as Ventolin.

W73; aspirin   W74; salbutamol (GSK's Ventolin)

Answer 7.11: The received wisdom about the Fries rearrangement on phenyl esters (see chapter 7 in the textbook) is that they give predominantly *ortho* products in nonpolar solvents and mainly *para* products in polar solvents. This simple argument would predict **W75** from aspirin in nonpolar and **W76** in polar solvents.

W73; aspirin   W75; *ortho*-product   W76; *para*-product

However, we should also consider the effect of the carboxylic acid group on the reaction. The acidic proton might interfere with the reaction by, for example, destroying the catalyst and we might suggest protection as an ester. It will also affect the regioselectivity of the reaction. It is a *meta*-director and both **W75** and **W76** have the new ketone group *meta* to CO_2H, so that should be all right. It looks as though we should be able to control the reaction. Glaxo-Wellcome certainly can in their salbutamol synthesis.

Analysis: The idea of a giant pharmaceutical company using the cheap product of another (Baeyer) in the synthesis of one of their own best sellers is charming, and that is what happens here. The intermediate must clearly be **W76** rather than **W75**. Disconnection of *t*-BuNH$_2$ from the ketone **W77** allows us to make **W78** by bromination of **W76**.

W74 salbutamol (GSK's Ventolin)   W77   W78

We might have preferred to make the amine by reductive amination or reaction of *t*-BuNH$_2$ with an epoxide, but neither of these approaches is compatible with the use of aspirin as starting material. We might also be concerned that the bromination of a ketone such as **W76** could also occur on the benzene ring, but it is doubly deactivated by two carbonyl groups.

Synthesis: The Fries rearrangement gives the *para* product **W76** in nitrobenzene as the polar solvent. The remainder is straightforward though the order of events, and the point at which the reductions should be done, need some thought. It turned out that the carboxylic acid did not interfere with the Fries rearrangement but did interfere with the bromination so it was protected as a methyl ester **W79**.

W73 aspirin $\xrightarrow[\text{PhNO}_2]{\text{AlCl}_3}$ W76 $\xrightarrow[\text{HCl}]{\text{MeOH}}$ (MeO$_2$C ... HO **W79**) $\xrightarrow[\text{CHCl}_3]{\text{Br}_2}$ (MeO$_2$C ... HO **W80** Br)

Direct displacement on **W80** with *t*-BuNH$_2$ was not satisfactory – we should not be surprised at this since reactions of primary amines with organic halides often result in multiple substitution. It was necessary to use an *N*-benzyl derivative **W81**, prepared by reductive amination, to control this step. The hindered tertiary amine **W82** does not react further.

t-BuNH$_2$ $\xrightarrow[\text{NaCNBH}_3]{\text{PhCHO}}$ (HN **W81** Ph) $\xrightarrow{\text{W80}}$ (MeO$_2$C ... HO **W82** N Ph)

Reduction of both carbonyl groups occurred with LiAlH$_4$ and the benzyl group was finally removed by catalytic hydrogenation. It is not very likely that GSK use two protecting groups and a LiAlH$_4$ reduction in a manufacturing process for a highly successful drug, but they are understandably coy about revealing what they actually do. This is a published lab synthesis from GSK so at least it works.

W82 $\xrightarrow{\text{LiAlH}_4}$ (HO ... HO OH N Ph **W83**) $\xrightarrow{\text{H}_2, \text{Pd/C}}$ (HO ... HO OH N H **W74; salbutamol (GSK's Ventolin)**)

In addition, the presence of a carboxylic acid group in the intermediates gave GSK a chance to resolve the compound and show that one enantiomer is more active than the other. This resolution approach to single enantiomers is discussed in chapter 22. **Problem 7.12**: Suggest a synthesis for the lactone **W84**, using some method to make a new C–C bond *ortho* to the oxygen substituent.

W84

Answer 7.12: *Analysis:* Opening the lactone **W84a** reveals the problem: we must introduce a new C–C bond *ortho* to the OH group **W85**.

An *ortho*-lithiation approach would require an S_N2 displacement by **W86** on an allylic halide **W87** at an unlikely α position in the presence of a carboxylic acid derivative. It is much more likely that any attempt to prepare such a derivative **W87** would give instead the more stable conjugated **W88** and, even if it were possible to prepare **W87**, that it would react at the γ position.

The fact that we are discussing controlling the regioselectivity of allylic halides should suggest a Claisen rearrangement approach. Reversing the Claisen rearrangement on **W85**, redrawn and 'ketonised' as **W89**, reveals **W90** as the starting material.

Starting material **W90** is a simple allyl phenyl ether and would come from the reaction of phenol with stable allylic isomer **W88** at the favourable γ position.

Synthesis: It is necessary to use the ester **W89**; R = Me for three reasons: the ester can be made by direct bromination of the methyl ester **W91**, the free carboxylic acid would be deprotonated during the alkylation step, and the acid might interfere with the Claisen rearrangement. Alkylation then requires only a weak base (K_2CO_3) because of the acidity of the phenol and occurs regioselectively at the γ position of **W88**; R = Me to preserve conjugation.

The Claisen rearrangement goes in good yield and, as a bonus, the closure of the five-membered lactone occurs under the same conditions to give **W84** directly.[7]

Problem 7.13: 2-Aryl propionic acids **W93** are much in demand[8] as nonsteroidal anti-inflammatory drugs such as the best selling ibuprofen **W94** (in Britain) and naproxen **W95** (in the USA). However, they are awkward to make. Simple alkylation of aromatic rings under Friedel-Crafts conditions fails because the intermediate cation is destabilised by the carbonyl group.

Your task is to propose a synthesis for an *ortho*-substituted aryl propionic acid for evaluation as an anti-inflammatory agent. You might consider how you would do this for an *ortho* group based on oxygen such as **W96**; X = OH or OMe and an *ortho* group based on a carboxylic acid such as **W96**; X = CONMe$_2$.

Answer 7.13: Two strategies seem promising: the Claisen rearrangement for X = OR and the *ortho*-lithiation strategy for both types of compounds. Consider the Claisen strategy first.

Claisen analysis: We cannot put the side chain in by a simple Claisen rearrangement but we can put the carbon skeleton in place and make the carboxylic acid by oxidative cleavage of the alkene **W97**. In reversing the Claisen, we must remember to invert the allylic side chain to get **W98**.

Claisen synthesis: Though each part of this synthesis is known separately, we are not aware of any attempts to put the whole scheme together for the synthesis of anti-inflammatory drugs. The first two stages are quite general.

An example is the dimethyl substituted compound **W99** which gives **W100** in 78% yield.[9]

The oxidation of the alkene has been performed on at least one compound and that gave ibuprofen in a patent.[10] There is every expectation that other methods too would give the acids in good yield.

ortho-*Lithiation analysis:* We might here look back to compound **71** in the chapter and consider a similar reagent for the electrophilic addition of the propionic acid side chain. Alkylation of the lithium derivative must be carried out in such a way that there is a methyl group in the right place.

The allylic bromide **W104** will probably react at the wrong end (the primary rather than the secondary end). Since we have no interest in what is on the end of the double bond we can either make the allylic system symmetrical **W105** or put a more double bond stabilising group than a methyl group on the other end. Possibilities include **W106–7**.

ortho-*Lithiation synthesis:* Choosing an oxazoline as the *ortho*-director **W108** and the symmetrical allylic halide **W105** as the electrophile, the synthesis looks straightforward, though we know of no examples.

The oxazoline can be converted into a number of other acid derivatives as explained in the chapter. You might well have preferred to use a tertiary amide instead. In making compounds such as **W96**; X = OR, you might have preferred a carbamate as *ortho*-director.

Double *Ortho*-Lithiations

We have seen several syntheses that have used *sequential* lithiations but we have not looked at situations where more than one lithiation is done *at the same time*. The use of dilithium derivatives is actually very rare in the realm of *ortho*-lithiation but mono *ortho*-lithiation of two aromatic rings has been used as a synthetic strategy.

The Synthesis of a Phosphepin

Analysis: The phosphepin **W111** was required in enantiomerically pure form. We disconnect between the benzene ring and the phosphorus atom on *both* sides. We will need a double phosphorus electrophile and a doubly lithiated species **W112**. Species **W112** looks viable – the lithium atoms are next to one good *ortho*-director (fluorine) and one mediocre director (the oxygen atom of the benzylic ether).

There are two chiral centres in the molecule. Notice that the first disconnections avoided bonds which involve the chiral centres. Breaking a bond to a chiral centre in a retrosynthesis demands that we consider the stereochemical control of that centre. Lithiated diether **W112** could be made from diether **W113** and the diether from diol **W114**.

Diol **W114** can be made from difluorostilbene **W115** in enantiomerically pure form, using the Sharpless asymmetric dihydroxylation reaction (chapter 25). The double bond of difluoro-stilbene could be synthesised in a variety of ways but a symmetrical disconnection across the double bond leads to two molecules of *meta*-fluorobenzaldehyde **W116**.

Synthesis: The synthesis[11] starts with the formation of the stilbene **W116** using a McMurry coupling (chapter 15). This stilbene is dihydroxylated enantioselectively to give the diol **W114** in a 97% yield and with >99% ee. The diol was deprotonated with NaH and then alkylated with MeI to give the diether **W113**.

A double lithiation is then achieved using *sec*-BuLi (the authors do not find it necessary to use TMEDA with their *sec*-BuLi). There are two positions on each ring where lithiation could occur. Since the fluorine atoms presumably acidify both of the protons *ortho* to themselves we might expect to see some lithiation at position 4 as well as position 2. Lithiation at position 4 is not observed. The oxygen atom, which is ineffective at directing a lithiation on its own, works together with the fluorine atom to achieve very good regioselection. A double nucleophile will need a double electrophile and PhPCl$_2$ is used and the phosphorus atom in the new seven-membered ring is oxidised using H$_2$O$_2$.

References

1. H. Reuman and A. I. Meyers, *Tetrahedron*, 1985, **41**, 837.
2. M. P. Sibi and V. Snieckus, *J. Org. Chem.*, 1983, **48**, 1937; K. Shankaran and V. Snieckus, *Tetrahedron Lett.*, 1984, **25**, 2827; V. Snieckus, *Chem. Rev.*, 1990, **90**, 898 (see scheme 20).

3. M. P. Sibi, S. Chattopadhyay, J. W. Dankwardt and V. Snieckus, *J. Am. Chem. Soc.,* 1985, **107**, 6312.
4. H. Reumann and A. I. Meyers, *Tetrahedron*, 1985, **41**, 854.
5. K. Shankaran and V. Snieckus, *Tetrahedron Lett.*, 1984, **25**, 2827.
6. P. A. Patil and V. Snieckus, *Tetrahedron Lett.*, 1998, **39**, 1325.
7. T. Padmanathan and M. U. S. Sultanbawa, *J. Chem. Soc.,* 1963, 4210.
8. J.-P. Rieu, A. Boucherle, H. Cousse and G. Mouzin, *Tetrahedron*, 1986, **42**, 4095.
9. D. McHale, S. Marcinkiewicz and J. Green, *J. Chem. Soc. (C)*, 1966, 1427.
10. J.-P. Rieu, A. Boucherle, H. Cousse and G. Mouzin, *Tetrahedron*, 1986, **42**, 4095.
11. P. Wyatt, S. Warren, M. McPartlin and T. Woodroffe, *J. Chem. Soc., Perkin Trans. 1*, 2001, 279.

8

σ-Complexes of Metals

The most striking example from the start of the chapter in the textbook was the preparation[1] and reaction of enantiomerically pure Grignard reagent **30**. The first few questions in the workbook consider the details of the reactions. The first stage involves the reasonably diastereoselective chlorination of the sulfoxide **27** by NCS **W2** to give a (6.4:1) mixture of **28** and **W1**. **Problem 8.1**: Explain this reaction as far as you can.

27; 99% ee
Ar = *p*-chlorophenyl

28; 97% ee   6.4:1   W1   W2; NCS

Answer 8.1: NCS can react as a source of radicals or electrophilic chlorine. The presence of base suggests an ionic reaction occurs here.[2] Only in the presence of K_2CO_3 was there some (about 6.5:1) diastereoselectivity. Crystallisation gave the pure major diastereoisomer. It is not easy to explain diastereoselectivities such as this. Perhaps chlorination at sulfur is followed by ylid formation and migration of Cl from S to C.

The next step is the creation, in turn, of two enantiomerically pure Grignard reagents, **30** and **31**, both stable at −78 °C. **Problem 8.2**: Suggest mechanisms for these reactions, defining and explaining the stereoselectivity.

28; 97% ee   29; 99% yield 96% ee   30; stable at −78 °C   31; not isolated stable at −78 °C

Workbook for Organic Synthesis: Strategy and Control   Paul Wyatt and Stuart Warren
© 2008 John Wiley & Sons, Ltd

Answer 8.2: As the sulfur atom is inverted in the first step, presumably nucleophilic attack occurs here. But how does the Mg atom get into position in **30**? Retention at the chiral C atom suggests concerted electrophilic substitution, something like **W5**.

Now the new Grignard reagent combines with five molecules of EtMgCl to give the simple enantiomerically pure secondary Grignard reagent **31**. **Problem 8.3**: Suggest reasons why they chose EtMgCl for this exchange.

Answer 8.3: Chloride exchanges more slowly than other halides and using the same reactive Grignard for both steps prevents the formation of other Grignards.[3]

Problem 8.4: Suggest details of how the trapping of this Grignard with $PhSCH_2N_3$ actually occurs. Other reagents for electrophilic nitrogen are described in chapter 33.

Answer 8.4: The azide is clearly an electrophile and attack by the Grignard must occur with retention of configuration on the terminal nitrogen atom. Something like this perhaps:

The Synthesis of B-Raf Kinase Inhibitors for the Treatment of Cancer

The Novartis company wished to make a series of heterocyclic compounds such as **W9** and **W10** as potential anti-cancer drugs.

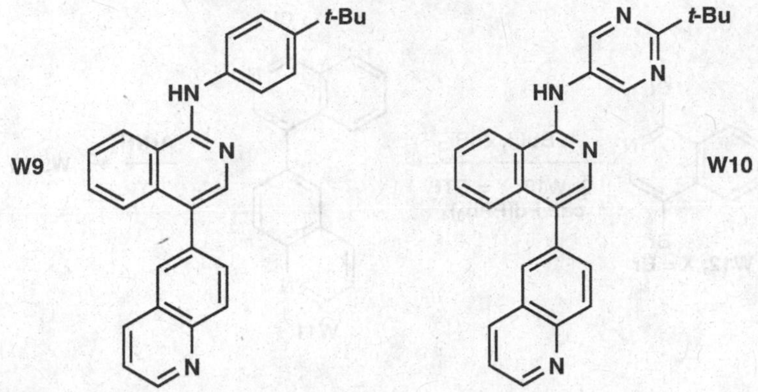

W9 W10

Since the lower two-thirds of the molecules is the same, it makes sense to disconnect that part **W9/10a** using the favourable nucleophilic substitution on an isoquinoline **W11**. The lower third could then also be disconnected using some kind of transition metal coupling. One of **W12** or **W13** must carry a Br or OTf and the other a metal.

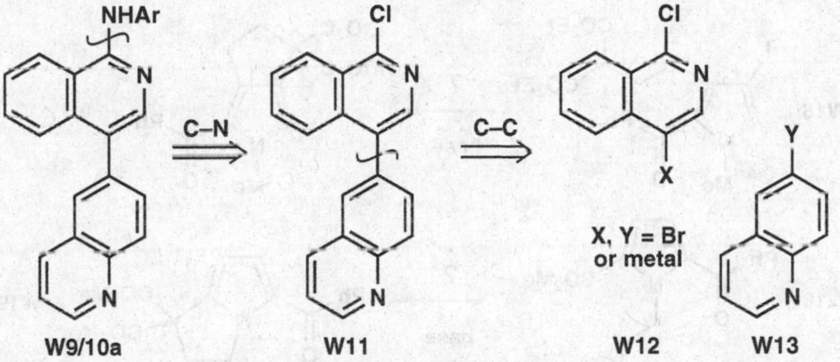

W9/10a **W11** **W12** **W13**

X, Y = Br or metal

It turns out that available **W14** can be converted into **W12** on a large scale via the N-oxide **W15**. This chemistry is explored in chapter 32.

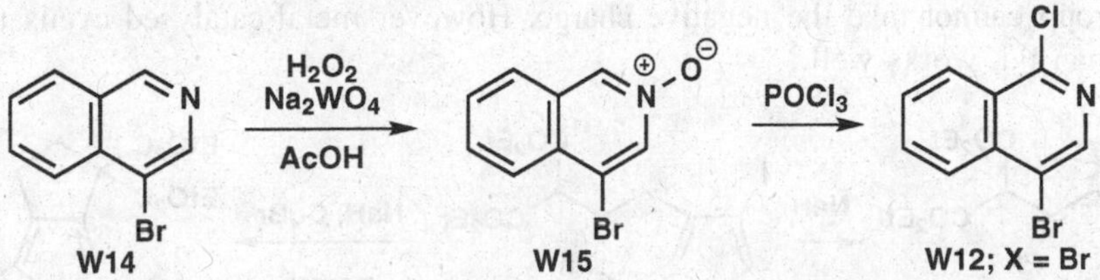

W14 **W15** **W12; X = Br**

Coupling to the triflate **W13**; Y = OTf also works well by the Negishi method. This requires Br/Li exchange on **W11**, exchange of the Li for Zn and Pd-catalysed coupling. Lithiation, either by Br/Li exchange or by directed lithiation, is an important preliminary to exchange for Li with other metals. All these reactions were carried out at Novartis on a multi-kilo scale.[4]

Contrasting Aromatic Nucleophilic Substitution with Palladium-Catalysed Couplings

Each of these reactions can be accomplished either by normal nucleophilic substitution or by metal-catalysed coupling. **Problem 8.5**: Assess the chances of success with each method.

Answer 8.5: The malonate enolate from **W16** cannot cyclise onto the pyrrole **W20** because the carbonyl group cannot take the negative charge. However metal-catalysed cyclisation, using Cu rather than Pd, works well.[5]

However, the ketone in **W18** is in exactly the right position to allow normal cyclisation **W21** and no transition metal is needed.[6] The sulfone is a poorish leaving group but this does not matter as the elimination **W22** is the fast step.

Lithiation of Heterocycles

This section uses material out of chapters 7 and 8. **Problem 8.6**: Which metal σ-complexes are formed in this sequence and why? Why does **W24** get brominated at that position (NBS supplies electrophilic Br rather than radicals in this case)? Give details of the conversion of **W25** into **W26**.

Answer 8.6: The first lithiation is on nitrogen as the NH proton is the most acidic. The product is **W27**. Lithiation of **W28** occurs by Br/Li exchange and the product is the aryl-Li **W28**.

Indoles generally react with electrophiles in the 3 position, especially here as the TBDMS substituent is so large. The intermediate cation retains the aromaticity of the benzene ring. Finally the lithium derivative **W28** reacts with EtI and TBAF removes the TBDMS group to give **W26**. This is a reliable *Organic Syntheses* procedure giving good yields.[7]

Finally the lithium derivative **W28** reacts with EtI and TBAF removes the TBDMS group to give **W26**.

A Synthesis of Pyralomycin

The naturally occurring antibiotic pyralomycin **W30** (R is a sugar derivative) can be made from the aglycone pyralomycinone **W31**. A synthesis by Tatsuta and his group[8] contains an interesting acylation of a lithium σ-complex.

W30; pyralomycin   W31; pyralomycinone

Double chlorination of pyrrole with NCS and protection gave the dichloro compound **W33** that could be brominated with NBS at one of the remaining sites to give **W34**, thus illustrating the high reactivity of pyrrole.

pyrrole   W32   W33; 90% yield   W34; 80% yield

The substituted benzene half **W38** started with the lithiation of **W35** between the two MeO directing groups (chapter 7) and carboxylation with dry ice to give the acid **W36**. Chlorination occurred at the position activated by the two MeO groups to give **W37**. Now the acid chloride **W38** was made. Note that SO_2Cl_2 is electrophilic at Cl and is a chlorinating agent while $SOCl_2$ is electrophilic at sulfur.

W35   W36; 72% yield   W37; 92% yield   W38; 90% yield

The two halves were coupled together by Br/Li exchange on **W34** and coupling with **W38**. It is unusual to make ketones this way: Weinreb amides are often used to prevent further acylation by the ketone product. However in this case the diaryl ketone is very unreactive as both aryl groups are electron-donating. One of the OH groups was freed by chemoselective deprotection with BBr_3: the low yield of **W40** is due to some deprotection of the other MeO group. Deprotection at N occurred under these conditions.

W39; 76% yield   W40; 56% yield

Treatment with base cyclised **W40** by displacement of one Cl from the pyrrole, activated by the new ketone. This was evident in the preference of the product **W41** for the enol form.

The mixture could be demethylated under more vigorous conditions with BBr$_3$ to give **W31** in 70% yield.

Problem 8.7: Are the two forms of **W41** aromatic? Draw a mechanism for the cyclisation of **W40** and explain the formation of the enol form of **W41**.

Answer 8.7: Yes, both forms are aromatic with $4n + 2$ delocalised electrons. There are various ways to count the electrons: for the enol it is probably best to count 10 electrons in rings A + B and six in ring C. For the ketoform, counting each ring separately is probably best. If you count all the electrons in both forms you get 14 so that is all right too.

You should of course cyclise **W42** onto the unsaturated ketone **W42** to give intermediate **W43**. Elimination of chloride from this enolate gives the keto form of **W41** directly, but equilibration to the rather stable nitrogen anion allows elimination **W44** to give the enol directly. There are other versions of this.

An Enantioselective Fluorescence Assay Using Zn, Cu and Pd

Three types of metal σ-complex are involved in this synthesis of a twisted diamine to be used in fluorescence determination of enantiomeric excess. One starting material was made from the simple bromobenzene **W45** by nitration and reduction to give the amine **W47**. The Br atom was used to direct the nitration and is removed by formation of a zinc σ-complex and reaction with acid.

A copper-catalysed coupling with *o*-chloro benzoic acid **W49**, resembling the Pd-catalysed Buchwald-Hartwig chemistry described in the textbook, joins the amine to the other aromatic ring **W50**. Cyclisation with POBr₃ gives the key intermediate **W51**.

A copper-catalysed coupling scheme:

HO₂C / Cl (W49) → W48, Cu, Cu₂O → *t*-Bu ... N-H ... HO₂C (W50; 95% yield) → POBr₃ → *t*-Bu ... N, Br (W51; 85% yield)

Two molecules of **W51** are joined to the dibromonaphthalene **W53** to give the fluorescent compound **W54** in a sequence involving σ-complexes of Li, Sn and Pd. Stille coupling via the tin compound **W52** is not very efficient as the yield is only 28% but the compound was made.[9]

W51 → 1. BuLi 2. Me₃SnCl → =Ar, SnMe₃, *t*-Bu, N (W52; 91% yield) + Br Br (W53) → Pd(PPh₃)₄, CuO → Ar Ar (W54)

Problem 8.8: Explain the regioselectivity on the formation of **W46** and **W51**. Attempt a drawing of the final product **W54**.

Answer 8.8: Both *t*-Bu and Br are *ortho*-directing but *t*-Bu is also much larger than Br so nitration of **W45** occurs *ortho* to Br and *meta* to *t*-Bu. In the cyclisation both N and *t*-Bu direct to the same sites but the site between the two is very hindered sterically. Drawing **W54** is very difficult as the two Ar groups cannot be in the same plane. As long as you can convey the twist in the molecule, your diagram is acceptable. The authors suggest something like this with the two Ar groups parallel and more-or-less orthogonal to the naphthalene.

3D shape of W54

Lithiated Alkynes

One type of metal σ-complex not discussed in the chapter is nevertheless very important. In the synthesis[10] of **W55**, Merck's reverse transcriptase inhibitor for treatment of HIV, disconnection of the heterocycle reveals and acetylenic alcohol **W56** most obviously made by addition of the cyclopropyl alkynyl anion synthon **W58** to the very electrophilic trifluoromethyl ketone **W57**.

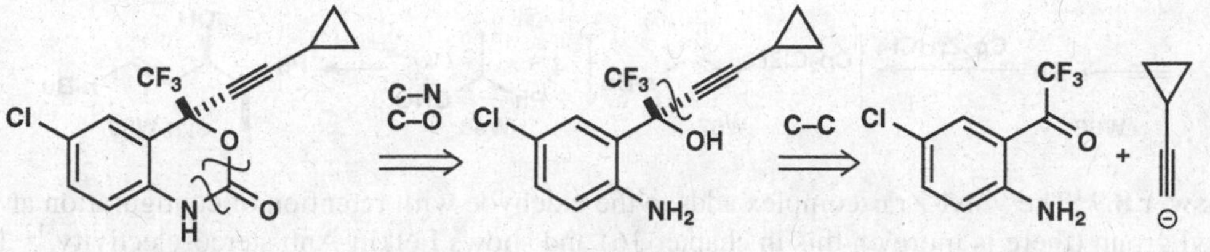

W55 W56 W57 W58

Acetylene gives an easily made σ-complex **W59** with BuLi that selectively reacts with the bromide end of 1-chloro-3-bromopropane to give the new acetylene **W60**. Notice that in both these reactions acetylenic σ-complexes are too stable to exchange Li with Cl or Br.

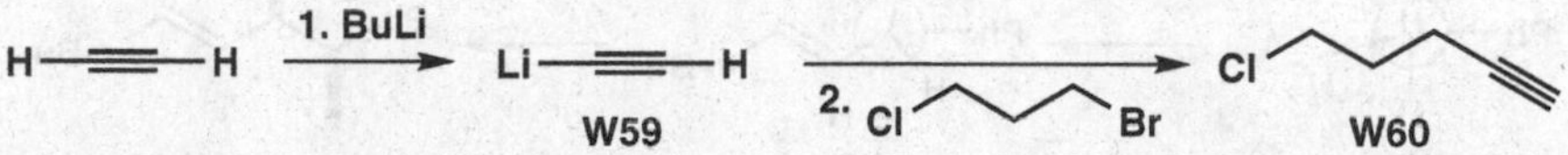

W59 W60

Reaction of **W60** with two molecules of BuLi generates first the Li-acetylide **W61** and then the propargyl-lithium **W62** that can be drawn as an allene derivative **W63**.

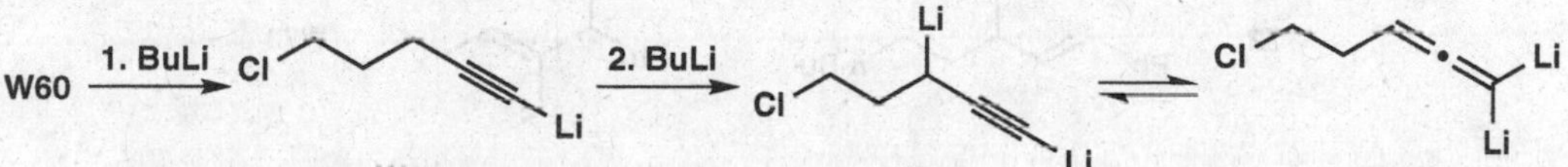

W61 W62 W63

However you draw it, this dilithium species now cyclises **W62a** to form, after quenching the intermediate **W58a** with aqueous ammonium chloride, the three-membered ring compound **W64** needed for the next reaction.

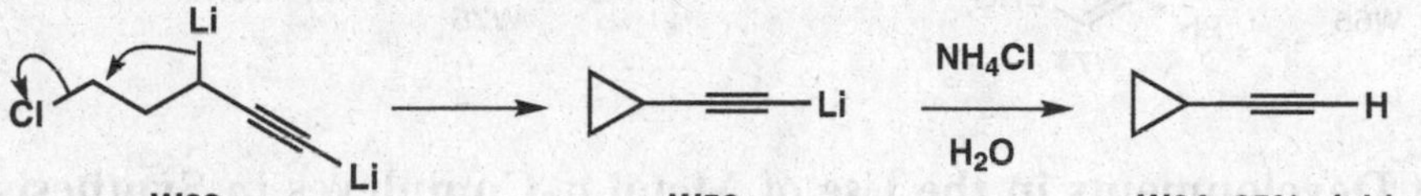

W62a W58a W64; 65% yield

However, if the reaction is quenched with ethyl trifluoroacetate, the acetylenic ketone **W65** is formed in 60% yield[11] from **W60** suggesting that a similar trap might lead directly to **W56**.

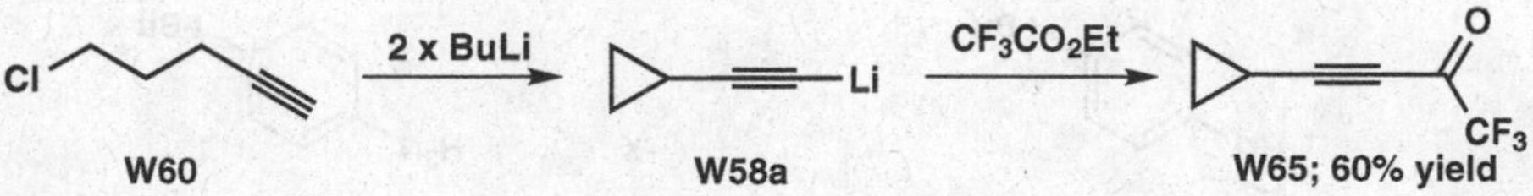

W60 W58a W65; 60% yield

Hydrozirconation in Synthesis

Acetylenes are also involved in the hydrozirconation route to vinyl σ-complexes as described in the textbook. They react well with aldehydes to give *E*-allylic alcohols such as **W69** in good yield (92%) and, in this case, with good selectivity (85:15 *syn:anti*). **Problem 8.9**: Suggest an explanation.

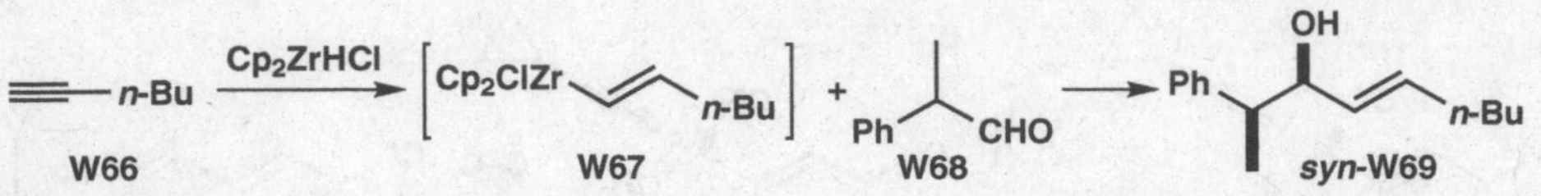

Answer 8.9: The vinyl-Zr σ-complex adds to the aldehyde with retention of configuration at the vinyl group (there is more on this in chapter 16) and shows Felkin-Anh stereoselectivity.[12] The best conformation for the addition **W70** has the largest group (Ph) orthogonal to the carbonyl group while the nucleophile (the vinyl-Zr σ-complex) approaches alongside the smallest group (H) as shown by the dashed arrow. This gives **W71** having OH and Me on the same side of the main chain.

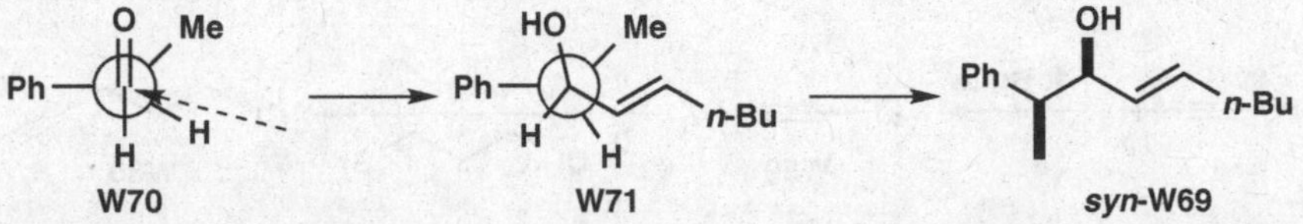

Problem 8.10: Suggest how this method could be used to make these allylic alcohols:

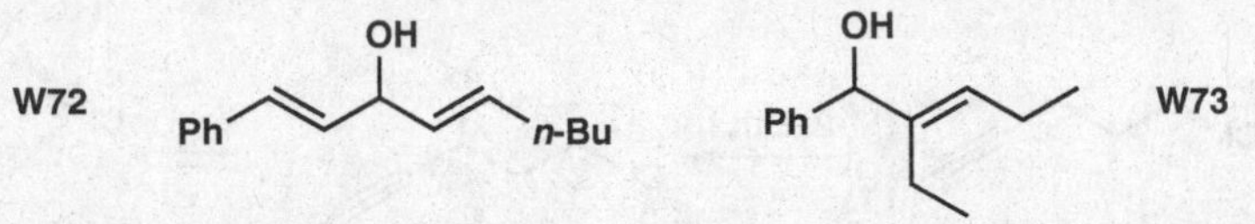

Answer 8.10: You have a choice in the first case **W72**: the method reported in the paper uses **W66** and **W74**. In the second case, symmetrical hex-3-yne **W75** must be used with benzaldehyde as the electrophile.[12]

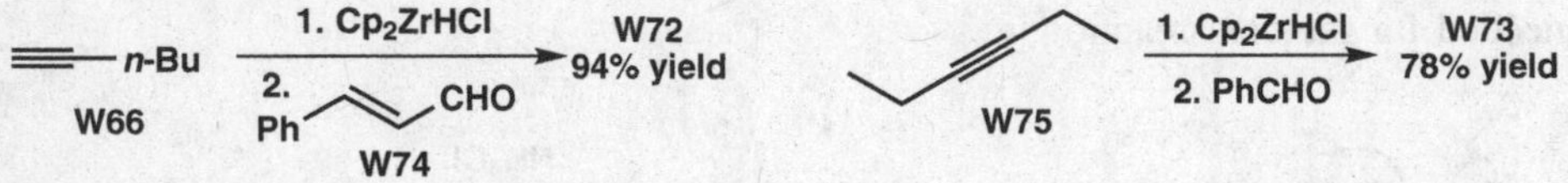

Recent Developments in the Use of Metal σ-Complexes in Synthesis

The most widespread developments are with Pd σ-complexes and a good illustration is the large scale synthesis of Novartis anti-cancer drugs[13] such as **W76**. Strategic bond disconnection suggests starting materials **W77–79** where the three substituents X, Y, and Z are different enough to allow for selectivity.

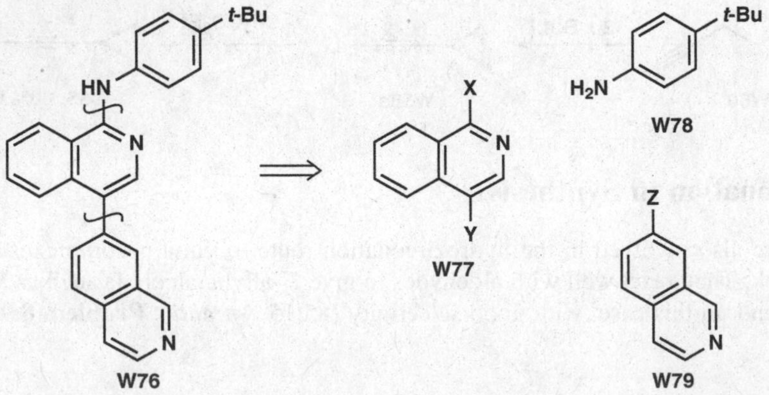

Both Suzuki and Negishi couplings were successful: the Negishi route involved selective formation of Li, Zn and Pd σ-complexes from **W77**; X = Cl and Y = Br followed by coupling with the triflate of **W79**. The aromatic amine was inserted by normal nucleophilic aromatic substitution activated by the isoquinoline nucleus.

However, another analogue demanded coupling of **W81** with the amino-pyrimidine **W84** that was synthesised[14] by Buchwald coupling of the bromo-pyrimidine **W82** with an ammonia equivalent. Pd catalysis is needed because the nitrogen atoms do not activate the ring in the right place for displacement of bromine. Benzophenone imine was used. Other effective ammonia equivalents include $LiN(SiMe_3)_2$, Ph_3SiNH_2 and even $LiNH_2$.

Even coupling of primary amines to aromatic rings has problems but the use of Cu can help.[15] The related bromide **W86** can be coupled to amine **W85** in 85% yield with Cu(I) and the simple ligand **W88**. The *p*-methoxybenzyl group can be removed to make the primary amine.

Coupling Grignard Reagents to Alkyl Halides

All organic chemists must have wished at some time that they could couple a Grignard reagent to an alkyl halide and make any C–C bond they wished. The problem is of course exchange of metal between the two species. With catalysis from another metal, Pd or Fe, the coupling is now possible.[16] Alkyl halides can be coupled with Grignards of the type RMgBr under catalysis by Pd with ligand tricyclohexylphosphine (Cy_3P) in *N*-methylpyrrolidinone (NMP) as solvent. The alkyl halide should be a chloride to minimise Grignard exchange. Benzyl chloride gives 100% yield, simple alkyl and aryl chlorides excellent yields and even alkyl chlorides with ester, nitrile or acetal groups work well.

W89 → W90 ; W91; NMP

RCl, NMP
4 mol% Pd(OAc)₂, 4 mol% Cy₃P

Catalysis by iron as the complex $[Li(TMEDA)]_2[Fe(CH_2CH_2)_4]$ also works well and the range of more functionalised examples is impressive. Bromides and iodides as well as chlorides can be used. Allylic halides may react with rearrangement (chapters 12 and 19). We give two examples: conditions are the same for both.[17]

W92 → W93; 87% yield   W94 → W95; 84% yield

5 mol%
catalyst
PhMgBr

References

1. R. W. Hoffmann, B. Hölzer and O. Knopff, *Org. Lett.*, 2001, **3**, 1945; R. W. Hoffmann and P. G. Nell, *Angew. Chem., Int. Ed.*, 1999, **38**, 338; R. W. Hoffmann, B. Hölzer, O. Knopff and K. Harms, *Ibid.*, 2000, **39**, 3072.
2. T. Satoh, T. Sato, T. Oohara and K. Yamakawa, *J. Org. Chem.*, 1989, **54**, 3130.
3. R. W. Hoffmann, B. Hölzer, O. Knopff and K. Harms, *Angew. Chem., Int. Ed.*, 2000, **39**, 3072.
4. M. Bänzinger, J. Cercus, H. Hirt, K. Laumen, C. Malan, F. Spindler, F. Struber and T. Troxler, *Org. Process Res. Dev.*, 2006, **10**, 70.
5. F. J. Lopez, M.-F. Jett, J. M. Muchowski, D. Nitzan and C. O'Yang, *Heterocycles*, 2002, **56**, 91.
6. F. Franco, R. Greenhouse and J. M. Muchowski, *J. Org. Chem.*, 1982, **47**, 1682.
7. M. Amat, S. Hadida, S. Sathyanarayana and J. Bosch, *Org. Synth.*, 1997, **74**, 248.
8. K. Tatsuta, M. Takahashi and N. Tanaka, *Tetrahedron Lett.*, 1999, **40**, 1929.
9. C. Wolf, S. Liu and B. C. Reinhardt, *Chem. Comm.*, 2006, 4242.
10. A. S. Thompson, E. G. Corley, M. F. Huntington and E. J. J. Grabowski, *Tetrahedron Lett.*, 1995, **36**, 8937.
11. E. G. Corley, A. S. Thompson and M. Huntington, *Org. Synth.*, 2000, **77**, 231.
12. P. Wipf and W. Xu, *Org. Synth.*, 1997, **74**, 205.
13. D. Denni-Dischert, W. Marterer, M. Bänziger, N. Yusuff, D. Batt, T. Ramsey, P. Geng, W. Michael, R.-M. B. Wang, F. Taplin, R. Versace, D. Cesarz and L. B. Perez, *Org. Process. Res. Dev.*, 2006, **10**, 70.
14. X. Huang and S. L. Buchwald, *Org. Lett.*, 2001, **3**, 3417.
15. F. Y. Kwong and S. L. Buchwald, *Org. Lett.*, 2003, **5**, 793.
16. A. C. Frisch, N. Shaikh, A. Zapf and M. Beller, *Angew. Chem., Int. Ed.*, 2002, **41**, 4056.
17. R. Martin and A. Fürstner, *Angew. Chem., Int. Ed.*, 2004, **43**, 3955.

9

Controlling the Michael Reaction

Problems from Chapter 9 in the Textbook

Problem 9.1: A revision question. Reagent **70** was used in the construction of the double electrophile[1] **W6**. Draw mechanisms for reactions **B**, **E** and **F**.

Answer 9.1: Step **B** is an unusual aldol between a malonic acid and formaldehyde. Evidently the stable enol of the malonic acid adds to very reactive formaldehyde **W7** and elimination and decarboxylation **W8** follow. Both reactions might involve cyclic proton transfer.

Step **E** is acetal hydrolysis and spontaneous hemiacetal formation. Cyclic hemiacetals are often more stable than the hydroxy aldehyde. The Wittig reaction uses one equivalent of the ylid to open the hemiacetal and the other to do the typically *trans*-selective Wittig with a stabilised ylid (chapter 15). **Problem 9.2**: Suggest how **80** might be elaborated into **72**.

Workbook for Organic Synthesis: Strategy and Control   Paul Wyatt and Stuart Warren
© 2008 John Wiley & Sons, Ltd

Answer 9.2: Various jobs have to be done. The exomethylene group must be oxidised away and the ketone reduced stereoselectively to an alcohol **W9**. The nitro group must be reduced to an amine and we might expect cyclisation to an amide **W10** to be spontaneous. The two-carbon bridge between the nitrogen atom and the benzene ring must be inserted and the amide reduced to the amine **72**.

Various orders of events are possible but the published synthesis[2] follows this route. The nitro group was partly reduced (to NHOH) by Zn dust and finally to the amide **W11** with TiCl₃. Conjugate addition to vinyl sulfoxide gave **W12** and a Pummerer rearrangement gave a mixture of diastereoisomers of the cyclised product **W13**. The sulfur could be removed in a radical reaction with Bu₃SnH and the alkene cleaved with OsO₄/NaIO₄ to give the amido ketone **W14**. All that remained was stereoselective Luche reduction of the ketone and complete reduction of the amide.

Problem 9.3: The explanation for the stereoselectivity in the formation of **90** is complex and you are referred to the paper by Martin and collegues[3] if you are interested. However, you can explain both of the first two selectivities, i.e. why only one of the two free OH groups reacts and only conjugate addition occurs.

Answer 9.3: The other OH group would form a four-membered ring if it did conjugate addition and, though a six-membered ring would be theoretically possible by direct attack on the carbonyl group, this is practically impossible across the *trans* alkene. By contrast, the OH group that does attack is perfectly placed to form a stable six-membered ring by conjugate addition in a favourable 6-*exo-trig* cyclisation.

Problem 9.4: Suggest reagents and more details for the formation of **99** from **98**. What must be done to get the purine in correctly to make neoplanocin A **92** if a Mitsunobu procedure is used?

Answer 9.4: The sulfur atom needs to be oxidised to the sulfoxide so that *syn*-elimination can occur regioselectively **W15** followed by selective deprotection to give **99**. Formation of **92** requires reduction of the ester group to the primary alcohol, controlled protection and inversion of the OH by oxidation to the enone and stereoselective reduction. Then the Mitsunobu gives the right stereochemistry.[4]

Problem 9.5: The stereoselectivity observed in this aldol reaction to give **102** is related to those we discussed in chapter 4. Can you explain the relative stereochemistry of the aldol? You are not expected to explain the absolute control. This will be discussed in chapter 28.

Answer 9.5: Aldol stereochemistry is related to the geometry of the enol(ate), here the silyl enol ether **101**. The problem suggests that you ignore the absolute control so we simply need to explain the *anti*-aldol stereochemistry (i.e. Me and OH *anti* in **102**). The propenyl side chain *chooses* to go equatorial in the Zimmermann-Traxler transition state **W17** and the methyl group on the enol must be equatorial as it is *anti* to the OSiMe$_3$ group in the enolate.[5] Hence the *anti* relative stereochemistry of **102**.

Problem Not Directly Derived from Chemistry in the Textbook

Problem 9.6: Deduce the structures of **W19** and of the intermediates in the rest of the scheme. Explain any regio- and stereoselectivity.

Answer 9.6: Conjugate addition of the vinyl copper and trapping of the resulting enolate gives the silyl enol ether **W19**. Methyl lithium releases the lithium enolate, alkylated by **W21** to give **W22** hydrolysed in aqueous acid to the keto-phosphonate **W23**. An intramolecular Horner-Wadsworth-Emmons reaction (chapter 15) completes the synthesis[6] of **W20**.

Prostaglandin Synthesis by Conjugate Addition and Enolate Trapping

The classic examples of three-component synthesis with enones are in the field of prostaglandins.[7] Assembly of three consecutive *anti* chains (as in **W27**) can be accomplished by adding a cuprate to a suitably functionalised cyclopentenone **W24** and trapping the resulting enolate with a suitable electrophile. Stork[8] used an optically active cuprate to resolve the enone and added the top chain as CH_2O in the three-component step. A more modern version,[9] appropriately described as 'an extremely short way to prostaglandins' uses the optically active cyclopentenone **W24** and the optically active copper complex **W25** followed by the addition of the whole of the top chain to the tin enolate to give **W26** only a step away from PGE$_2$ **W27**.

Posner's synthesis[10] of guaiene analogues **W28** illustrates these methods dramatically as two copper-catalysed additions appear. This group of natural products has a five- and a seven-membered ring fused together with several branches. The obvious aldol disconnection removes the seven-membered ring and suggest that the specific enolate required for the aldol reaction might be made by a cuprate addition to the enone **W30**.

Disconnection of **W30** at the branchpoint suggest two possible cuprate approaches. Addition of a d^3 reagent **W35** to the trienone **W34** would require selective attack at the middle double bond of the three and we know of no way to do this. The alternative **W30b** requires addition of a vinyl cuprate to the terminus of the dienone **W33** and looks more promising. This strategy needs the cyclopentenone **W31** and we can use a method developed in chapter 6 for this.

The synthesis is shown in detail on the chart. The first Michael addition is accomplished with a copper-catalysed Grignard reaction with the protected vinyl halide and does indeed go at the most remote position. The second is a cuprate addition and the developing enolate is trapped by silylation. The final aldol can be carried out on the acetal **100**, providing $TiCl_4$ is used as a Lewis acid catalyst, to give **W29**.

Nor does this preclude three-component reactions, as in the addition of cyanide-stabilised carbanions, e.g. from **W39**, followed by alkylation to give mixtures of diastereoisomers of **W40** in good yield.[11]

An Unusual Acetal Disconnection

The polycyclic compound **W41** was needed as an intermediate in the synthesis of the even more complex 'azaspiracid' natural products.[12] The obvious acetal disconnection to give the diketo-diol **W42** is not very appealing because there are two ketones that might make acetals and because there is a sensitive N, O-acetal at the other end of the molecule that might unravel under the acidic conditions required.

A successful alternative was based on a double Michael disconnection of the same two C–O bonds hoping that the two OH groups in **W43** would add Michael-fashion to the ynone. This turned out to be very simple.

The silyl protected molecule **W44** was simply treated with fluoride and cyclisation occurred to give **W41** in 85% yield. Draw a mechanism for the two Michael additions: we suggest you ignore the left hand half of the molecule and represent it as **W43a**.

The fluoride removes both silyl groups, maybe one at a time, to give an intermediate such as **W45** that can do conjugate addition to the ynone. After protonation of the enolate a second conjugate addition of the other oxygen anion to the enone **W46** gives the skeleton of **W41**. Addition of either oxygen first gives the same product but they suggest that this is the order of events.

A Double Michael Addition Where Stereochemistry Does Not Matter

Problem 9.7: In a partial synthesis of the taxane skeleton, Nah and colleagues[13] added cyanide twice to the dienone **W47** and developed the product into the polycyclic dione **W52**. Suggest mechanisms for all the reactions. Compound **W48** is formed as a mixture of diastereoisomers. What stereochemistry is necessary for conversion of **W50** to **W51**? Why does it not matter that **W48**, **W49**, and **W50** do not exist solely as the 'right' isomer?

Answer 9.7: For cyclisation to occur the two carboxylic acids must be *cis* but the *trans* compound can equilibrate *via* enols. The conformations of **W51** and **W52** will have chair six-membered rings with a 1,3-diaxial bridge.

cis-W50   conformation of W51   conformation of W52

Formation of the taxane ring system involves interesting chemistry. An enol ester **W53** derived from **W52** cyclises photochemically with cyclohexene to give **W54** which fragments by retroaldol reaction with KOH to give the required eight-membered ring **W55**.

W53; R = CO₂Bn   W54; R = CO₂Bn   W55

Michael Reactions with Complex Enolates: The Synthesis of Vertinolide

Vertinolide **W56** is a naturally occurring tetronic acid.[14] Among other possible disconnections, the dienone portion suggests an aldol reaction from an enolate of **W57** and crotonaldehyde. Further disconnection of protected **W57** reveals a possible Michael reaction between **W58** and some reagent for the extended enolate synthon **W59**.

vertinolide *Analysis*

W56; vertinolide   crotonaldehyde   W57

Protected W57   W58   W59; d³ synthon
                                    γ-extended enolate

The initial Michael reaction worked rather well but subsequent aldols were low-yielding.

W60   LDA   →   W61   →   W58   protected W57; 75% yield

1. LDA
2. croton-aldehyde   poor yield of W56

Problem 9.8: What are the structures of **W61** and the intermediate in the attempted aldol reaction? Why is it surprising that an intermediate of this kind undergoes a successful Michael addition but not a successful aldol reaction with crotonaldehyde? What special property does **W61** have that explains why Michael addition is, after all, successful?

Answer 9.8: Removal of the γ-proton from **W60** by LDA gives an extended lithium enolate **W61** while kinetic deprotonation of the methyl group of protected **W57** gives a simple lithium enolate **W62**. Lithium enolates are usually bad at Michael additions as they are highly reactive and basic but for the same reasons are usually good at aldol reactions. Intermediate **W61** is unusual as it is aromatic (furan) making it less basic, less reactive and better at Michael reactions.

The best synthesis[14] so far reported is even more strange. The lithium enolate **W61** from **W60** does a superb Michael reaction on a trienone **W63 to** give a near-quantitative yield of vertinolide. **Problem 9.9**: Define the selectivity involved in this reaction and offer an explanation.

Answer 9.9: The trienone **W63** has four electrophilic sites: the carbonyl group, two a³ sites and an a⁵ site. It prefers one of the a³ sites as we might have expected from the previous result. This could not easily be predicted but it looks as though steric hindrance plays a part.

Remarkable Stereochemical Control in Michael Additions

A recent report[15] reveals that the lithium enolate **W65** from methyl dithioacetate **W64** can be persuaded to add to 4-silylated cyclohexenone **W66** with either *cis* or *trans* selectivity according to the conditions. In this way, either lactone *cis*-**W67** or *trans*-**W67** can be made at will.

Clearly some explanation is needed! It is the temperature of the Michael addition that makes the difference. At $-78\,^\circ$C the expected *trans* addition occurs to give the dithioester **W68** in good yield. Removal of sulfur using Hg(II) gives the normal methyl ester **W69** and cyclisation in acid (HCl in aqueous MeCN) gives *trans*-**W67** in 73% yield.

If the reaction is allowed to warm up to room temperature before quenching with ammonium chloride solution, a totally different product is formed: the bicyclic compound **W70**. A great deal has happened here – conjugate addition with the opposite stereoselectivity, cyclisation onto the carbonyl group and, most mysteriously, migration of MeS also onto the carbonyl group.

Spivey suggest that the Michael addition is reversible at higher temperatures and that the *cis* adduct **W71** can form a new lithium dithioenolate **W72** that eliminates MeSLi to form a thioketene **W73** that cyclises with addition of MeS– to give **W70**. This is all remarkable stuff and we quote it mainly to show you that new discoveries are being made all the time even in such well established reactions.

References

1. A. Marfat and P. Helquist, *Tetrahedron Lett.*, 1978, 4217.
2. C. Jousse-Karinthi, C. Riche, A. Chiaroni and D. Desmaële, *Eur. J. Org. Chem.*, 2001, 3631.
3. H. J. Martin, M. Drescher, H. Kählig, S. Schneider and J. Mulzer, *Angew. Chem., Int. Ed.*, 2001, **40**, 3186.
4. M. Ono, K. Nishimura, H. Tsubouchi, Y. Nagaoka and K. Tomioka, *J. Org. Chem.*, 2001, **66**, 8199.
5. M. Braun, B. Mai and D. Ridder, *Eur. J. Org. Chem.*, 2001, 3155.
6. W. K. Bornack, S. S. Bhagwat, J. Ponton and P. Helquist, *J. Am. Chem. Soc.*, 1981, **103**, 4647.
7. J. S. Bindra and R. Bindra, *Prostaglandin Synthesis*, Academic Press, New York, 1977, pp. 99–144; A. Mitra, *The Synthesis of Prostaglandins*, John Wiley & Sons, Ltd, New York, 1977, pp. 247–277.
8. G. Stork and M. Isobe, *J. Am. Chem. Soc.*, 1975, **97**, 6260; see also *t*-BuMe$_2$Si protected version by T. Tanaka, S. Kurozumi, T. Toru, M. Kobayashi, S. Miura and S. Ishimoto, *Tetrahedron*, 1977, **33**, 1105.
9. M. Suzuki, A. Yamasigawa and R. Noyori, *J. Am. Chem. Soc.*, 1985, **107**, 3348.

10. G. Posner, *Tetrahedron Lett.*, 1978, 4205.

11. E. Hatzigrigoriou, M.-C. Roux-Schmitt, L. Watski and J. Seyden-Penne, *Tetrahedron*, 1983, **39**, 3415.

12. C. J. Forsyth, J. Hao and J. Aiguade, *Angew. Chem., Int. Ed.*, 2001, **40**, 3663.

13. H. Nah, S. Blechert, W. Schnick and H. Jansen, *Angew. Chem., Int. Ed.*, 1984, **23**, 905.

14. K. Takabi, N. Mase, M. Nomoto, M. Daicho, T. Tauchi and H. Yoda, *J. Chem. Soc., Perkin Trans. 1,* 2002, 500.

15. A. C. Spivey, L. J. Martin, D. M. Grainger, J. Ortner and A. J. P. White, *Org. Lett.*, 2006, **8**, 3891.

10

Specific Enol Equivalents

Problems from Chapter 10 in the Textbook

You will need chapter 9 of the textbook for the first question and chapter 10 for the next few questions. **Problem 10.1**: Look through the reactions in chapter 9 of the textbook that involve a specific enol or enolate and say what controls the enol(ate) component in each reaction.

Answer 10.1: In most cases the specific enol(ate) was formed by conjugate addition. Diagrams in chapter 9 that illustrate this include: **12**, **17**, **49**, **63**, **64**, **68**, **69**, **71**, **86**, and **97** but you may have found more. Other examples include the lithium enolates and silyl enol ethers in **100** to **101**.

If you extend your study to chapter 10 you will find a greater variety such as the lithium enolates and silyl enol ethers **134** to **135** and **146**, **147** and **163**, the enamine **127**, the malonate (twice) in the synthesis of **143**, and the boron enolate **150**. **Problem 10.2**: Under equilibrating conditions, ketoester **7** forms the stable enolate **8**. With an excess of NaH, which enolate will be formed from **7**? Where will this enolate be methylated? What happens then?

Answer 10.2: The first equivalent of NaH will give **8** but the second will create a dianion, best represented as **W1**. Methylation occurs on the last formed enolate to preserve the stable enolate **W2**. You might have done many things afterwards, such as react with a second electrophile at the remaining enolate to form **W3**.

Problem 10.3: Compound **51**; X = OTs was needed for the synthesis of multistriatin **48**. Suggest what happens in detail in the formation of **51** from **52**.

Answer 10.3: There is clearly an allylic rearrangement as the CH_2OTs group must come from reduction of the newly introduced CO_2 unit. This chemistry is explored in more detail in chapter 12 where we reveal that allyl Grignard reagents normally react with electrophiles at the far end of the allylic system **W4**. Reduction to **W6** and tosylation complete the synthesis.[1]

Problem 10.4: In the asymmetric synthesis of multistriatin, which other stereoisomer was present in the final mixture with (−)-**48**?

Answer 10.4: In the textbook we pointed out that 'It emerged from this synthesis that the chiral centre next to the ketone could be epimerised to the correct configuration during acid-catalysed cyclisation of the final intermediates.' In fact rearrangement of the mixture of diastereoisomers of **56b** gave a mixture of all four diastereoisomers of **48** known as α-δ-multistriatin.[2] Only the centres fixed by the 1,3-diaxial bridge cannot epimerise. Natural (−)-α-multistriatin **48a** was the major isomer and this allowed assignment of the absolute stereochemistry. Problems are the low enantiomeric purity of the natural citronellol starting material and the poor stereoselectivity of the epoxidation.

(−)-**48a**; α-multistriatin   (−)-**48b**; β-multistriatin   (−)-**48c**; γ-multistriatin   (−)-**48d**; δ-multistriatin

Problem 10.5: What happens in these two methods[3] to get **75** from **74**? Draw the intermediates and the mechanisms. What is the other product in each case?

Answer 10.5: Sodium periodate oxidises the imine *via* the diol to the ketone **75** and the nitroso compound Me_2N-NO. The cyclic iodate **W8** decomposes by a reverse 1,3-dipolar cycloaddition as the organic compound is oxidised and I(V) is reduced to I(III). The methylation route is easier to see: reaction at nitrogen gives an imine salt that is hydrolysed to the ketone **75** and Me_2NHMe.

Corey's Ginkgolide Synthesis

Compounds **117** to **130** in the textbook chapter 10 are the first stages of this synthesis. You might like to read more in *Corey and Cheng* pages 221–226. Among other interesting chemistry there are some specific enolates used. **Problem 10.6**: What is the role of each reagent in the transformation **138** to **132**? Why was a thioester chosen for this synthesis?

Answer 10.6: The acid in step 1 hydrolyses the THP derivative, hydrogenation over Lindlar's catalyst gives the *cis*-alkene from the alkyne, and Hg(II), being a good Lewis acid for sulfur, catalyses the hydrolysis of the *t*-butyl thioester. Protic acid might have promoted reactions of the alkene and base might have enolised and epimerised the ketone though in fact the authors used the dithioester as they were keen to demonstrate efficient conjugate addition of such enolates.[4]

Problem 10.7: Suggest a synthesis for **148** needed to make the specific enolate **147** by olefin metathesis.

Answer 10.7: The obvious route is by double alkylation of malonate.[5] The allyl group should go in very easily so it would be good strategy to add the other alkyl halide **W9** first. This compound will have to be made from the γ-bromo ketone **W10**.

Specific Enolate Examples not from the Chapter in the Textbook

Enamines and Aza-Enolates

The cyclic ketone **W11** was needed by Schmid and colleagues for a study of Cope [3,3]sigma-tropic rearrangements.[6] Disconnection by the Robinson annelation strategy gives simple starting materials **W13** and **W14**. At every stage except for the cyclisation of **W12**, specific enol(ate)s will be needed.

They decided to alkylate the available nitrile with allyl bromide and to reduce the product **W16** to **W14** with DIBAL. The pyrrolidine enamine **W17** prevented self-condensation of the aldehyde and promoted conjugate addition. Cyclisation to **W11** was spontaneous.

Sometimes a choice of specific enolates is crucial to success. Neither the enamine **W18** nor the enolate from **W19** would react with the tosylate **W20** (made from cyclopentadiene by hydroboration and tosylation) to give **W22** but the magnesium aza-enolate **W21** was successful.

Aza-enolates are also involved in a successful way to avoid self-condensation with aldehydes during attempted aldol reaction.[7] A cunning variant is preliminary silylation of an aza-enolate before the aldol reaction.[8] The aza-enolate from **W24** will react cleanly even with cyclohexanone to give the crowded enal **W25**. **Problem 10.8**: Draw a detailed mechanism for the last step: the formation of **W25** from **W24**.

Answer 10.8: The lithium aza-enolate **W26** is formed with LDA and then adds to cyclohexanone either by simple nucleophilic addition **W27** or through a cyclic mechanism. The product **W28** loses Me_3SiO^- in a Peterson reaction (chapter 15) by attack of the oxyanion on silicon. Hydrolysis of the imine gives **W25**.

Enolates of 1,3-Dicarbonyl Compounds

With reactive electrophiles, simple enolates of 1,3-dicarbonyl compounds are good reagents as they are too stable to do self-condensation and can be made with weak bases. The keto-lactone **W29** is prepared[9] from ethyl acetoacetate and ethylene oxide with NaOH as base in aqueous solution. Further transformations via **W30** give the cyclopropyl ketone **W31**. **Problem 10.9**: Suggest mechanisms for these reactions commenting on the enolates used and the selectivity in the last step.

Answer 10.9: The stable enolate reacts with ethylene oxide **W32** to give and adduct that can cyclise **W33** by ester exchange to **W29**. Reaction with HCl by S_N2 at the ester carbon atom gives a keto acid that decarboxylates by a cyclic mechanism **W35** to give the enol of **W30**.

The last step is more interesting. Either enolate could be formed: one can cyclise **W37** in a superficially attractive way to cyclopentanone while the other can form **W31**. Cyclisations to five- and three-membered rings are both kinetically favoured but **W37** has the enolate double bond *endo* to the mechanism ('*enolendo*') and is disfavoured by Baldwin's rules. The alternative **W36** has the enol double bond *exo* to the new ring ('*enolexo*') and is favoured by Baldwin's rules.

Lithium Enolates and Silyl Enol Ethers

You will be familiar with these important intermediates so one example of each will do. The cyclopentenone **W38**, made by a Pauson-Khand reaction (chapter 6) needed alkylation with one extra methyl group. The lithium enolate, made with LiHMDS, gave **W39** in 94% yield. The product was used to make asteriscanolide.[10]

The synthesis of the ant pheromone mannicone required an aldol reaction between pentan-3-one and the enantiomerically enriched aldehyde **W41**. The ketone is symmetrical but a specific enol is need to avoid self-condensation of the aldehyde. The silyl enol ether **W40** was chosen and reacted cleanly with **W41** with TiCl$_4$ as catalyst (a Mukaiyama aldol) to give a mixture of diastereoisomers of **W42** in 92% yield.

The mixture of diastereoisomers is not a problem as both dehydrate with catalytic TsOH in refluxing benzene to give mannicone in 83% yield. The product is entirely *E*-alkene: no *Z*-alkene could be detected.[11] This is clearly not an E2 elimination as it is stereo*selective* rather than stereo*specific*. Either the enol is an intermediate or else the *Z*-alkene may equilibrate with the *E*-alkene under the reaction conditions.

Other Methods

Problem 10.10: Workers at Hoffmann-La Roche made bicyclic compounds, e.g. **W44**, by a simple one-step reaction. Draw a detailed mechanism and explain any selectivity, particularly identifying the ways they used to prepare specific enolates.

Answer 10.10: Conjugate addition of the anion of the thiol to cyclohexanone **W45** creates a specific enolate that cyclises **W46** onto the ester to give the product.[12] Under the reaction conditions, the methoxide eliminated in the cyclisation converts **W44** into its stable enolate **W47**. The stereochemical control (*cis* ring fusion) is therefore thermodynamic: fused 6/5 rings prefer to be *cis*. The first step shows regioselectivity in preferring conjugate to direct addition either because sulfur is a soft (orbital controlled) nucleophile or because addition to the carbonyl group is reversible and conjugate addition gives the thermodynamic product. The specific enolate **W46** would be difficult to make from its parent ketone.

Specific Enolates by Reduction of Enones

The only examples we gave in the textbook chapter of reduction of enones to simple specific enolates were on cyclic compounds where the stereochemistry of the resulting lithium enolate was inevitable. If the enolates are required for stereoselective aldol reactions (chapter 4), the geometry of the enolate is very important. Chamberlin and Reich[13] have reported that some enones have predominant conformations about the single bond in the middle of the enone and that this becomes the configuration of the enolate on reduction. Thus enone **W48** exists as >100:1 s-*trans*: s-*cis* (the s-*trans* form is shown) to keep the large groups apart: this is the normally preferred conformation for enones. Reduction with lithium in ammonia fixes the configuration of the lithium enolate, best trapped as the silyl enol ether **W49**. The yield is good and the other geometrical isomer undetectable – they claim >300:1.

By contrast, *exo*-methylene enones such as **W50**, exist mainly (>20:1) in the s-*trans* conformation and reduction with L-selectride gives the silyl enol ether **W51** in poorer yield but a 170:1 ratio of geometrical isomers.

W50 → W51; 69% yield

1. Li B(s-Et₃)H, THF, −78 °C
2. Me₃SiCl, Et₃N, −10 °C

References

1. W. E. Gore, G. T. Pearce and R. M. Silverstein, *J. Org. Chem.*, 1976, **41**, 2797.
2. G. J. Cernigliaro and P. J. Kocienski, *J. Org. Chem.*, 1977, **42**, 3622.
3. E. J. Corey and D. Enders, *Tetrahedron Lett.*, 1976, **3**, 11; E. J. Corey, D. Enders and M. G. Bock, *Tetrahedron Lett.*, 1976, 7; E. J. Corey and D. L. Boger, *Tetrahedron Lett.*, 1978, 4597; D. Enders and P. Wenster, *Tetrahedron Lett.*, 1978, 2853.
4. K. Gerlach and P. Kunzler, *Helv. Chim. Acta*, 1978, **61**, 2505.
5. A. Okada, T. Ohshima and M. Shibasaki, *Tetrahedron Lett.*, 2001, **42**, 8023.
6. P. Vittorelli, J. Peter-Katalinic, G. Mukherjee-Müller, H.-J. Hansen and H. Schmid, *Helv. Chim. Acta*, 1975, **58**, 1379.
7. G. Wittig and H. Reiff, *Angew. Chem., Int. Ed.*, 1968, **7**, 7.
8. E. J. Corey, D. Enders and M. C. Bock, *Tetrahedron Lett.*, 1976, 7.
9. W. L. Johnson, U.S. Patent 2,443,829, *Chem. Abstr.*, 1949, **43**, 677.
10. M. E. Krafft, Y. Y. Cheung and K. A. Abboud, *J. Org. Chem.*, 2001, **66**, 7443.
11. K. Banno and T. Mukaiyama, *Chem. Lett.*, 1976, 279.
12. P. N. Confalone, E. Baggiolini, B. Hennessy, G. Pizzolato and M. R. Uskokovic, *J. Org. Chem.*, 1981, **46**, 4923.
13. A. R. Chamberlin and S. H. Reich, *J. Am. Chem. Soc.*, 1985, **107**, 1440.

11

Extended Enolates

Problems from Chapter 11 in the Textbook

You will need chapter 11 of the textbook for the first few questions.

Wittig Approaches to γ-Aldols

In the textbook we pointed out that γ-bromoesters can be made by radical bromination with NBS and converted into phosphorus compounds such as **29** or **33** for Wittig or Horner-Wadsworth-Emmons reactions to give compounds such as **34**. It does not matter if some reaction occurs at the γ position (to phosphorus now!) of **29** or **33** as the Wittig elimination cannot occur on such an intermediate and it reverses. **Problem 11.1**: Draw this intermediate from **29** or **33** and explain.

Answer 11.1: The intermediates from **33** are the extended enolate **W1** and then the aldol adduct **W2**. The oxyanion is too far away from the phosphorus atom for the elimination to occur and **W2** reverts to **W1** and adds RCHO α to phosphorus (i.e. γ to the carbonyl).

Problem 11.2: In the textbook we said: 'Extended lithium enolates of conjugated aldehydes (such as **51**) are more stable than those of simple unconjugated aldehydes, and can be used particularly where there is a branch at the α-position as in the vernolepin intermediate **53** which can be made by alkylation of the lithium derivative **52**'. Why are lithium extended enolates of conjugated aldehydes more stable than lithium enolates of simple aldehydes? How does a branch in the α-position help? Give details of the potential danger of fragmentation of **52**.

Workbook for Organic Synthesis: Strategy and Control   Paul Wyatt and Stuart Warren
© 2008 John Wiley & Sons, Ltd

Answer 11.2: Even one extra alkene stabilises an enolate and this and the branch in the α-position makes self condensation less likely. Compound **52** also contains a ketone protected as an acetal. This makes it vulnerable to fragmentation **W3** that would destroy the molecule and probably produce an aromatic compound such as **W4**.

Using Chloro-Sulfides in γ-Alkylation

The sequence **62** to **67** from the textbook illustrates the preparation and use of **63** in γ-alkylation of extended enolates but leaves a number of questions unanswered. **Problem 11.3**: In the formation of **62** with PhSH and NaOH, why is the alcohol RCH$_2$OH not formed?

Answer 11.3: There are two reasons. Like all thiols, PhSH is more acidic than water or alcohols. So fast proton transfer converts it into PhS$^-$ and H$_2$O. Further, sulfur nucleophiles, having higher energy lone pairs, are softer and more effective at S$_N$2 reactions because S$_N$2 reactions are orbital controlled.

The second step, the conversion of **62** into **63**, is essential to the whole operation as nobody would use **63** unless they were easily made. This a type of Pummerer rearrangement in which functionality initially put on sulfur is transferred to carbon. Chlorination occurs on sulfur to give a sulfonium salt **W6** that can lose a proton to form the ylid **W7**. Chloride now leaves **W7** but immediately returns to carbon to give **63**.

Problem 11.4: Fill in the gaps in the formation of **66** from **65**. We show only the initial γ-alkylation **64**.

Answer 11.4: First drawing the result of the arrows in **64** we get the stable cation **W9** from which chloride (or some other nucleophile) removes the SiMe$_3$ group.

Problem 11.5: This chemistry is used in the textbook to make retinal.[1] Give detailed mechanisms for these first two stages.

Answer 11.5: The first step is acetal formation from natural cyclocitral **76** with methyl orthoformate. Methanol can add to the aldehyde but the reaction stops there unless something removes the OH group. The orthoester can do this and, unlike acetals, can react by an S$_N$2 reaction **W10**. If you prefer, you can lose MeOH first to make a much more electrophilic cation. In either case the product can lose HCO$_2$Me to give the essential intermediate for the second addition of methanol.

The Mukaiyama aldol used the Lewis acid TiCl$_4$ to create the same cation **W12** that reacts in the γ-position **W13** with the silyl enol ether **79**. The product is desilylated, probably with methanol, to give **80**. **Problem 11.6**: How would you make **79**?

Answer 11.6: This is just the silyl enol ether from either the conjugated **W15** or non-conjugated **W16** enals. The conjugated enal can easily be made by an aldol reaction but is readily available. It reacts in the γ-position as it is a weak nucleophile under orbital control.

Problem 11.7: Why do you think the aldol dimerisation of cyclopentanone gives **103**? What other product might you have expected? Does it matter which is formed?

Answer 11.7: The intermediate aldol **W17** evidently prefers to dehydrate to the non-conjugated enone **103** rather than the conjugated enone **W18** in contrast to the enal we have just considered.[2] This may be because conjugation requires **W18** to be planar around the enone system and there is a steric clash between the carbonyl group and at least one H atom. It does not matter which we use – treatment of either with $NaNH_2$ gives the same extended enolate anion **101**.

Problem 11.8: The starting material **104** is described as a 'Robinson annelation product'. Explain what this means. Why does enamine formation occur in the way it does? Give a full mechanism.

Answer 11.8: You do not have to remember the Robinson annelation as normal enone **104a** followed by 1,5-diCO **W19** disconnection reveals the starting materials **W20** and cyclo-hexanone for the annelation sequence. As **W20** can enolise and self-condense, a specific enol(ate) from cyclohexanone (enamine or silyl enol ether) will be needed.[3]

The initial product from **104** and pyrrolidine is the imine salt **W21** that can form an enamine by loss of H^A or H^B. Loss of H^A gives the less stable cross-conjugated enamine **W22** but loss of H^B gives the more stable linearly conjugated **105** that is actually formed. Enamine formation is reversible and under thermodynamic control.

Problem 11.9: Cycloaddition of the silyloxydiene **120** was used[4] in the synthesis of senepoxide **123**. Explain the regiochemistry of the Diels-Alder adduct **122**.

Answer 11.9: The 'nucleophilic' end **120a** of the diene, i.e. the end with the largest coefficient in the HOMO, and the 'electrophilic' end **121a** of the alkyne, i.e. the end with the largest coefficient in the LUMO are circled **120b** and **121b** and will combine. It is not 'cheating' to use arguments like this as bond formation between the circled atoms will be well ahead of formation of the other bond in the rather unsymmetrical transition state **W23**.

Synthesis of Senepoxide 117

The synthesis of senepoxide **123** from intermediate **122** is completed in the following steps.[5]
Problem 11.10: State the selectivity (or selectivities, remember there might be three kinds) in each step and offer a reasonable explanation. Comment on the order of events in the conversion of **W28** into **123**.

Answer 11.10: The three kinds of selectivity are *chemo*, *regio* and *stereo*. Epoxidation of **122**: *chemo*: the more nucleophilic alkene reacts with *m*CPBA; *stereo*: addition occurs on the face of the ring opposite the large OSiEt₃ group. Reduction of **W24**: *chemo*: epoxide is not reduced, *regio: reduction* of C=O and not conjugate reduction – DIBAL goes for most *nucleo*philic site. Addition of PhSe group to **W25**: *regio* and *stereo*: nucleophiles normally add to the end of the epoxide that leads to the *trans*-di-axial product. Here that is rather unconvincing as the large OSiEt₃ group would have to be pseudo-axial **W30**. In any case, the product is *not* a chair and it looks better to draw the starting material in a folded conformation and have the nucleophile add to end away from the large *exo* OSiEt₃ group.

Oxidation of **W26** with *m*CPBA: *chemo*: no epoxidation of alkene as Se is very easily oxidised to the selenoxide. Elimination step: *regio* and *stereo*: the elimination is stereospecifically *cis* **W32** but this does not control the regiochemistry as both Hs (marked in **W32**) are *cis* to PhSeO. The hydrogen that is lost is allylic (weakens C–H bond and activates towards elimination) and not next to oxygen (deactivates towards elimination). Note the low temperature needed for these very easy selenoxide eliminations.

Oxidation of **W27** with *m*CPBA: *regio:* only one alkene is epoxidised. Both are allylic alcohols so this is not important. Epoxidation is in essence an electrophilic attack on the alkene and the one that reacts is trisubstituted. This shows that it is the high reactivity of Se that controls the previous step and not the unreactivity of the alkenes. *Stereo:* reaction *anti* to the large OSiEt₃ group (a Houk-style argument) but just might be directed by the OH group on the other side of the molecule if it is axial. Benzoylation of **W28**: *chemo:* only the primary alcohol reacts with this fairly large acylating reagent at low temperature. The OSiEt₃ group is removed by HCl in MeOH – nucleophilic attack on silicon by Cl⁻ or MeOH. Acetylation shows that the secondary alcohols can be esterified at higher (room) temperature.

Birch Reduction of Pyridines

In the textbook we said: 'The pyridine diester **138** is reduced to an intermediate that could be drawn as **139**. Both anions are extended enolates but one has the charge delocalised onto the nitrogen atom and so is less reactive than the other.' **Problem 11.11**: Draw another delocalised form of **139** showing that both anions are extended enolates and convince yourself that only one anion is delocalised onto nitrogen. How would you use this chemistry to make **W33**?

Answer 11.11: You could draw various delocalised versions of **139**: perhaps the most obvious is **139a** giving **139b** with α and γ carbons easy to see. Attempts to delocalise the other anion onto nitrogen are doomed: one attempt is **139c**.

The Synthesis of Mniopetal F

We left unfinished the synthesis of the vital diene **159**; R = $SiPh_2t$-Bu. **Problem 11.12**: Suggest how this might be carried out from **158** easily prepared by a Horner-Wadsworth-Emmons reaction from **157**.

Answer 11.12: The ester needs to be reduced and protected and the isolated alkene needs to be converted into an aldehyde with the same number of carbon atoms. This must be done after the reduction or the aldehyde will be reduced too. Obvious solutions are hydroboration and oxidation. We give the method used by Jauch[6] but you may have other solutions. You might have used DIBAL to reduce the ester but $LiAlH_4$ is a more obvious choice and you are most unlikely to use hypervalent iodine and a radical source as the oxidising agent. The exposed monosubstituted alkene in the triene **W34** is more susceptible to reaction with the large borane 9-BBN than the diene.

In the textbook we said: 'The product **172** is clearly destined for a Diels-Alder reaction and the product **173** is nearly ideal for conversion into mniopetal F **174**.' **Problem 11.13**: Why only 'nearly' ideal? What are you going to do to correct that small problem?

172; 88% yield   173; 68% yield   174; mniopetal F

Answer 11.13: The problem is that the OH group has the wrong configuration. Mitsunobu inversion might be a good answer, as it happens they invert a triflate with KNO_2 as nucleophile but the yield was less that 50%.

Problem 11.14: Bromination of **179** initially occurs between the two carbonyl groups. How does that product rearrange to **180**? Suggest a mechanism for the formation of the tetronic acid **181** from **180**. Why are compounds like **181** acids?

179   180   181; 67% from 167   182; 99% yield

Answer 11.14: The initial bromination product is **W36**. Air sets off a radical chain reaction: bromine radicals add to the enol **W37** and the intermediate **W38** loses then other Br as a radical to continue the chain and give the more stable conjugated enol **W39** of **180**.

W36   W37   W38   W39   180

The enol of **180** can cyclise by a favourable 5-*exo-tet* reaction **W40** and the bromide released can dealkylate the (probably protonated) intermediate **W41**. The anion **W42** of the tetronic acid **181** is stable and delocalised hence the 'acid' in the name.[7]

W40   W41   W42   181

The Synthesis of Cholesterol and Cortisone

Problem 11.15: Explain the control exercised in the following sequence of reactions to create the tricyclic A/B/C ring system for the steroids.[8]

Answer 11.15: Though **W43** is an enone it cannot form an extended enolate because of the quaternary ring junction and must react in the α′ position with electrophilic but unenolisable ethyl formate to give the keto-aldehyde that prefers to exist in the enol form **W46**. Base creates the stable enolate that carries out conjugate addition to the enone **W47** as expected (chapter 9).

Now the Robinson annelation is completed by cyclisation **W48** to the only intermediate that can form a stable six-membered cyclic enone. Deformylation by hydroxide **W49** may occur before or after cyclisation. The last intermediate **W50** is (at last!) an extended enolate and is protonated in the γ position. The protonation is under thermodynamic control and also fixes the stereochemistry of the new centre: we get the more favourable *trans, trans* ring junction.

Minor adjustments of functionality gave the next intermediate **W51**. **Problem 11.16**: Suggest how these might be carried out and explain the stereochemistry of the acetal. The formation of ring A required some interesting chemistry outlined below. Give mechanisms for the reactions, explain the regiochemistry, particularly of the extended enolate reaction, and justify the stereoselectivity.

Answer 11.16: We cannot hope to match the explanation of these reactions given in Fleming's *Selected Organic Syntheses* but, as that is out of print, we offer a brief summary. *Synthesis of* **W51**: One alkene must be removed and one dihydroxylated. The alkene not conjugated with the ketone is the most nucleophilic so dihydroxylation must come first. There is no question of *trans*-di-axial here as dihydroxylation is *cis* specifc so reaction occurs on the face opposite the axial methyl group.

Formation of W52: The first step is the same as the formation of **W46** and the amine replaces OEt by conjugate substitution (addition–elimination). However, the enamine is now both a blocking and an activating group and reaction occurs on the extended enolate side. The base 'Triton B' – a tetra-alkyl ammonium hydroxide – creates the extended enolate that adds Michael-fashion to acrylonitrile in the α-position **W57**. This is kinetic control. Stereochemical control in this step was not very good but regio-control was total (no γ-product). The immediate product **W58** loses the imine during hydrolysis of the nitrile.

The remaining steps are among the more puzzling. Acetic anhydride forms an enol lactone **W59** from the ketone, no doubt via the mixed anhydride, and this reacts with MeMgBr by nucleophilic substitution with an enolate as the leaving group to give a new ketone **W60** that can cyclise with KOH to give the ring A enone.

The Birch Reduction Route to Extended Enolates: The Synthesis of Anisatin

The initial stages in a synthesis of the toxic compound anisatin from the spice star anise have examples of material from earlier chapters. **Problem 11.17**: Comment on the reagents for the first step, on the stereochemistry of the reduction of **W63**, and on the regiochemistry of the synthesis of **W65**. Explain the formation of **W66** by Birch reduction which was intended as a preliminary for extended enolate chemistry.

Answer 11.17: Copper was needed to ensure conjugate addition in the first step (chapter 9). The reduction occurs simply by addition to the other side of the flat ring to the methyl group. The regiochemistry of the carboxylation reveals that it is an *ortho*-lithiation directed by the OLi group (chapter 7).

However, direct alkylation of the extended enolate derived from **W66** failed as the compound re-aromatised rather easily by loss of water. Instead the sequence below gave the required compound.[9] **Problem 11.18**: Explain the mechanism, regio- and stereoselectivity in these two steps.

Answer 11.18: The first reaction is an aliphatic Claisen rearrangement: acetal exchange gives **W72** and elimination of methanol gives the allyl vinyl ether **W73**. The [3,3]-sigmatropic rearrangement seems to need a long stretch, but, providing the side chain is passed across the top surface of the molecule through a boat transition state, it is all right. This Claisen rearrangement is constrained to give a *cis* alkene unlike the usual *trans* alkene in open chain compounds (chapter 15).

Finally, in the second step we have some extended enolate chemistry as **W70** has no leaving group and can no longer easily aromatise. The extended enolate could form into the ring by loss of H^A in **W70a** or *exo* to the ring (H^B) to form **W74**. Presumably with LDA the more acidic proton (H^B) is preferred. Alkylation occurs α, as expected, and even shows some stereoselectivity in favour of addition from the opposite side of the large and pseudo-axial amide.

References

1. T. Mukaiyama and A. Ishida, *Chem. Lett.*, 1975, 1201.
2. A. T. Nielsen and W. J. Houlihan, *Org. React.*, 1968, **16**, 1: see table on page 114; H. Ueberwasser, CIBA, Ger. Pat., 1,059,901, *Chem. Abstr.*, 1962, **56**, 355f.
3. G. Stork, A. Brizzolara, H. Landesman and J. Szmuszkovicz, *J. Am. Chem. Soc.*, 1963, **85**, 207.
4. G. A. Berchtold, J. Ciabattoni and A. A. Tunick, *J. Org. Chem.*, 1965, **30**, 3679.
5. R. H. Schlessinger and A. Lopes, *J. Org. Chem.*, 1981, **46**, 5252.
6. J. Jauch, *Eur. J. Org. Chem.*, 2001, 473.
7. K. Takabi, N. Mase, M. Nomoto, M. Daicho, T. Tauchi and H. Yoda, *J. Chem. Soc., Perkin Trans. 1*, 2002, 500.
8. I. Fleming, *Selected Organic Syntheses*, John Wiley & Sons, Ltd, 1973, pp. 57–65.
9. T.-P. Loh and Q.-Y. Hu, *Org. Lett.*, 2001, **3**, 279.

12

Allyl Anions

Material from Chapter 12 in the Textbook

It will help you if you have chapter 12 in the textbook open for the first few questions. **Problem 12.1**: Cyclisation of the epoxide **107** gives a seven-membered ring **109** via a cation **108**. Draw two mechanisms for the formation of **108** from **107**, one based on an S_N1 opening of the epoxide and the other on an S_N2. Which do you prefer?

Answer 12.1: The Lewis acid binds to the epoxide and could open it by an S_N1 reaction **W1** that would give the more highly substituted (tertiary) cation and this would cyclise[1] **W2** to give **108** and hence **109**.

By contrast, opening by the S_N2 mechanism would give a six-membered ring by attack at the less substituted end of the epoxide. This gives the wrong product **W5** and we prefer **W1**.

Problem 12.2: We made artemisia ketone in the textbook from the allyl silane **117**. How would you make **117**?

Answer 12.2: The cyclohexene **117** is an obvious Diels-Alder adduct[2] from the silylated butadiene **W6** and dimethyl fumarate **W7**. As the allyl group (top half) in **W6** is symmetrical, it could be made by reaction of the allyl lithium with Me$_3$SiCl. Hosami and collegues[3] made it by the nickel-catalysed reaction between chloroprene and Me$_3$SiCH$_2$Cl.

Reactions of Substituted Cyclopentadienes

We used the sequence **90** to **93** in the textbook chapter. The intermediate **92** is a stable compound and ideal for both $2 + 2$ and Diels-Alder reactions. By contrast, attempts to use alkyl cyclobutadienes in Diels-Alder reactions lead to mixtures of products. **Problem 12.3**: Explain why, say, methyl cyclopentadiene is useless but **92** useful.

Answer 12.3: Diels-Alder reactions with methyl cyclobutadiene **W8** with, say, maleic anhydride **W9** give a mixture of adducts **W10**, **W11**, and **W12**. The 'expected' adduct **W10** is a minor product.

The reason is that [1,5]-sigmatropic H shifts equilibrate **W8** with the more stable isomers (more highly substituted alkenes) **W13** (giving **W11**) and **W14** (giving **W12**). The Me$_3$Si group migrates around the ring by [1,5] sigmatropic shifts about 10^6 times faster than does the proton, thus maintaining structure **92** at low temperature.[4] This is an observation of wider importance but for this chapter we should note that, if the substituent migrates, it remains allylic while if hydrogen migrates the previously allylic substituent becomes vinylic.

In the textbook we said: 'Reaction with Me₃SiCl and careful acidic work-up gives the allyl silane **126**.' **Problem 12.4**: Why is care needed in the acid work-up to give **126**? What might happen?

Answer 12.4: Any allyl silane may lose silicon by protodesilylation.[5] It is not always easy to know how fragile a given compound might be, but one often finds out by experience. As in the last example, it all depends on whether H or Me₃Si is lost from the intermediate cation **W16**.

Problem 12.5: Why is only one C=O group in **128** reduced by NaBH₄? What kind of selectivity is this? Draw a complete mechanism for the cyclisation to **130**.

Answer 12.5: The imide in **128** is very electrophilic as the lone pair on nitrogen is delocalised into two carbonyl groups. The first reduction product **129** is a much less electrophilic simple amide and is not reduced by NaBH₄. This is rather like comparing anhydrides and esters. This is chemoselectivity. The cyclisation probably proceeds via an acyl iminium cation **W19**, cyclisation and desilylation **W20**.

Problem 12.6: The absolute stereochemistry of **134** was shown by comparison with the saturated compound **W21** also produced[6] from the imine **W22**. What is the mechanism of the second reaction?

Answer 12.6: Lithiation must occur at the benzylic position and, though there is little mechanistic detail in the otherwise comprehensive published method,[7] we can assume that the imine nitrogen directs lithiation to the benzylic position **W23** but would it be to the top face so that alkylation occurs with retention? Whether the lithium derivative is **W24** or **W24a** it is not easy to see the origin of the stereochemical induction.

Problem 12.7: Draw mechanisms for the formation of **146** and its reaction with **147**. What is the other isomer?

Answer 12.7: BuLi attacks the tin atom, presumably directed by one of the nitrogen atoms **W25**. The aza-allyl anion does a 1,3-dipolar cycloaddition **W26** onto the alkene with regioselectivity suggesting that the HOMO of the allyl anion combines with the LUMO of the styrene.[8] The other isomer is the regioisomer of **149**.

Problem 12.8: Suggest a mechanism for the reaction **72 + 150** giving **151** on the way to indium allyls **153**.

Answer 12.8: Presumably acylation at oxygen provides an electrophilic species (an allylic cation rather than anion) for addition of bromide.[9] The Lewis acid may coordinate to oxygen or to bromine.

This same paper[9] gives a display of various allyl metal derivatives with two heteroatoms, one at each end, and lists their stereoselectivities in reactions with aldehydes. Their discussion of the [1,3]-sigmatropic shifts of the various methods is also worth a look. **Problem 12.9**: Why is this elimination on **184** so easy? Draw a mechanism.

Answer 12.9: Removal of a marked γ-proton **W30** gives the extended enolate (another kind of allyl anion–see chapter 11) that eliminates sulfinate anion by the E1cB mechanism. The product is *E*-**185** as the side chain can orient itself in the more favourable way to give the more stable diene.[10]

Problems and Further Examples from Other Sources

A selection of (fairly) recent results. The simplest extension of an allyl metal is a crotyl metal and a great deal of work has been done in this area particularly on the stereochemistry of their addition to aldehydes. Nokami[11] analyses and summarises results in a useful way. Addition to aldehydes is rather similar to the Zimmerman-Traxler analysis of aldol reactions (chapter 4). *Trans* allyl metals *E*-**W32** give *anti*-γ-adducts *anti*-**W35** while *cis* allyl metals give *syn* adducts. The cyclic mechanism **W33** directs the aldehyde to the γ-carbon and the chair transition state **W34** shows the methyl group in its *enforced* equatorial position while R *chooses* the equatorial conformation.

The metal can be a wide variety of elements such as Ti, Cr, Al and B. The less reactive allyl stannanes and some of the *cis* allyl compounds need Lewis acid catalysis. A remarkable development is the transformation of the normal γ-adducts into α-adducts **W41** by reversing the reaction with Sn(OTf)$_2$. The trick is to do the first reaction with acetone to give γ-adduct **W36** and the rearrangement with the aldehyde you actually want. Two hemiacetals **W37** and **W40** are involved but the key step is the [3,3]-sigmatropic rearrangement **W38**. There are tables of results in the paper.

An Asymmetric Synthesis of Spirocyclic Ketal Ionophores

The synthesis of these complex molecules (see the papers and a comprehensive review[12]) uses two allyl anion additions to aldehydes. One needs to set up the *syn* relationship between OH and Me in **W42** and therefore a Z-allyl metal **W44** is required.

No doubt Z-crotyl stannane would do the trick but this is an asymmetric synthesis and so an enantiomerically pure 'metal' is needed and a suitable boron compound is the answer. You will meet di-isocampheyl boranes in chapter 24: for the moment we shall use R*$_2$B as the substituent. Stereospecifically *syn*-**W42** is formed as a single diastereoisomer in 80% yield.

The second uses an allyl stannane addition to the enantiomerically pure aldehyde **W46** made from natural malic acid (chiral pool strategy, chapter 23). Since the aldehyde is one enantiomer we need only diastereoselectivity and you would expect that Z-crotyl stannane would be the answer.

Unfortunately the relationship between the OH and the other Me group needs Felkin-style control. The best Felkin arrangement is probably **W48** though Me and CH$_2$OMOM are of similar size. In any case it gives the wrong result **W49**. Chelation is the answer and a Lewis acid is needed such as Mg(II). Now there is a clear distinction **W50** and it gives the right answer.

Now a further and happy complication arises. It was already known that Lewis acid catalysed crotyl stannane additions to aldehydes reverse the normal stereospecificity. now *E*-crotyl stannanes give *syn*-addition products.[13] So all we need is a chelating protecting group on the aldehyde (MOM, MeOCH$_2$O–, is ideal) and a Lewis acid (MgBr$_2$.Et$_2$O was chosen) to get the right diastereoisomer

Problem 12.10: Further development used the start of a hydroboration reaction to give **W51** and an interesting transformation into **W52**. See if you can unravel what is going on in this last step.

Answer 12.10: The chemistry of the last step will probably be new to you.[14] The nucleophile adds to the boron and a migration of the whole molecule from B to C occurs **W53** as the SPh group is removed by Hg(II). The product finally loses boron by oxidation with H$_2$O$_2$ in the usual way and releases the aldehyde **W55** which cyclises to the hemiacetal **W52**.

Allylic or Vinylic Anions Stabilised by Sulfur and/or Silicon

Alkenes **W56** having allylic sulfones at one end and allylic silanes at the other end can be deprotonated at the sulfur end **W57** by BuLi. You might have expected that a sulfone is more anion stabilising that a silicon atom. Reaction with an alkyl halide occurs at the sulfone end[15] (α-selectivity) of the allylic anion to give **W58**.

This is a useful reaction because the two activating groups can be removed in one frag- mentation step **W59** with TBAF giving a substituted butadiene **W60**. Dialkylation and TBAF treatment gives a 1,1-disubstituted butadiene. In other words **W57** is a reagent for the butadiene anion **W61**–a vinyl anion.

If the alkene is just one position nearer the sulfone **W62**, the vinyl-lithium **W63** is formed rather than the allyl lithium and can be alkylated to give **W64**. Now a second lithium must give the allyl lithium **W65** and alkylation followed by fragmentation gives the disubstituted butadiene **W67** but without much control over alkene geometry.

Reaction of **W62** with 1,3-dichloropropane and cyclisation with BuLi gives the disubstituted **W69** without any need to control stereochemistry and the fragmentation gives the interesting butadiene **W70** in 84% yield.

Allyl Anions Derived from Dithians

Hepialone is a pheromone of a moth and has been synthesised by various routes by Schaumann.[16] The strategy that interests us in this chapter starts with a Peterson reaction between silylated dithian **W71** and an aldehyde to give **W72**. The allylic anion (or Li derivative) **W73** can easily be made from **W72** but the question is: at which end does it react?

In fact oxygen tends to direct γ and sulfur tends to direct α so the two cooperate in giving alcohols **W74** on reaction with epoxides. Treatment with catalytic acid cyclises these adducts to tetra-hydro pyrans **W75** close in structure to hepialone.

Allyl Anions Derived from Phosphonate Esters

Reactions of anions with epoxides are commonplace but azetidinium cations are rarely used as electrophiles. One study[17] reports on their reactions with anions derived from allylic phosphonate esters **W75**. They make the salutary observation that the sense of α/γ selectivity depends on a number of factors particularly the electrophile and the solvent. Reaction with an epoxide in THF, gave only the α-product **W77** though the same reagents gave an α/γ mixture in toluene.

Proving their point, reaction with the azetidinium ion **W78** gave only the γ-adduct **W79** in a slightly different solvent mix. The warning here is clear for all to see—α/γ selectivity is delicately balanced and depends on many factors: the metal, the stabilising group, the electrophile, the solvent and the conditions.

Reactions of Allyl Silanes with Cobalt-Stabilised Cations

We said in the textbook: 'More exotic electrophiles include cobalt-stabilised cations derived the alcohol (*S*)-**138** made by a sequence of reactions that shows the stability of allyl silanes to bases. The cuprate from *Z*-**135** adds to a single enantiomer of the epoxide (*S*)-**136** and the tosylate in the product (*S*)-**137** is displaced by a Co(I) anion to give the intermediate (*S*)-**138** as a stable orange solid.'[18]

The group 'Co(dmgH)₂py' is made from the complex of Co(II) with the monoanion of dimethylglyoxime **W81**. Two molecules of the oxime monocation from a square coplanar complex[19] made into an octahedron with axial H and pyridine ligands **W82**. The base NaOH removes this proton to create a Co(I) anion **W83**.

In acid solution, cobalt creates a π-stabilised cation at the site of the OH group that cyclises onto the allyl silane with the expected regioselectivity and excellent enantioselectivity. The cobalt is removed by photochemical oxidation with the stable radical TEMPO **W84**.

References

1. D. Wang and T.-H. Chan, *J. Chem. Soc., Chem. Commun.*, 1984, 1273.
2. A. Hosomi, M. Saito and H. Sakurai, *Tetrahedron Lett.*, 1980, **21**, 355.
3. A. Hosomi, M. Saito and H. Sakurai, *Tetrahedron Lett.*, 1979, 429.
4. A. Bonny and S. R. Stobart, *J. Am. Chem. Soc.*, 1979, **101**, 2247; C. W. Spangler, *Chem. Rev.*, 1976, **76**, 187.

5. J.-C. Gramain and R. Remuson, *Tetrahedron Lett.*, 1985, **26**, 327.

6. R. Yamaguchi, M. Tanaka, T. Matsuda and K. Fujita, *Chem. Commun.*, 1999, 2213.

7. A. I. Meyers, M. Bóes and D. A. Dickman, *Org. Synth.*, 1989, **67**, 60.

8. W. H. Pearson, E. P. Stevens and A. Aponick, *Tetrahedron Lett.*, 2001, **42**, 7361.

9. M. Lombardo, R. Girotti, S. Morganti and C. Trombini, *Org. Lett.*, 2001, **3**, 2981.

10. K. Uneyama and S. Torii, *Tetrahedron Lett.*, 1976, 443.

11. J. Nokami, L. Anthony and S. Sumida, *Chem. Eur. J.*, 2000, **6**, 2909.

12. M. M. Faul and B. E. Huff, *Chem. Rev.*, 2000, **100**, 2407.

13. G. E. Keck and D. E. Abbott, *Tetrahedron Lett.*, 1984, **25**, 1883.

14. R. K. Boeckman, A. B. Charette, T. Asberom and B. H. Johnston, *J. Am. Chem. Soc.*, 1987, **109**, 7553; *J. Am. Chem. Sec.*, 1991, **113**, 5337.

15. T. B. Meagher and H. W. Schechter, *J. Org. Chem.*, 1998, **63**, 4193.

16. S. Dreessen, S. Schabbert and E. Schaumann, *Eur. J. Org. Chem.*, 2001, 245.

17. A. Bakalarz-Jeziorna, J. Helínski and B. Krawiecka, *J. Chem. Soc., Perkin Trans. 1*, 2001, 1086.

18. G. Kettschau and G. Pattenden, *Tetrahedron Lett.*, 1998, **39**, 2027.

19. G. N. Schrauzer, *Inorg. Synth.*, 1968, **11**, 61.

13

Homoenolates

Material from Chapter 13 in the Textbook

It will help you if you have chapter 13 in the textbook open for the first section.

Cyclopropanes as Homoenolates

In the textbook chapter we used sodium, lithium and silicon derivatives of cyclopropanols as homoenolates in homoaldol reactions with aldehydes.

The preliminary trapping of the magnesium equivalent **W3** of **21** or **22** in an aldol-like reaction followed by a homoaldol reaction to give five-membered rings has been reported.[1] **Problem 13.1**: Explain what is happening.

Answer 13.1: At first sight two nucleophiles (**W2** and **W3**) seem to be reacting together but the magnesium 'homoenolate' **W3** must be an electrophile as it is in equilibrium with cyclopropanone. Base (NaH) opens the cyclopropanol in **W4** to release a true homoenolate **W8** that reacts intramolecularly with the ketone. The stereochemistry comes from the tether. Whether cyclopropanone or **W8** are really free is debatable.

W6 → W4 — NaH, Et₂O → W7 → W8 → W5

(Scheme structures W6, W7, W8)

Problem 13.2: Explain the stereoselectivity of the formation of **24**. Removal of the Cbz protecting group leads to spontaneous cyclisation to give **26**, an intermediate on the way[2] to the neurotoxin pumiliotoxin 251D.

23 → (OSiMe₃, OEt **21**, TiCl₄) → **24; 49% yield** → (H₂, Pd/C) → **25; 100% yield**

Answer 13.2: Felkin control **W9** would put NCbz orthogonal to the carbonyl group but predicts the wrong diastereoisomer **W10**. If the Ti atom bonds to both the carbonyl oxygen and to the NCbz group (through N or O) that gives **W12** that looks the same as **25** to us.

W9 → **W10**   **W11** → **W12**

There is another reason for rejecting simple Felkin control: addition of MeMgBr to the aldehyde **W14**; R = H related to **23** is not stereoselective. It was easier to make **23** from proline by direct reaction on the proline ester **W14**; X = OMe using the Tebbe reagent **W15** to get the enol ether **W16** and hence **23** by acid hydrolysis.

W13 ← (MeMgBr, X = H) **W14** → (Cp₂Ti–CH₂–AlMe₂Cl, **W15**, X = OMe) **W16** → (H⊕, H₂O) **23**

Problem 13.3: Suggest details for the conversion of **35** to **36** and **37** (X is a halogen).

X–CH₂–CH(CO₂Me) **(+)-35** → (Zn) → **36** (Zn–O–C(OMe)) → (RHal) → R–CH₂–CH(CO₂Me) **37**

Answer 13.3: Oxidative insertion of Zn into the C–X bond would give an organo-zinc compound **W17** that could be stabilised by chelation with carbonyl oxygen as shown in **36**. Alternatively cyclisation onto the carbonyl group would give a cyclopropane **W18** that could be in equilibrium with **36**. This is rather like the titanium homoenolates already described. Reaction[3] of either **36** or **W18** with an alkylating agent would give **37** (arrows on **W18**).

Later in the chapter we give details of a zinc homoenolate **57** made from a serine derivative **56** and drawn like **W17**. It could also be drawn as **W19** or **W20** but the authors[4] prefer **57**. The synthesis of this homoenolate is greatly assisted by ultrasound, a technique often used to assist heterogeneous reactions. Ultrasound may help to break up the solid zinc and make it more reactive or it may produce 'hot spots' in the solution.

Another application of zinc homoenolates is in the synthesis of ginkgolides. The first couple of paragraphs of the paper by Crimmins and colleagues[5] make interesting reading. You may also be interested to see that the authors do not draw their simple ester homoenolate as we do **39** nor in the style of **W17** nor **W18**. There is scope for individual choice here.

The synthesis of **39** has an interesting stereoselective step. The lithiated alkyne **W22** adds to the aldehyde **W21** with 3.3:1 selectivity in favour of the right (*anti* as drawn) diastereoisomer *anti*-**W23**. This is remarkable as it is 1,3-control. No explanation is offered but fortune was with them.

The next step is also interesting–a photochemical 2 + 2 cycloaddition creating a bridging four-membered ring. Two diastereoisomers were formed: **W25** was wanted and fortunately the other cyclised rapidly to the lactone **W26** so they were easily separated. Ginkgolide is a complex structure and you are advised to read the paper for more details.[5]

Homoenolate Equivalents from Three-Membered Rings

This simple lactone synthesis uses a cyclopropyl sulfonium ylid **63** as a rather indirect homoenolate equivalent. **Problem 13.4**: Draw mechanisms for the last three reactions and explain any selectivity.

The ylid adds to the ketone and cyclisation of the betaine intermediate **W27** gives the remarkably strained oxa-spiro-pentane **64**. Protonation obviously occurs on oxygen but which C–O bond will break **W28**? Each would give a tertiary cation but only one breaks: **a** or **b**?

Answer 13.4: Only bond **a** breaks. Cyclopropanes are strained but cyclopropyl *cations* are more strained because the cationic carbon would like a 120° bond angle. By contrast cyclopropyl methyl cations, such as **W29**, are stabilised by conjugation with the π-like σ-bonds in the three-membered ring. This also helps the rearrangement **W29**. The final step is a Baeyer-Villiger rearrangement and the more substituted carbon migrates to oxygen.

The cyclopropane **69** provides[6] another homoenolate equivalent in the formation of **72** by a homo-aldol reaction. **Problem 13.5**: Draw the mechanism for the step **70** to **71**.

Answer 13.5: This time the cyclopropane fragments **W31** with push from the MeO group and pull from the Lewis acid. Acetal exchange with the thioalcohol gives **71**.

Problem 13.6: Compound **149** was made in the textbook chapter by enantioselective homo-enolate alkylation. It was then used to make the simple natural product **W35** in enantiomerically pure form as summarised in the chart. Explain the reactions (including selectivity) and give the structure of the intermediate in the two-stage conversion of **W34** to **W35** Suggest an alternative synthesis of this intermediate.

Answer 13.6: Double alkylation of **149** at the only position where an enolate can form gives **W33** and reduction gives a mixture of diastereoisomers of **W34**. The interesting steps come next.[7] There are two acetal-like functional groups in **W34** and aqueous acid hydrolyses both of them. Our answer starts with protonation at nitrogen and cleavage of one of these groups **W36**. Completion of the hydrolysis gives the hydroxyketone **W38**.

You could equally well have hydrolysed the other **W39** first. Whichever way you do it, hydrolysis of both groups gives the intermediate keto-aldehyde **W41** that can cyclise in only one way by an equilibrium-controlled aldol reaction to give the final product **W35**.

The most obvious alternative synthesis of the 1,5-diCO compound **W41** is by conjugate addition but that might not be asymmetric. Another would be kinetically controlled α′ alkylation (chapter 12) of the parent cyclohexenone. Any homoenolate alkylation from this textbook chapter would also be all right.

Further Examples and Problems Not Directly Drawn from the Textbook Chapter

Titanium Homoenolates from Unsaturated Acetals

Divalent titanium alkoxides **W43** react with acetals **W42** of α,β-unsaturated aldehydes to form titanium homoenolates[8] **W44**. The problem is that they need to react at the α-carbon to become equivalent to homoenolate synthon **W45**.

The solution is curious. If R = Et reaction occurs predominantly at the γ-carbon but if cyclic acetals **W46** are used, reaction occurs mostly at the α-carbon. The products **W47** are *E*:*Z* mixtures but this does not matter if they are to be hydrolysed to aldehydes **W48**.

This discovery opens the way to asymmetric homoenolates if C_2-symmetric diols are used to make the acetals **W49**. Reactions with aldehydes give low enantioselectivities but reactions with imines were excellent. The Ti-homoenolate initially forms with a structure like **W44** but soon equilibrates to **W50**. It is not unusual for transition metals to move from one end of an allylic system to the other, presumably via an η^3 allyl complex. Now reaction is required at the γ-carbon (with respect to Ti) and this is what occurs.

Asymmetric Hoppe Homoenolates

We discussed some of the famous Hoppe homoenolates in the textbook chapter but this seems an appropriate moment to introduce his recent work[9] on asymmetric lithium homoenolates. The starting material **W52** is an allylic carbonate that is also a vinyl silane. Reaction with BuLi catalysed by sparteine removes H^A selectively to give chelated allyl lithium **W53**. Sparteine is a natural product from lupins that is an excellent asymmetric ligand for lithium.

Direct reaction with aldehydes gives homoaldol products with low stereoselectivity (1:1 *syn*:*anti*-**W56**). Transmetallation with titanium goes with inversion to give reactive homoenolate equivalents **W55** (compare with **W50**) that react reliably in the γ-position with aldehydes to give *anti* homoaldol products **W56** in excellent yield and near perfect ee. Removal of silicon with TBAF occurs with retention of configuration at the vinyl ether.

Tandem Cyclopropanation and Homoenolate Formation

Modern developments may not necessarily fit any of the classes of homoenolates we have described. One such example[10] is the cyclopropanation of the stable zinc enolates **W59** of 1,3-dicarbonyl compounds such as **W58** with Simmons-Smith (Furukawa) reagent. The cyclopropanes **W60** are in equilibrium with the zinc enolate **W61** of a 1,4-dicarbonyl compound (hence chain extension). Alternatively silylation to **W64** and a second reaction with the Simmons-Smith reagent adds another carbon to give **W63** and hence **W62**.

Trost's Ruthenium Complexes in Vinylation of Homoenolates Using Alkynes

Even more remote from the material in the textbook is the fascinating use of alkynes and allylic alcohols to produce what looks like a homoenolate vinylation.[11] A silylated alkyne, preferably bearing a functional group **W66** based on O or N, combines with a silylated allylic alcohol catalysed by a ruthenium cation complex. The product **W67** is a diene in reasonable yield and with full control over alkene geometry.

One would like to think that the Ru allyl complex reacts at the γ-position with the middle of the alkyne **W68** (the curly arrows are pure conjecture) but the published suggestion gives only the product of such a reaction **W69**. The future surely promises reagents like this, remotely connected with homoenolates, that produce products difficult to make from conventional homoenolate equivalents.

References

1. M. J. Bradlee and P. Helquist, *Org. Synth.*, 1997, **74**, 137.
2. A. G. M. Barrett and F. Damiani, *J. Org. Chem.*, 1999, **64**, 1410.
3. E. Nakamura, K. Sekiya and I. Kuwajima, *Tetrahedron Lett.*, 1987, **28**, 337.
4. R. F. W. Jackson, N. Wishart, A. Wood, K. James and M. J. Wythes, *J. Org. Chem.*, 1992, **57**, 3397; R. F. W. Jackson and M. Perez-Gonzalez, *Org. Synth.*, 2005, **81**, 77.
5. M. T. Crimmins, J. M. Pace, P. G. Nantermet, A. S. Kim-Meade, J. B. Thomas, S. H. Watterson and A. S. Wagman, *J. Am. Chem. Soc.*, 2000, **122**, 8453.
6. E. J. Corey and P. Ulrich, *Tetrahedron Lett.*, 1975, 3685.
7. J. B. Schwarz and A. I. Meyers, *J. Org. Chem.*, 1998, **63**, 1732; A. G. Waterson and A. I. Meyers, *J. Org. Chem.*, 2000, **65**, 7240.
8. X. Teng, Y. Takayama, S. Okamoto and F. Sato, *J. Am. Chem. Soc.*, 1999, **121**, 11916; S. Okamoto, X. Teng, S. Fujii, Y. Takayama and F. Sato, *J. Am. Chem. Soc.*, 2001, **123**, 3462.
9. J. Reuber, R. Fröhlich and D. Hoppe, *Org. Lett.*, 2004, **6**, 783.
10. R. Hilgenkamp and C. K. Zercher, *Org. Lett.*, 2001, **3**, 3037.
11. B. M. Trost, J.-P. Surivet and F. D. Toste, *J. Am. Chem. Soc.*, 2001, **123**, 2897.

14

Acyl Anion Equivalents

Material from Chapter 14 in the Textbook

It will help if you have chapter 14 in the textbook open for the first section. **Problem 14.1**: In the textbook chapter 14 we considered three main types of acyl anion equivalent: (a) reagents which can be considered as modified acetals, that is *protected* aldehydes, (b) *masked* carbonyl compounds such a nitroalkanes, and (c) substituted vinyl-lithiums. Compare these three types of reagents for the d^1 synthon with the three strategies for the d^3 synthon (homoenolates) presented in chapter 13. You might say, for example, that 'acetal' is purely protective for d^3 but must be anion-stabilising for d^1.

Answer 14.1: (a) Acetals: The two synthons are both nucleophilic and both exhibit *umpolung* so we might expect some similarities. In this chapter we saw acetal-style d^1 reagents based on dithians **21**, monosulfoxides of dithioacetals **W1**, silylated cyanohydrins **69** and amino-nitriles **W2**. The first two use sulfur atoms to stabilise the 'anion' while the last two use the electron-withdrawing cyanide group. Structures **68** and **W2** could be drawn with lithium on nitrogen rather than on carbon.

By contrast in chapter 13 (compound numbers are from chapter 13) we used simple acetals of aldehydes **78**, alcohols **84**, sulfones **W3** or phosphonium ylids **W4** where the role of the acetal seemed purely protective and allowed the formation of a Grignard reagent or lithium derivative at the other end of the molecule. You may object (rightly) that chelation of Mg or Li by an oxygen atom of the acetal does contribute to the stability of the reagent as in **78**.

Workbook for Organic Synthesis: Strategy and Control   Paul Wyatt and Stuart Warren
© 2008 John Wiley & Sons, Ltd

(b) Masked carbonyl compounds: In this chapter we met phosphonium ylids **W5**, lithium derivatives of phosphine oxides **121** (and other phosphorus compounds), and of course anions **W7** of nitroalkanes **W6**. In each case a non-carbonyl functional group stabilises the anion.

W5: ylid from 116   **121**   **W6**   **W7**

We suggest (though you may disagree) that the homoenolates nearest to this idea are the allylic anions bearing silicon **110**, oxygen **125**, or nitrogen **131** functionality (numbers again from chapter 13) and most similar of all, the amino nitrile derivatives **138**. None of these contains a carbonyl group which appears only after hydrolysis of an enol ether or enamine or oxidation of a vinyl silane. You might also have mentioned allyl carbamates.

110   **125**   **131a**   or   **131b**   **138**

(c) Substituted vinyl lithiums: In this chapter the vinyl ether could be simple **81** and used as its lithium derivative **82**. In the homoenolate chapter the directly homologous cyclopropyl ethers (numbers from chapter 13) **69** were used as their lithium derivatives **W9**. The two chapters meet in the lithiation of vinyl ether **W8**. No doubt the proton next to OMe is initially removed but the stable lithium derivative is **85**, identical to the derivative **126** we have just mentioned.

81   **82**   **W8**   **85**   **69**   **W9**

Dithians as Acyl Anion Equivalents

Alkylation of dithian itself **27** with the iodide **32** establishes the 1,4-diO relationship in **34** and the lithium derivative of this new dithian can be acylated with DMF (Me_2NCHO). **Problem 14.2**: Why would, say, HCO_2Et not be a suitable electrophile for this formylation and how does DMF overcome the problem?

32   **27**   1. BuLi / 2. 32   **33**   1. BuLi / 2. DMF Me_2NCHO **34**

Answer 14.2: Acylation at carbon is a classic problem of organic synthesis and is explained in chapter 2. In this case, ethyl formate would react to give the aldehyde **34** in the presence of the lithium derivative **32** and the aldehyde is more electrophilic than ethyl formate so a second addition would occur. By contrast, DMF gives a tetrahedral intermediate **W10** that does not decompose to the aldehyde under the conditions of the reaction but only on work-up **W11** in aqueous acid.

Pyrenophorin

Problem 14.3: When the dimer **36** was formed, Seebach and his co-workers were relieved to find only eleven signals in the ^{13}C NMR spectrum. Why were they relieved? Identify the eleven carbon atoms. Draw a mechanism for the Mitsunobu reaction showing how inversion occurs.

Answer 14.3: The natural diastereoisomer has C_2 symmetry and the two halves are identical. The C_2 axis rises vertically out of the plane of the molecule through the circled dot **36a**. If stereochemical control had been imperfect, either because **24** had low ee or because the Mitsunobu reaction was not completely stereospecific, some of the other (centrosymmetric and therefore achiral) diastereoisomer **W12** would also have been formed and there would have been 22 signals. The eleven carbon atoms are marked on **36a**.

36a C_2 axis of symmetry   W12; centre of symmetry

There is incomplete agreement about the detailed mechanism of the Mitsunobu reaction, but agreement that it goes with inversion. The simplest version of the reaction, using RCO_2H and an alcohol adds Ph_3P to DEAD **W13**, the alcohol to the phosphorus atom, and ensures inversion with an S_N2 reaction by the carboxylate ion **W15**. A proton must be added to each nitrogen atom of DEAD and Ph_3P takes the oxygen atom away to make Ph_3PO.

Synthesis of Acifran

The carbon framework of **38**, the last intermediate in the synthesis of acifran **37**, contains a remarkable number of relationships: two 1,2- and one 1,3-diCO relationships **38a** as well as two 1,4-diCO and a 1,5-diCO. **Problem 14.4**: Identify these last three relationships. Why is a 1,5-diCO disconnection unlikely to be helpful?

37; Acifran   **38**   **38a**

Answer 14.4: The two 1,4 relationships are between C4 and C1 and between C2 and C5 in **38b**. The 1,5 relationship is between C1 and C5. One disconnection (**b** in **38c**) is impossible as we cannot put in the necessary alkene. The other (**a** in **38c**) requires conjugate addition of a d^2 reagent **W18** to a strange alkene **W19** that is not very electrophilic.

38b   **38c**   **W18**   **W19**

Problem 14.5: We prefer odd-numbered to even-numbered disconnections and, of the two possible 1,3-diCO disconnections, **38b** is better as it gives an enolisable ketone **39** and the unenolisable, symmetrical and very electrophilic diethyl oxalate **41**. What would be the starting materials for disconnection **a**? Could this reaction be made to work? How would you prepare the starting materials?

Answer 14.5: An acylating agent **W20** would be required to react with the enolate of pyruvic acid **W21**. This might be possible if both the OH (silyl ether?) in **W20** and the CO_2H (ester?) in **W21** were protected. Pyruvic acid is available and **W20** could in principle be made by addition of a one-carbon d^1 reagent (cyanide?) to acetophenone PhCO.Me.

Dithioacetal Monoxides

Problem 14.6: In the textbook chapter the mono-sulfoxide **51** was used in the synthesis of the tricyclic triether **65**. We can now explore this chemistry a bit further.[1] Comment on the stereochemistry of **51**. Why is it not worthwhile to make one stereoisomer of **51**?

Answer 14.6: There are two chiral centres in **51**: a carbon atom and a sulfur atom marked with circles in **51a** There are two diastereoisomers **51b** and **51c**. As each is chiral each also has two enantiomers. It is not worth separating the diastereoisomers of **51**, nor resolving either into enantiomers as both centres disappear in the formation of **54**. As all isomers of **53** give the same product – a mixture of epimers of **54** – purification of **63** would be a mistake.

Vinyl Ethers as d^1 Reagents

This reagent **75** (R = EthoxyEthoxy) is easily prepared and used in the synthesis of amides **77** by alkylation at what is to become the carbonyl group. **Problem 14.7**: Suggest a synthesis of **75** and a mechanism for the last step – the conversion of **76** into **77**.

Answer 14.7: Since **74** is a derivative of malononitrile, the obvious starting point is some derivative of malonic acid. As it happens, diethyl bromo-malonate is available, so the chemists started from there and used a series of remarkably simple reactions. One problem is finding a non-basic oxygen nucleophile – acetate ion is a good answer – and another is to dehydrate the amides. The Burgess reagent, an inner salt **W26** works best.[2]

They suggest that an acyl cyanide **W28**, formed by elimination **W27** on the cyanohydrin **76**, is an intermediate in the formation of **77**.

Problem 14.8: The starting material **90**, used in the textbook for the synthesis of phyllanthoside is made as follows. Explain the bromination of the acetal **W31**.

Answer 14.8: At first sight this bromination is surprising but reflect on the behaviour of acetals in acid solution. The reaction begins as if the acetal were to be hydrolysed by protonation of **W31** to give **W33** but there is no water present so the acetal forms the enol ether that can react

W35 with bromine from Br$_3^-$. The product can reclose the acetal **W36** to give **90**. Normally acetals protect against unwanted reactions: here the acetal activates towards a wanted reaction.

Synthesis of Pederin

Problem 14.9: The reagents for the conversion of **102** to **96** were 1. NaIO$_4$, 2. heat, 3. adjustment of protecting groups on nitrogen. Explain the action of the first two reagents.

Answer 14.9: Oxidation of the selenide to the selenoxide prepares the way for a concerted elimination by a cyclic mechanism. As the side chain is free to rotate, both diastereoisomers (Se is chiral in the selenoxide) give the same alkene.[3]

Problem 14.10: The stannane **100** was prepared by the sequence below. Explain the various steps.

Answer 14.10: The selenide anion is very good at S$_N$2 and poor at attack on a carbonyl group so it opens the lactone **W41**. The ester in **W38** is reductively cleaved by the unusual combination of BuLi and DIBAL. This gives *i*-Bu$_2$AlHBu.Li and prevents epimerisation by strong bases. Finally the enol triflate couples with Me$_3$Sn–SnMe$_3$ with palladium catalysis. If you drew a full mechanism for this you should have started by oxidative insertion of Pd(0) into the C−O bond of the vinyl triflate (chapter 8).

Lithium Derivatives of Allenyl Ethers

In the textbook we used the allenyl lithium **105** as a reagent for the synthon **108**. **Problem 14.11**: Draw the orbital of the allene that is protonated during the hydrolysis and convince yourself that this will give the Z-alkene in **107**.

Answer 14.11: The key point is that the two alkenes of the allene are orthogonal. Protonation **W43** must therefore occur in the plane of the terminal alkene.[4] Diagram **W44** shows as circles the p-orbitals that overlap to form the terminal alkene and the p-orbital that is being protonated in the plane of the paper (THF refers to the rest of the molecule – the tetrahydrofuran). Protonation occurs away from R to give initially **W45**.

Synthesis of *O*-Methyljoubertiamine

A brief discussion in the textbook focused on enamines prepared from aminomethyl-phosphorus reagents such as **121**. This is the further discussion on strategy promised there. The alkaloid is an amino enone **W46**. Removal of the amine reveals an aldehyde **W47** and further aldol disconnection of the α,β-unsaturated ketone reveals a ketodialdehyde **W48** with 1,4- and 1,5-dicarbonyl relationships.

W46: *O*-methyljoubertiamine   W47   W48

Clearly **W48** is an unstable intermediate whose cyclisation might go in a number of possible ways so the decision was taken to tackle the 1,4-dicarbonyl relationship first by a reconnection strategy. This allowed the double disconnection **W49** back to the simple 1,4-diketone **W50** available by acylation of the Grignard reagent from **W51** with a protected ketonitrile **W52**. This is an unusual strategy. Normally a 1,5-dicarbonyl compound would be made by conjugate addition but here the quaternary centre dominates and the key reaction is **W50** to **W49**.

W49   W50   W51   W52

Problem 14.12: The keto-group in **W49** was protected as the acetal seen in compound **130** in the main text. Suggest how the final stages of the synthesis from **131** to **W46** might be performed.

131   W46: *O*-methyljoubertiamine

Answer 14.12: Oxidative cleavage (ozone) gives the aldehyde **W47** and then reductive amination with $Me_2NH.HCl$ and $NaB(CN)H_3$ in *t*-BuOH gives the alkaloid itself.[5]

Nitroalkanes as d^1 Reagents

Problem 14.13: Tetrahydropyran **152** is in equilibrium in base with an open chain compound. What is the structure of this compound, how is **153** formed, and why is it immediately converted into **154**?

152   +   NO2   base   153 NO2   154 NO2

Answer 14.13: Tetrahydropyran **152** is a hemiacetal so it is in equilibrium with the hydroxy-aldehyde **152a**. The aldehyde condenses with nitroethane in a 'nitro-aldol' reaction[6] to give **W53**. The very powerful electron-withdrawing effect of the nitro group now ensures dehydration to **153** by the E1cB mechanism and rapid conjugate addition to give **154**.

152   152a   +   NO2   base   W53   base   153   → **154**

In Raphael's strigol synthesis, the hydroxyketone **162**; R=H was the originally proposed intermediate. When the reaction of the nitroketone with $TiCl_3$ was carried out, the diketone **169** was actually formed.

160; strigol   161   162

166   165 / base   167 $H^{\oplus}$ 168 $TiCl_3$ / H_2O 169

Problem 14.14: Do you think reduction of **169** to **162** is reasonable, under the conditions of the reaction, given the successful conversion of **139** to **W54**? Explain.

162; R = H   ? $TiCl_3$ / H_2O 169   HO   OH / $H^{\oplus}$ W54

Answer 14.14: Entirely reasonable as $TiCl_3$ is a reducing agent. The difference is presumably that acetal formation is under thermodynamic control so the conjugated enone is retained.[7]

Further Examples and Problems not Directly Drawn from the Textbook

Problem 14.15: Yes, this reaction is in the textbook but now we ask you to draw a detailed mechanism for the conversion of **113** into **114** with propargyl silane.

Answer 14.15: Propargyl silanes behave like allyl silanes (chapter 12) and react with electrophiles at the end remote from silicon[8] **W55** so that a β-silyl cation **W56** is an intermediate. The silyl triflate removes the OMe group from the acetals to give the cation. The Me$_3$Si group is probably removed by MeOH. We draw only one of the reactions and delay a discussion of stereochemistry until the next problem.

Problem 14.16: The starting material **113** was made from the simple enone **W57** as shown. Account for the reactions. Do not get too involved with the stereochemistry. Nelson warns; 'The cyclisation of the epoxydiketone...(**W59**)... was central to the success of our synthesis...the transformation...exhibits an exquisite combination of kinetic and thermodynamic control.'

Answer 14.16: The first step is a self metathesis between two molecules of **W57** to give **W58** with the loss of ethylene gas. It is most simply seen by the diagram **W60** where the dotted double arrows show the atoms that join and the dotted line shows the bonds that split. Metathesis is discussed in detail in chapter 15. The *E*-alkene **W58** is expected. The epoxide **W59** must be *trans*. Acid protonates the epoxide which is opened by the ketone **W61** (only half the molecule is shown) and addition of MeOH to the cation **W62** gives (half of) **113**.

In fact things are much more complicated. The kinetic product of the cyclisation is the *bis*-THF **W63** rather than **113**. This equilibrates to **113** preserving the stereospecific opening of the epoxide but needing the anomeric effect (OMe prefers to be axial if next to a ring oxygen) **113a** to get the stereochemistry at the two acetals. The stereochemistry of the acetals in **W63** is not controlled as there is no anomeric effect in five-membered rings.

The rearrangement presumably occurs by protonation and loss of methanol from **W63** and participation by the ether oxygen atom **W64** to give the 6/6 fused system **W66** that can pick up methanol axially and, by equilibration at the other acetal centre, give **113a**.

Problem 14.17: Suggest how the aldehyde **52** might be made by the scheme outlined here. Notice that a *Z*-alkene is required. The protecting group R could be benzyl or a hindered silyl group.

Answer 14.17: The first two steps are the Corey-Fuchs method of making substituted alkynes from aldehydes.[9] The first reaction is a Wittig with the ylid **W74** made in an unusual way from Ph₃P and CBr₄ without base. Nucleophilic attack at the carbon atom of CBr₄ is impossible because the four Br atoms surround it with electrons. Attack at Br **W70** is possible and nucleophilic displacement at phosphorus now leads to **W73**. A second molecule of Ph₃P also attacks at bromine to form the ylid **W74** and a second molecule of **W71**.

Treatment of the dibromoalkene **W68** with a two-fold excess of BuLi gives the lithium derivative **W76** of the alkyne. Elimination of HBr with the first molecule of BuLi gives the bromo-alkyne **W75** and Br/Li exchange gives **W76** that can be trapped with electrophiles: formaldehyde gives the alcohol **W69**.

All that remains is Lindlar reduction to give the *cis*-alkene and oxidation. Swern with $(COCl)_2$ and DMSO followed by Et_3N is best for allylic alcohols as equilibration of **52** to the *trans* alkene is all too easy. This sequence has been followed[10] with R = Bn and with R = $SiMe_2t$-Bu.

Other d¹ Reagents: TosMIC

There are now many other d¹ reagents based on the symmetrical or unsymmetrical acetal model but we have space for only one more. This is the TosMIC reagent of van Leusen.[11] This belongs to the *iso*cyanides,[12] with a sulfone to help stabilise the anion, so that the vital carbon atom is trapped between a sulfur and a nitrogen atom, and the basic reagent **W79** is made from formaldehyde in two simple steps.[13] Metallation with BuLi followed by alkylation and then acylation gives **W81** which can be hydrolysed to the 1,2-dione **W82**. This is simple d¹ + a¹ synthesis.

Further TosMIC examples include the synthesis of heterocycles.[14] The synthesis of imidazoles **W86** involves addition of **W79** to imines followed by a *5-endo-dig* cyclisation to give the anion **W84** that transfers a proton and eliminates **W85** to give the imidazole.

The synthesis of pyrroles **W90** from enones **W87** by conjugate addition of the lithium derivative of TosMIC **W79** and the same style of cyclisation **W88** leads to 3,4-disubstituted pyrroles that are tricky to make by electrophilic substitution.

Problem 14.18: Complete the mechanism for the transformation of **W89** into **W90** and give the disconnections for this pyrrole synthesis.

Answer 14.18: This time it is the enolate proton that is exchanged allowing the elimination of *p*-toluene sulfinate **W91** giving a compound **W92** that needs only a proton shift (could be a [1,5]-sigmatropic H shift) to give the pyrrole **W90**. The disconnections of the pyrrole synthesis are shown on **W90a**: the 1,3-dipole **W93** may or may not be an intermediate in the synthesis.

References

1. K. Fujiwara, K. Saka, D. Takaoka and A. Murai, *Synlett.*, 1999, 1037; K. Fujiwara, D. Takaoka, K. Kusumi, K. Kawai and A. Murai, *Synlett.*, 2001, 691.
2. H. Nemoto, Y. Kubota and Y. Yamamoto, *J. Org. Chem.*, 1990, **55**, 4515.
3. P. Kocienski, R. Narquizian, P. Raubo, C. Smith, L. J. Farrugia, K. Muir and F. T. Boyle, *J. Chem. Soc., Perkin Trans. 1*, 2000, 2357.
4. P. Kocienski, R. C. D. Brown, A. Pommier, M. Procter and B. Schmidt, *J. Chem. Soc., Perkin Trans. 1*, 1998, 9.
5. S. F. Martin and T. A. Puckette, *Tetrahedron Lett.*, 1978, 4229; S. F. Martin, G. W. Phillips, T. A. Puckette and J. A. Colapret, *J. Am. Chem. Soc.*, 1980, **102**, 5866.
6. R. Ballini, M. Petrini, E. Marcantoni and G. Rosini, *Synthesis*, 1988, 231.
7. G. MacAlpine, R. A. Raphael, A. Shaw, A. W. Taylor and H.-J. Wild, *J. Chem. Soc., Chem. Commun.*, 1974, 834; *J. Chem. Soc., Perkin Trans. 1,* 1976, 410.
8. J. M. Holland, M. Lewis and A. Nelson, *Angew. Chem., Int. Ed.*, 2001, **40**, 4082.
9. E. J. Corey and P. L. Fuchs, *Tetrahedron Lett.*, 1972, 3769.
10. K. Fujiwara, D. Takaoka, K. Kusumi, K. Kawai and A. Murai, *Synlett.*, 2001, 691.
11. A. M. van Leusen, *Tetrahedron Lett.*, 1977, 4233.
12. D. Hoppe, *Angew. Chem., Int. Ed. Engl.*, 1974, **13**, 789.
13. A. M. van Leusen, *Org. Synth.*, 1977, **57**, 8, 102.
14. A. M. van Leusen, *J. Org. Chem.*, 1977, **42**, 1153; A. M. van Leusen, *Helv. Chim. Acta*, 1976, **50**, 1698; A. M. van Leusen, *Tetrahedron Lett.*, 1975, 3487.

C
Carbon–Carbon Double Bonds

15

Synthesis of Double Bonds
of Defined Stereochemistry

E- and Z-Alkenes

We have the opportunity here to mention nomenclature. Alkenes are defined as E or Z according to whether the two higher ranking substituents are on the same side (Z for *zusammen* – German for together) or on opposite sides (E for *entgegen* – German for against). The nomenclature experts have thus chosen the only language in the world where the letter for *cis* looks like *trans* and the letter for *trans* looks like *cis*. You can remember this system by the malice of its inventors. Notice that both the enone **W1** and the enolate **W2** are alkenes with the same stereochemistry – that is they have a *cis* double bond inside the six-membered ring. However **W1** is a Z-alkene while **W2** is an E-alkene! **Problem 15.1**: Assign the stereochemistry for the *cis*-alkenes **W3–W6**.

Answer 15.1: Alkenes **W3** and **W5** are Z-alkenes while **W4** and **W6** are E-alkenes. So we may need two systems: E and Z for published papers and *cis* and *trans* for ordinary chemical discussion but they must be accompanied by a diagram.

The cis-Alkene is Formed Stereoselectively

Commercial 1-methoxy-but-1-en-3-yne **W8** is the Z compound because it is made by addition of methanol to butadiyne. **Problem 15.2**: Suggest a mechanism that explains the stereochemistry.

Answer 15.2: Addition of MeOH (or methoxide) to the terminus of one of the alkynes **W9** gives an anion delocalised into the other alkyne. The negative charge is in a p-orbital in the plane of the new alkene and will prefer to add a proton **W10** (from MeOH?) *anti* to OMe.

Only One Alkene is Possible. This is Usually a Cyclic cis *Alkene*

A new enone synthesis was developed in 2001 by Ballini and co-workers.[1] Oxidation of the furan **W11** with a peracid gives the *cis*-enedione **W12** in very high yield. The new enone synthesis is to treat **W12** with an aliphatic nitro compound and base. Enones **W13** are formed at room temperature in good yield. **Problem 15.3**: Explain the formation of **W12** and why it is formed as the less stable Z-isomer.

Answer 15.3: Epoxidation of either alkene gives a strained acetal that could open up in two ways. Only **W14** leads to the product by two successive ring openings. The alkene has to be *cis* as it is already there inside the ring when the ring is opened **W15**.

Problem 15.4: The enone **W13** was immediately hydrogenated and treated with acid to give a new furan. What are the structures of the compounds in this sequence?

Answer 15.4: The saturated 1,4-diketone **W16** is formed by hydrogenation of **W13**. These compounds are notoriously unstable in acid or base, giving cyclopentenones in base (chapter 6) and furans in acid. The final product **W17** looks as though it might be made directly from **W11** by electrophilic substitution but furan much prefers to react at C-2 or C-5.

The Wittig Approach to Alkenes

The insect pheromone **W18** is made almost totally stereoselectively using a lithium salt-free ylid made from the phosphonium salt with various bases such as sodium hexamethyldisilazide in THF.

The most completely salt-free ylids are made with sodium amide in benzene.[2] The sodium salt of the counterion of the phosphonium salt precipitates and a benzene solution of the ylid results. This gives almost pure Z-**W18**.

The Horner-Wadsworth-Emmons (HWE) Method

In the textbook we said 'The extra stabilisation (in conjugated ylids) makes the ylid rather unreactive and phosphonate esters **91** are often used instead of phosphonium salts in these reactions. Treatment with a base (NaH or RO⁻ is often used, BuLi will certainly not do) gives an inherently more reactive enolate anion **92** rather than an ylid. These Horner-Wadsworth-Emmons reagents[3] (HWE as we shall call them, though they go under many other names) react with ketones as well as aldehydes and the product is normally the E-alkene **93**.' **Problem 15.5**: Why will BuLi 'certainly not do'? How is it that EtO⁻ is a good enough base?

Answer 15.5: You will have realised that BuLi might attack the carbonyl group as a nucleophile rather than a base. It might not be so obvious that BuLi could also displace EtO⁻ (twice!) from phosphorus. Not only is EtO⁻ a 'good enough' base, it is the best base as it does not matter if it attacks C=O or P=O as a nucleophile since it would merely recreate **91**.

Polyzonimine

In the synthesis of polyzonimine described in the textbook, the HWE reaction between **91** and **96** actually gave a 3:2 mixture of **98:97** that was isomerised to E-**98** with acid in refluxing CH_2Cl_2. **Problem 15.6**: How and why did this happen? Comment on the stereochemistry of polyzonimine.

Answer 15.6: The product of the HWE reaction must initially be **98** but this will be in equilibrium with the extended enolate **W19** (chapter 11) that can be protonated at the α-position **W19** to give **97**. Acid evidently equilibrates these two products.[4] The *Z* isomer of **W98** has the carbonyl oxygen between the two methyl groups and is evidently too unstable. Polyzonimine is a *spiro* compound (two rings sharing one atom): each ring is orthogonal to the other. The plane of neither ring is a plane of symmetry and polyzonimine is chiral.

The Horner-Wittig Method

The essence of the Horner-Wittig is that lithium derivatives **120** of phosphine oxides **119** contain a genuine C–Li bond and react stereoselectively with aldehydes to give stable adducts **121** that in turn give the highly crystalline alcohols *syn*-**122**, easily purified by crystallisation and, if necessary, chromatography.[5]

In more detail, the lithium derivative **120** probably reacts with the aldehyde by first forming a Li–O bond. Only then does the C–Li bond attack the carbonyl group **W21** forming a folded transition state **W21a** with large substituents preferring the *exo* (or convex) face (below, as drawn).

Strategy of Synthesis of Alkenes by Wittig-Style Methods

A particularly interesting example comes from the world of macrolides – large ring lactones – where a synthesis of the natural product vermiculine **W22** (an antibiotic from a *Penicillium* mould) uses an *E*-selective Wittig reaction to control the regio (chemo?) selectivity of a cyclisation reaction. Vermiculine has a 16-membered ring containing two lactones, two ketone groups and two *E*-alkenes inside the ring. Because the whole structure is C_2 symmetrical it makes sense to disconnect both alkenes to give two identical starting materials **W23** or **W24**. There is a choice of Wittig reagents – we can put the Ph_3P^+ group at either end because it will give a stabilised ylid in both cases and hence an *E*-selective reaction. Notice that the starting materials **W23** are different from **W24**, but that *each set of two starting materials are the same*. Each proposed reaction is a dimerisation.

W22; vermiculine   **W23**   **W24**

Though we are happy about the phosphonium end of both starting materials, we are less happy about the other end where we have a rather unstable 1,2-dicarbonyl compound. Also we should rather not have free ketones in the molecule when we are going to do a Wittig reaction elsewhere. For both reasons, Burri and his group[6] at Hoffmann-La Roche decided to mask both ketones, one as an acetal and one as a dithiane, and to use a protected version of starting material **W23**. Both problems are solved by this strategy.

(half of) W23   **W25**   **W26**

Now another problem arises: will the Wittig reaction occur intramolecularly to give an eight-membered ring, **W26** or intermolecularly to give vermiculine? Normally intramolecular reactions are preferred but not when they give medium rings (8 to 14-membered). In practice they used the HWE reagent **W27** to get a 50% yield of the dimeric macrolide **W28** and none of the eight-membered ring **W26** at high dilution. One factor must be that the molecule can easily have two *E*-alkenes in the 16-membered macrolide, but will have to accept a *Z*-alkene in the eight-membered ring (see this chapter in the textbook). **Problem 15.7**: How might you deprotect **W28** to make vermiculine **W22**? And – rather more difficult – how would you make the starting material **W27**?

Answer 15.7: *Deprotection*: There are many ways to remove a dithian. You might have tried acid- or Hg(II)-catalysed hydrolysis, methylation, or oxidative methods. There is no certain way of forecasting which will work. In practice reaction with *N*-iodosuccinimide removed the dithians in 90% yield and the easier removal of the acetals went best with acetone (i.e. transacetalisation) and BF$_3$ etherate.

W27

W28; protected vermiculine

Preparation of the Starting Material W27: The obvious strategy is to keep the dithian, to get more value from it in the synthesis, but to remove the acetal and the ester. This reveals an obvious aldol product from acetone and the dial **W30** that can be made from dithian itself. There are problems of selectivity here as we need to react one aldehyde and not the other but no doubt these could be overcome. There are of course many other possibilities.

W27a   W29   W30
some protection
needed for CHO

In fact the chemists used a rather different strategy because they wanted to resolve an intermediate to make enantiomerically pure vermiculine. So, having added one CHO group to the simple alkylated dithian **W31**, they carried out a 1,3-dipolar cycloaddition **W32** of acetonitrile *N*-oxide onto the isolated alkene to form an isoxazoline **W33**. Though it may not be obvious, the correct side chain is there.

W31   W32; 75% yield   W33; 90% yield

Reductive cleavage of the isoxazoline with DIBAL gave one diastereoisomer of the amino-alcohol but annoyingly also reduced the aldehyde. Choosing to regard this as a kind of protection, they resolved the amino diol **W34** (chapter 22) before oxidising the amine with 3,5-di-*t*-butyl *ortho*quinone to reveal the side chain in **W35**. All compounds are now single enantiomers. Protection of the ketone by acetal exchange, acylation of both OH groups with chloroacetic anhydride and deacylation of the less hindered primary alcohol gave **W37**. All that remains is Swern-style oxidation of the primary alcohol and conversion of the primary chloride to the phosphonate in an Arbuzov reaction.

Crossing the Stereochemical Divide in the Wittig Reaction

The Schlosser Modification

An ingenious extension of Schlosser's ideas demonstrates that extra base is needed for the equilibration. The 3-hydroxy phosphonium salt **W38** is easily made from the simple ylid **63** and an epoxide and can be protected as a silyl ether **W39**.

A Wittig reaction with the corresponding silyl ether **W39** then gives normal stereo-selectivity – a predominance of the *Z*-alkene **W40**. The free alcohol **W38** gives exclusively *E*-alkene **W41** with the same aldehyde.[7] **Problem 15.8**: Give details of this process.

Answer 15.8: The first molecule of BuLi removes the proton from the OH group **W42** but the second forms the ylid **W43** This alkoxide acts as a base to equilibrate the stereochemistry of the Wittig intermediate **W44** via **W45** along these lines:

More HWEs and the Charette Version of the Julia Method

This chemistry was used in the asymmetric synthesis of the natural product **W56** (+)-U-106305, a compound that inhibits cholesterol metabolism. **Problem 15.9**: Explain the stereoselectivity of each olefination step and suggest why that method was chosen.

Answer 15.9: The formation of **W47**, **W49** and **W56** are HWE reactions, the last showing how versatile this method is as the reagent **W55** has an acidic amide proton. The formation of **W53** is the Charette modification of the Julia method.[8] The benzothiazole **164** starts a normal Julia reaction but the thiazole is electrophilic enough to cause automatic cleavage of the sulfone (**158** to **160**). These diagrams are from the chapter in the textbook where there is more explanation. The Julia reaction was chosen because the great anion-stabilising ability of the thiazole sulfone minimises opening of the cyclopropane ring. The Charette version was chosen as merely using the appropriate solvent produces good *E*-selectivity.

The Kocienski Modification of the Julia Reaction

The reaction uses tetrazole sulfones **167**, hexamethyldisilazide bases, and is again sensitive to solvent. Polar solvents (dimethoxyethane, DME, is the best) and $(Me_3Si)_2NK$ give high *E*-selectivity (usually >95:5 *E:Z*) while non-polar solvents (toluene is the best) with $(Me_3Si)_2NLi$ give moderate *Z*-selectivity. **Problem 15.10**: Draw a mechanism to show the details of this modified Julia reaction.

Answer 15.10: Tetrazoles are even more electron-withdrawing than thiazoles so the spontaneous loss of the sulfone is even easier (**W57** and **W58**) with the alkene being *E*- or *Z*-**150** depending on the conditions.[9]

The Synthesis of Nafuredin

We describe in the textbook chapter how the central strategic *E*-alkene in nafuredin **168** was put in by a Kocienski-modified Julia reaction between the sulfone **169** and the appropriate aldehyde **170**.

Problem 15.11: The Julia reaction gave only the *E*-alkene **W60**. What might have gone wrong with this reaction, or any other method involving strong base? What is the structure of **W60** and how might it be converted into nafuredin **168**?

Answer 15.11: The structure of **W60** is shown below. It is amazing that the mesylate in **172** survives this reaction and that the chiral centre next to the aldehyde in **170** does not epimerise. All that remains is to remove the protecting groups, oxidise the OTIPS centre to the lactone, and close the epoxide. As so often happens, this proved annoyingly difficult. The epoxide was closed with NaH, the benzyl group was replaced by an allyl-oxy-carbonyl (Alloc), the TIPS group removed with HF in pyridine, the lactol oxidised to the lactone with the Dess-Martin periodinane, and finally the Alloc group was removed.[10]

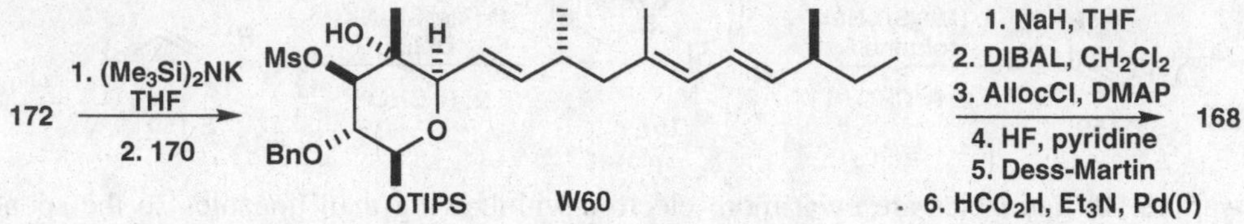

Problem 15.12: The sulfone **172** for the Julia reaction was prepared from **171**, in turn prepared from glucose (chiral pool strategy – chapter 23) *via* the protected intermediate **W61**. Suggest how the alkene in **171** might be introduced and explain what is going on in the formation of **172**.

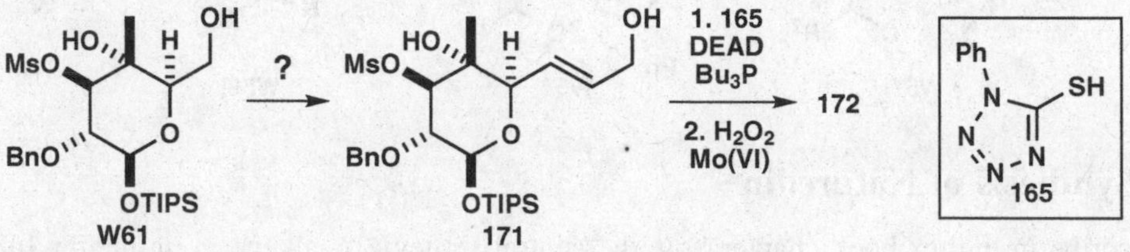

Answer 15.12: The simplest way is to oxidise the primary alcohol to the aldehyde **W63** (they used their favourite Dess-Martin periodinane) and carry out some Wittig-style olefination. They used HWE with the allyl ester so that they could remove the allyl group with Pd(0), activate the acid as the anhydride with $ClCO_2Me$ and reduce it to the alcohol **171**. The other two alkenes in **170** were also made by *E*-selective Wittigs with stabilised ylids or HWE reactions.

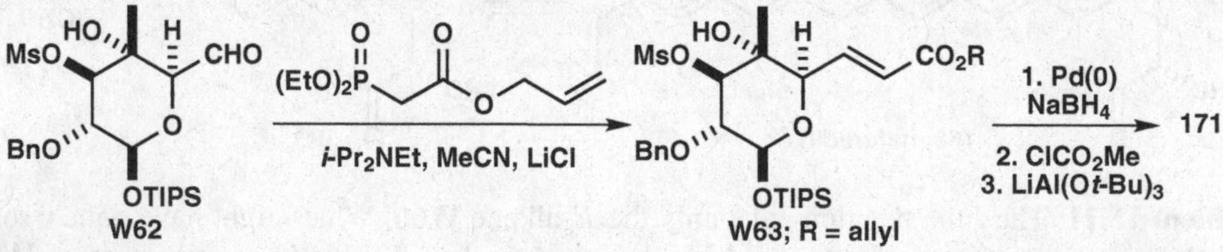

The McMurry Reaction

We described but did not explain this useful reaction in the textbook. Two molecules of the same aldehyde are coupled with low-valent titanium to give an alkene. There is no general agreement about the details of the mechanism but it seems clear that titanium forms Ti−O bonds by donating one or more electrons to the carbonyl group **W64** and then forming a C−C single bond, rather in the style of the pinacol reaction. The next step is different. Titanium has a strong affinity for oxygen and breaks both C−O bonds, perhaps in a process like **W65** to leave the alkene. The preferred *E* stereochemistry of the alkene suggests that **W65** has *anti* stereochemistry.

We gave an example of an indole **178** being formed from a keto-diamide **177** in an intramolecular McMurry. **Problem 15.13**: Comment on the chemoselectivity of the McMurry step. Explain where the chirality comes from in **177**.

Answer 15.13: The ketone is the most electrophilic of the three carbonyl groups so it will couple with one of the less electrophilic amides. An intramolecular reaction will be favourable if a five-membered ring **178** is formed but not if the other amide reacts as an eight-membered ring would result.[11] The starting material **177** is clearly made from proline (chiral pool strategy – chapter 23).

A Synthesis of Tamoxifen Using the McMurry Reaction

Tamoxifen, an antiestrogen, is a leading breast cancer drug and is metabolised *in vivo* into 4-hydroxytamoxifen **W66**. There is obviously a serious stereochemistry problem in making the active *Z*-isomers of these drugs as the difference between the two aromatic rings at the southern end of the molecule is slight and distant. Amazingly, one successful approach[12] uses a McMurry reaction from starting materials such as **W67** and **W68**.

The obvious starting material is 4,4′-dihydroxybenzophenone **W69** which could be monoalkylated in around 50% yield in base. But then a new problem emerged. The McMurry reaction gave a good yield of **W66** but the compound was mostly the inactive *E*-isomer. It turned out that *E*/*Z* isomerisation was exceptionally easy, especially on chromatography, presumably because the electron-rich alkene is easily protonated.

The deduction from many experiments was that a free hydroxyl group on a benzene ring prefers to be *trans* to the ethyl chain. As they say: 'Current knowledge of mechanisms does not provide an explanation.' The solution was easy: protect **W69** with something removable (a pivaloyl group was chosen) and transform it into the amine side chain later. This way they got a 99:1 *Z*:*E* ratio.

Alkene Synthesis by Olefin Metathesis

When we introduced metathesis in the textbook we said that: 'The first cycle is different from the main reaction in that the other product is styrene and the catalyst then changes to the methylene complex **187**. The reaction goes from left to right as the other product is gaseous ethylene.' **Problem 15.14**: Draw a mechanism for the second cycle in which **179** combines with **187**.

Answer 15.14: The initial sequence of 2 + 2 cycloadditions and reversions is the same but without the phenyl group. This means that the conversion of **W71** to **W72** drives the reaction as ethylene gas is lost in this step.

Problem 15.15: In the second cycle, what happens if the catalyst **187** adds the 'wrong way round' to the next molecule of **179**?

Answer 15.15: It does not matter as the cyclobutane is then symmetrical and metathesises **W74** back to starting materials.

In the textbook we gave a simple example: the synthesis of analogues of the antibiotic and antifungal streptazolin by Cossy.[13] Enantiomerically pure diene carbamate **188**, prepared from the chiral pool, was treated with the Grubbs catalyst to form the six-membered ring **189** required for (−)-4,5-dihydrostreptazolin **190**. This metathesis product inevitably contains a Z-alkene. **Problem 15.16**: Suggest how **188** might be made from protected glyceraldehyde **W75** and how **189** might be converted into **190**.

Answer 15.16: The right-hand part of **188** obviously comes from glyceraldehyde so we can disconnect the carbamate and put in the acetal. Now disconnection of the butenyl side chain from nitrogen **W76** and a reagent for the vinyl anion synthon (chapter 14) from **W77** reveals the certainly unstable imine of protected glyceraldehyde **W75**.

The chemists chose to carry out the synthesis with an easily removed *p*-methoxybenzyl (PMB) group on nitrogen so that a stable imine would be formed, and to use vinyl Grignard for the addition. The rest is straightforward except that methyl chloroformate debenzylated the amine and gave the carbamate **W83** in one step.

Alkenes by Sigmatropic Rearrangement

Problem 15.17: In the textbook we explained why this Cope rearrangement *E*-**216** was successful but eight-membered rings are not normally very stable: what makes **217** much more stable than **216**?

Answer 15.17: Eight-membered rings are unstable because of transannular interactions between substituents, even H atoms on CH_2 groups. The two *Z*-alkenes in **217** have no inward-pointing substituents and so avoid transannular interactions.[14]

[2,3]-Sigmatropic Rearrangements: The Wittig Rearrangement

In the textbook we gave details of the [2,3]-Wittig rearrangement (**219** to **221**) and mentioned the [1,2]-Wittig rearrangement (**219** to **218**). The stereochemistry of the alkene in the final product **222** comes from a chair-like transition state with the substituent R taking a pseudo-equatorial position by choice. **Problem 15.18**: Show how this works with a conformational drawing.

Answer 15.18: There are various ways to do this all something like **W84**. If R is pseudo-equatorial the *E* stereochemistry of the new alkene is already in place. This may be easier to see with the H atoms inserted.

The [2,3]-Wittig is a normal sigmatropic rearrangement but the [1,2]-Wittig is something else. The mechanism **W85** cannot be taken seriously as it would requite inversion (it is very like an S_N2 reaction). There is evidence that a radical intermediate **W87** is involved. This can also be drawn as **W88** and can recombine to give **218**.

Reduction of Alkynes

We described the synthesis of the acetal **235** in the textbook. This was used to make a trisub-stituted alkene **W91**. **Problem 15.19**: Explain the stereoselectivity.

Answer 15.19: The reaction is an aliphatic Claisen rearrangement and takes place by a [3,3]-sigmatropic shift **W92** controlled by the preference for the large substituent to occupy an equatorial position **W93**. Note that the OEt and Me substituents remain on trigonal carbon throughout.[15]

Using Cyclic Compounds

We stated in the textbook that simple pyridines, such as 2-methylpyridine **248**, can be alkylated and reduced to leave just one necessarily Z-alkene not conjugated with nitrogen **251**. **Problem 15.20**: Suggest an explanation for the reduction of the pyridinium salt.

248
2-methylpyridine   249

Answer 15.20: The first reduction is easy: the imine salt is reduced. The next step is claimed to be the protonation of the extended enamine **W94** by the solvent (water or an alcohol) to give a new iminium salt **W95** that is again reduced next to nitrogen. Only when the remaining alkene is not an enamine **251** does this sequence stop.[16]

Problem 15.21: We described the cyclisation of the hydroxy-ester **258** so that the alkene formed by elimination on the lactone **259** was forced to be Z. But how might **258** be made?

Answer 15.21: The obvious way is a controlled aldol reaction. As both components can give enolates, the lithium enolate **W96** of EtOAc was formed first.[17]

This synthesis was done partly to find out the stereochemistry of the natural product (verrucarin) and partly to be able to make analogues as verrucarin is too toxic to be useful as a drug. It turned out that Z,E,Z-**255** was not the same as the natural product so they made E,E,Z-**255** which was.

Stereospecific Interconversion of *E*- and *Z*-Isomers

In describing a good method for switching from *Z* to *E* and vice versa, we said that the epoxide **285** is made without change of stereochemistry and opening this epoxide with the very nucleophilic Ph$_2$PLi ensures one inversion in the formation of the phosphine **286**.

The synthesis of Ph$_2$PLi is interesting. The very cheap Ph$_3$P is treated with lithium metal: this adds two electrons to Ph$_3$P causing the loss of a phenyl anion so that the immediate products are PhLi and Ph$_2$PLi that cannot be separated physically. Addition of *t*-BuCl decomposes PhLi by elimination leaving a solution of Ph$_2$PLi ready for use.[18] The other products are benzene, LiCl and isobutene. You will notice that PhLi is the stronger base but Ph$_2$PLi is the better nucleophile for saturated carbon.

Further Examples of the HWE Reaction

Formyl phosphonates **W99** for the synthesis of enals **W100** by the HWE reaction can be made by the formylation of lithium derivatives of alkylphosphonates prepared directly from triethyl phosphate and alkyl lithiums.[19]

The weakly basic nature of the anions formed from phosphonates such as **91** is well demonstrated by the lack of any racemisation[20] in the HWE reaction with enolisable aldehydes such as (+)-2-*O*-benzylthreitol **W102** giving **W103** in >97:3 *E*:*Z*.

Reduction of Acetylenes

In his exploration of stereo- and regioselectivity in a Prins cyclisation to give tetrahydropyrans (THPs) with three chiral centres, Rychnowsky needed both *E*- and *Z*-isomers of the homoallylic alcohol **104**. The reduction of the alkyne **105** is the obvious route.[21]

Lewis acid catalysed addition of lithiated propyne **107** (propyne and BuLi) to the epoxide **106** gave the alkyne **105** in 96% yield and reduction with Raney nickel followed by acylation gave the Z-acetate **108**. Alternatively, reduction with sodium in liquid ammonia and acylation gave E-**108** in 77% yield.

Cyclisation to THPs showed this stereochemical control to be worthwhile as DIBAL reduction of **108** gave the E- and Z-isomers of **109** that each cyclised to a different stereoisomer of the THP: all *syn*-**110** from Z-**109** and **111** from E-**109**.

The most likely explanation is that concerted cyclisation of the alkene onto the oxonium ion (whose methyl group chooses to go equatorial) **112** preserves the stereochemistry of the alkene and that *trans* addition to the alkyne establishes the final centre. This neatly leads us to the subjects of chapters 16 and 17.

References

1. R. Ballini, G. Bosica, D. Fiorini and G. Giarlo, *Synthesis*, 2001, 2003.
2. H. J. Bestmann, P. Range and R. Kunstmann, *Chem. Ber.*, 1971, **104**, 65; H. J. Bestmann, W. Stransky, O. Vostrowsky and P. Range, *Chem. Ber.*, 1975, **108**, 3582.
3. W. S. Wadsworth and W. D. Emmons, *J. Am. Chem. Soc.*, 1961, **83**, 1733.
4. T. Sugahara, Y. Komatsu and S. Takano, *J. Chem. Soc., Chem. Commun.*, 1984, 214.

5. J. Clayden and S. Warren, *Angew. Chem., Int. Ed.*, 1996, **35**, 241.

6. K. F. Burri, R. A. Cardone, W. Y. Chen and P. Rosen, *J. Am. Chem. Soc.*, 1978, **100**, 7069.

7. W. G. Salmond, M. A. Barta and J. L. Havens, *J. Org. Chem.*, 1978, **43**, 790.

8. A. B. Charette and H. Lebel, *J. Am. Chem. Soc.*, 1996, **118**, 10327.

9. P. R. Blakemore, W. J. Cole, P. J. Kocienski and A. Morley, *Synlett*, 1998, 26.

10. D. Takano, T. Nagamitsu, H. Ui, K. Shiomi, Y. Yamaguchi, R. Masuma, I. Kuwajima and S. Omura, *Org. Lett.*, 2001, **3**, 2289.

11. A. Fürstner, A. Hupperts and G. Seidel, *Org. Synth.*, 1999, **76**, 142; A. Fürstner, A. Hupperts, A. Ptock and E. Janssen, *J. Org. Chem.*, 1994, **59**, 5215.

12. S. Gauthier, J. Mailhot and F. Labrie, *J. Org. Chem.*, 1996, **61**, 3890.

13. J. Cossy, I. Pévet and C. Meyer, *Synlett*, 2000, 122.

14. B. Alcaide, C. Rodríguez-Ranera and A. Rodríguez-Vicente, *Tetrahedron Lett.*, 2001, **42**, 3081.

15. W. S. Johnson, Y.-Q. Chen and M. S. Kellogg, *J. Am. Chem. Soc.*, 1983, **105**, 6653.

16. G. Dressaire and Y. Langlois, *Tetrahedron Lett.*, 1980, **21**, 67.

17. J. D. White, J. P. Carter and H. S. Kezar, *J. Org. Chem.*, 1982, **47**, 929.

18. G. W. Luther and G. Beyerle, *Inorg. Synth.*, 1977, **17**, 186; A. M. Aguiar, J. Beisler and A. Mills, *J. Org. Chem.*, 1962, **27**, 1001.

19. P. Savignac and C. Patois, *Org. Synth.*, 1995, **72**, 241.

20. B. Steuer, V. Wehner, A. Lieberknecht and V. Jäger, *Org. Synth.*, 1997, **74**, 1.

21. J. J. Jaber, K. Mitsui and S. D. Rychnovsky, *J. Org. Chem.*, 2001, **66**, 4679.

16

Vinyl Anions

Problems from Chapter 16 in the Textbook

It will help if you have chapter 16 in the textbook open for the first section.

A Stereospecific Elimination to Give a Vinylic Bromide

In the textbook we said: 'One explanation is an intramolecular elimination through an *anti-peri*-planar transition state in a chair-like conformation using OLi as an internal base **26**. It can reach H-3 in a five-membered cyclic array.'

Another explanation is a conventional E2 reaction with the OLi group making H-2 less acidic. The cyclic mechanism **26** is intellectually more appealing but that does not mean that it is right.[1]

A Vinyl Stannane from a Shapiro Reaction

In the textbook we described a Shapiro reaction that gave Z-**48** trapped as the vinyl stannane E-**49** and hence the vinyl lithium Z-**W1** for combination with the epoxy-ketone **W2** to give a 17:1 mixture of E-**W3** and its diastereoisomer.[2]

Workbook for Organic Synthesis: Strategy and Control   Paul Wyatt and Stuart Warren
© 2008 John Wiley & Sons, Ltd

E-49   Z-W1   W2   E-W3

Some aspects of this sequence are worth further discussion. Trivially, the configuration of the alkenes in the formation of **W1** and **W3** appears to have changed. In fact both reactions occur with retention but the priority of the groups has changed! The synthesis of **W2** is more interesting. The benzoic acid was partly methylated and then reduced to the alcohol **W6**. RedAl is $Na(MeOCH_2CH_2O)_2AlH_2$. The most interesting step is the formation of **W7** by oxidation of **W6**. Finally reduction with diimide generated from potassium azodicarboxylate (the dipotassium salt of the DEAD acid $KO_2CN = NCO_2K$) in acetic acid, gave **W2** in 77% yield.

W4   W5; 83% yield   W6; 97% yield   W7; 61% yield

The oxidative cyclisation of **W6** is intriguing. The original report[3] offers no mechanism but remarks that a bulky substituent is necessary to prevent dimerisation of dienones like **W7** by the Diels-Alder reaction and that sodium iodate precipitates from the reaction mixture. We may suppose that a periodate ester cyclises with loss of iodate ion **W8**.

W6; 97% yield   W8   W7; 61% yield

Asymmetric Synthesis of Grandisol

In the textbook we revealed how *cis* but racemic allylic alcohol **61** could be made and that kinetic resolution using the Sharpless asymmetric epoxidation with L-(+)-di-isopropyl tartrate (see chapter 27) removed the unwanted enantiomer as the epoxide **62** and left the required enantiomer (−)-**61** for transformation into (+)-grandisol **63**. **Problem 16.1**: Suggest how (−)-**61** might be transformed into (+)-grandisol **63**.

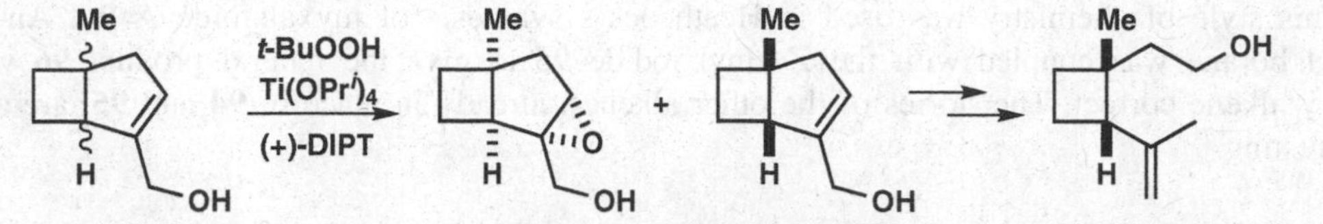

(±)-*cis*-61 62 (–)-*cis*-61; 94% ee 31% yield 63; (+)-grandisol

Answer 16.1: The hydroxyl group must be removed, the alkene split open and a carbon atom added to make the alkene. There are various ways to do this; the published[4] way is:

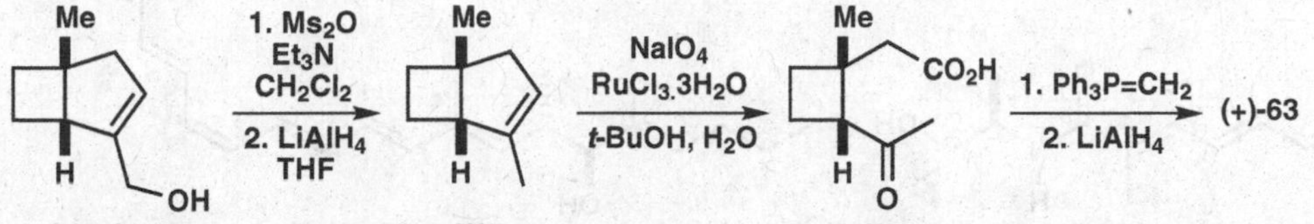

(–)-*cis*-61 W9 W10

The Synthesis of Myxalamide A

We described in the textbook a stereospecific conversion of vinyl boronic acids **91** into two kinds of halide **92** and **93**. **Problem 16.2**: Suggest a mechanism for both these reactions of vinyl boronic acids

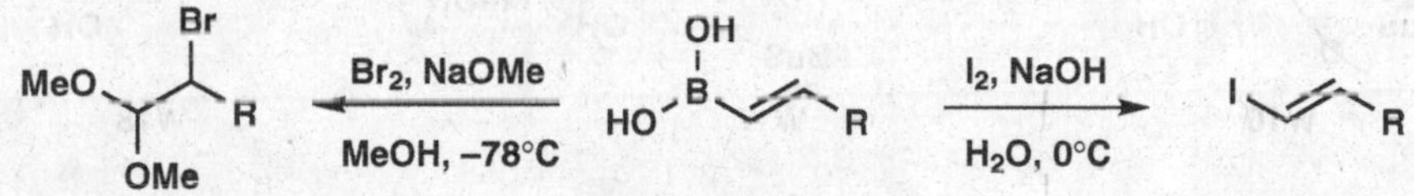

92; 90-94% yield *E*-91; boronic acid *E*-93; 100% yield

Answer 16.2: As iodine alone does not react with *E*-**91**, the iodination presumably occurs on the ate complex **W11** to give **W12** or possibly the *anti*-di-iodide derived from it.[5] E2 Elimination gives *E*-**93**.

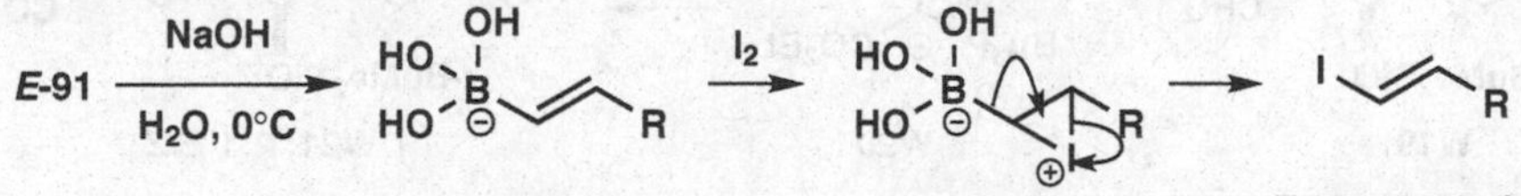

W11 W12 *E*-93; 100% yield

The bromination is more interesting. Reaction with Br_2 and NaOH gave mixtures of products including both *E*- and *Z*-vinyl bromides.[6] Addition of NaOMe gave a clean reaction to give **92**. The suggested mechanism involves *trans* addition of Br and MeO to the ate complex of **91** to give **W13** that eliminates stereospecifically **W14** to give **W15** and hence, by a second addition, **92**.

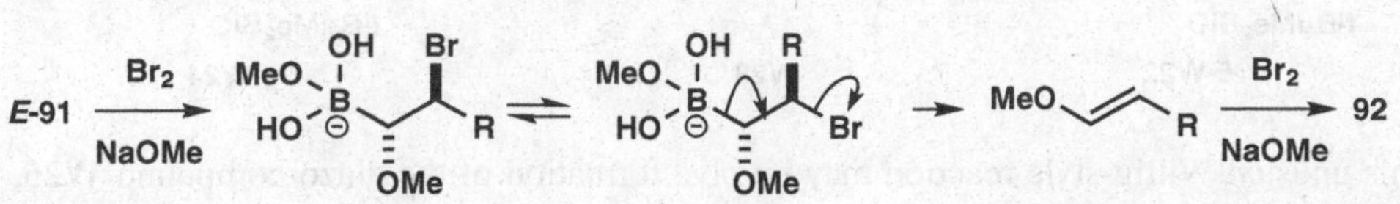

W13 W14 W15

This style of chemistry was used in Heathcock's synthesis of myxalamide A **96**. An *E*-vinyl borane was coupled with the *Z*-vinyl iodide **95** to give the natural product **96** with every alkene correct. The stories of the other alkenes, already in place in **94** and **95**, are also interesting.[7]

The first alkene came from a [2,3]-sigmatropic shift that also transferred chirality **W16**. Removal of the sulfur by a thiophile (chapter 15) gave one enantiomer of one diastereoisomer of the *E*-alkene **W18**.

The protected aldehyde **W19**, derived from **W18**, gives the next alkene **W21** by an unusual Wittig reaction. The Bu_3P^+ group is less anion stabilising than the more usual Ph_3P^+ group and gives a good yield (92%) but, as expected for a trisubstituted alkene, poor stereoselectivity.

The alkyne in **94** finally comes from a most unusual Horner-Wadsworth-Emmons reaction with a diazophosphonate **W23** on the protected aldehyde **W22** derived from **W21**. The details are in papers by Seyferth and collegues[8] and Gilbert and Weerasooriya.[9]

This unusual Wittig-style reaction may involve formation of the diazo-compound **W25**, loss of nitrogen to give a vinyl carbene that rearranges by hydrogen migration **W26** to give the alkyne **W24**. The protecting group is removed with TBAF.[9]

E-W22 → W23 → *E*-W25 → −N$_2$ → *E*-W26 → W24

The synthesis of the other half of myxalamide starts with the vinyl bromide **W27** but first attempts were eventually abandoned. Coupling to allylic alcohol in a Heck reaction gave **W28** in good yield but as a mixture of isomers. Hydroboration of the alkyne **W29** gave the vinyl boronate and Suzuki coupling with **W27** worked well but conversion to the vinyl iodide went with poor stereoselectivity.

W27 + HO → Pd(OAc)$_2$, Ar$_3$P, Et$_3$N → W28

W29 → 1. catechol borane 2. W27 Pd(OAc)$_2$ → W30 → W31

The successful method used a vinyl stannane **W34** made from the enyne **W32** by stannylation, hydrozirconation and removal of Zr and Si. The final alkyne was put in with a Horner-Wadsworth-Emmons reaction, the tin replaced by iodine with retention and the amide coupled to give **95**.

W32; R = *t*-BuMe$_2$Si → 1. KHMDS 2. Bu$_3$SnCl → W33; R = *t*-BuMe$_2$Si → 1. Cp$_2$ZrHCl 2. H+, H$_2$O 3. TBAF → W34

W35 + (EtO)$_2$P O ⊖ CO$_2$Et → W36 → 95

Hydrosilylation

In the textbook we discussed the radical reaction of alkynes with R$_3$SiH to give vinyl stannanes **112** with typical *Z:E* selectivity of >20:1. The products **112** can be converted into *Z*-vinyl bromides by reaction with bromine.

Et$_3$B → O$_2$ → Et· + H—Si(SiMe$_3$)$_3$ → EtH

·Si(SiMe$_3$)$_3$ → (Me$_3$Si)$_3$Si—H → 111 → *Z*-112

110

The stereochemistry of the replacement of silicon with hydrogen was explored in two reactions. The *Z*-vinyl silane **W37** was combined with Me$_3$SiCl and D$_2$O and the deuterated compound Z-**W39** was treated with HCl. Each gave the vinyl chloride with retention of configuration.[10]

Z-W37   Z-W38   Z-W39   E-W38

Hydrozirconation and Carbometallation

In the textbook we stated that during hydrozirconation of alkynes **126** the zirconium atom transfers the least stable anion among its ligands. This is obviously H⁻ as Cl⁻ is much more stable. The metal prefers to take the terminal position because the resulting σ-complex is more stable as it is a σ-complex of the less-substituted carbanion.[11] Now please note that this regioselectivity can be reversed with other metals such as copper and in other reactions. Sometimes the most stable anion is transferred to the terminus and the metal forms a π-complex with the more substituted carbon atom. (See chapters on Pd, especially the Heck reaction!)

122; Cp₂ZrHCl   **123**   **124; π-complex**   **125; hydro-**   **126; σ-complex**
16 electrons   **18 electrons**   **zirconation**   **16 electrons**

We gave a recent application of carbometallation to the synthesis of the macrolide antibiotic concanamycin using carbometallation of the alkyne **133** to give the vinyl iodide E-**134** followed by a palladium-catalysed coupling with a vinyl stannane, also created from an alkyne, to give the diene E,E-**135**. **Problem 16.3**: Which vinyl stannane would be needed and how would it be coupled to **142**?

141   E-142; 88% yield   E,E-143; 72% yield

Answer 16.3: The vinyl stannane **W40** is needed and coupling by the Stille procedure (chapter 18) using Pd(0) to make a σ-complex from the vinyl iodide **142**. The main point is that the configuration of both the vinyl iodide and the vinyl stannane is retained in the product[12] E,E-**142**.

E-142   W40   E,E-143; 72% yield

It is fairly easy to see how **142** might be transformed into the required macrocycle **W41** by building on a few extra carbon atoms and closing the ring by lactone formation. Macrolides were always and often still are made this way. However the Stille coupling could also be used to close the macrocycle by making a C–C bond between a vinyl stannane and a vinyl iodide: cyclisation of **W42** gives **W41** in 72% yield. Vinyl anion and vinyl cation equivalents add flexibility to the design of syntheses.

$$Pd_2(dba)_3$$
$$i\text{-}Pr_2NEt$$
$$Ph_3As$$

W41   **W42**

Preparation of Vinyl Silanes

The Peterson reaction with two SiMe$_3$ groups on the same carbon atom **183** was among the methods in the textbook for the preparation of vinyl silanes **185**.

183   **184**   **185**

However, *not* the Wittig reaction as the products from tempting Wittig reactions with ylids **W43** containing α-SiMe$_3$ groups as the Peterson reaction is preferred. **Problem 16.4**: Why?

W43   **W44**   **W45**

Answer 16.4: The Wittig would involve attack by the oxyanion on P$^+$ while the Peterson involves the same anion attacking Si.[13] The Wittig is tempting because it looks good to attack a cation rather than a neutral molecule, but Si is more electropositive than even P$^+$ and is very susceptible to attack by hard nucleophiles such as halide and oxy-anions. If we draw the reaction as a nucleophilic substitution at silicon **W44a**, the leaving group **W46** is the phosphorus ylid **W47** and the other product the stable anion **W48**.

W44a   **W46**   **W47**   **W48**

Reduction of Alkynyl Silanes

The stereoselective reduction of alkynyl silanes **W49** is a simple way to make either *E*- or *Z*-vinyl silanes: the alcohol **W51** is a good example.[14]

This product can be used in an interesting synthesis of a cyclohexene ring. **Problem 16.5**: Suggest a mechanism for the formation of **W53**.

Answer 16.5: A Mannich-like imine salt is formed from the amine and formaldehyde and cyclises in the normal vinyl silane manner **W54** to give the silicon-stabilised cation **W55** that loses silicon to some oxygen-based nucleophile to give **W53**. Notice that the new bond has to be *cis* so the vinyl silane reacts with retention, as it prefers.[15]

Reactions of Vinyl Silanes

In the textbook we showed that the anion of the allyl phosphonate **192** could be silylated to give the vinyl silane **193** and this reacts in turn with the classical combination of acetyl chloride and AlCl₃ to give the enone *E*-**194** that can be used in HWE reactions to make dienones (chapter 15). **Problem 16.6**: Suggest how **193** is formed.

Answer 16.6: The first lithiation must give **W56** that reacts in the γ-position to give the vinyl phosphonate (and allyl silane) **W57**. This is more acidic than **193** and the second molecule of LDA creates a new lithium derivative, either **W58** or **W59**, that is protonated on work-up to give[16] *E*-**193**.

The Synthesis of Epothilones A and B

The alkynyl silane **205**, used in the synthesis of epothilones, was made from a simpler aldehyde **W60** produced by asymmetric reduction. **Problem 16.7**: Suggest reagents for the conversion of **W60** into **W61** and explain the details of the formation of **205**.

Answer 16.7: The lithium derivative of trimethylsilyl acetylene is added to **W60** and the alcohol trapped with methyl chloroformate. The (lack of) stereoselectivity does not matter as the new centre is lost in the next step. This is a 'transfer hydrogenation' from the formate anion to the propargyl cation complex **W62** that could be drawn in other ways.[17] The carbonate leaving group and the formate anion each give a molecule of CO_2.

Reactions of Vinyl Silanes: The Nazarov Cyclisation

The Nazarov cyclisation (chapter 6) makes cyclopentenones by Lewis acid-catalysed cyclisation of dienones. There is a regioselectivity problem in knowing where the alkene will end up in the product. This problem is beautifully solved by allyl silane chemistry. The Shapiro reaction (this chapter) gives a vinyl-lithium **W65** that adds to the silyl propenal **W66** to give, after oxidation, the dienone **W68**. Cyclisation with $FeCl_3$ as the Lewis acid gives a high yield of one isomer of the cyclopentenone **W69**. **Problem 16.8**: Draw a mechanism for the formation of **W69** commenting on the stereo- and regiochemistry.

Answer 16.8: The Nazarov cyclisation can be drawn **W70** to show the stabilisation of the resulting cation by silicon and hence the formation of one regio-isomer **W69** after removal of silicon by some nucleophile, probably chloride ion. The double bond in the product has to be *cis* and so the vinyl silane is forced to react by inversion.[18]

A Recent Application of Vinyl Anion Chemistry

Chemists at DSM Pharma[19] needed the amino acid derivative **W72** for the synthesis of anti-HIV drugs. A simple route is by S_N2 with a nitrogen nucleophile on the alcohol **W73** derived from the epoxide of **W74**.

The process was simplicity itself. They made disiamyl borane on a large scale by the hydro-boration of 2-methylbut-2-ene and used it to hydroborate the alkyne **W75**. Suzuki coupling with *m*-fluorobenzyl chloride worked very well on the anion of **W76** and the epoxidation, cyclisation and displacement were achieved. This paper is worth reading to see what problems have to be overcome in large scale synthesis.

References

1. E. J. Corey, H. F. Wetter, A. P. Kozikowski and A. V. Rama Rao, *Tetrahedron Lett.*, 1977, 777; M. G. Bock, A. P. Kozikowski, A. V. Rama Rao, D. Floyd and B. Lipshutz, *Tetrahedron Lett.*, 1978, 1051.
2. E. J. Corey and J. P. Dittami, *J. Am. Chem. Soc*, 1985, **107**, 256.
3. H.-D. Becker, T. Bremholt and E. Adler, *Tetrahedron Lett.*, 1972, 4205.
4. D. P. G. Hamon and K. L. Tuck, *J. Org. Chem.*, 2000, **65**, 7839.
5. H. C. Brown, T. Hamaoka and N. Ravindran, *J. Am. Chem. Soc*, 1973, **95**, 5786.
6. T. Hamaoka and H. C. Brown, *J. Org. Chem.*, 1975, **40**, 1189.
7. A. K. Mapp and C. H. Heathcock, *J. Org. Chem.*, 1999, **64**, 23.
8. D. Seyferth, R. S. Marmor and P. Hilbert, *J. Org. Chem.*, 1971, **36**, 1379.
9. J. Gilbert and U. Weerasooriya, *J. Org. Chem.*, 1979, **44**, 4997.
10. K. Miura, K. Oshima and K. Utimoto, *Bull. Chem. Soc. Japan*, 1993, **66**, 2356.
11. J. Schwartz and J. A. Labinger, *Angew. Chem., Int. Ed. Engl.*, 1975, **15**, 333.
12. T. Jyojuma, M. Katahno, N. Miyamoto, M. Nakata, S. Matsumuna and K. Toshuna, *Tetrahedron Lett.*, 1998, **39**, 6003; T. Jyojuma, N. Miyamoto, M. Katahno, M. Nakata, S. Matsumuna and K. Toshuna *Tetrahedron Lett.*, 1998, **39**, 6007.
13. L. F. van Staden, D. Gravestock and D. J. Ager, *Chem. Soc. Rev.*, 2002, **31**, 195.
14. L. E. Overman, M. J. Brown and S. F. McCann, *Org. Synth.*, 1989, **68**, 182.
15. L. E. Overman, C. J. Flann and T. C. Malone, *Org. Synth.*, 1989, **68**, 188.
16. B. S. Lee, S. Y. Lee and D. Y. Oh, *J. Org. Chem.*, 2000, **65**, 4175.
17. D. Sawada, M. Kanai and M. Shibasaki, *J. Am. Chem. Soc.*, 2000, **122**, 10521.
18. T. K. Jones and S. E. Denmark, *Helv. Chim. Acta*, 1983, **66**, 2377.
19. D. J. Ager, K. Anderson, E. Oblinger, Y. Shi and J. VanderRoest, *Org. Process. Res. Dev.*, 2007, **11**, 44.

17

Electrophilic Attack on Alkenes

Problems and Further Examples from the Textbook

You will find it helpful for the first part of this chapter to have chapter 17 of the textbook open.

Chemoselective Epoxidation

In a synthesis of biologically active marine polycyclic ethers, McDonald and Wei[1] first reacted farnesyl acetate with NBS in aqueous THF. The three trisubstituted alkenes look about equally reactive but in aqueous solution the hydrocarbon core coils and protects the middle alkene. The allylic acetate is less nucleophilic because of the electron-withdrawing ester group and only the other terminal alkene reacts with the expected regioselectivity to give racemic bromohydrin.

Enantioselective epoxidation (chapter 27) was carried out using the Shi epoxidation catalyst[2] 'epoxone' **W4** prepared from fructose **W3** in straightforward steps.[3] The true epoxidation reagent is the trioxane **W5** prepared *in situ* by oxidation of **W4** with 'Oxone'. Practical details of the large scale use of the Shi epoxidation have appeared.[4]

Epoxidation occurs only on the middle alkene (chemoselectivity) to give one enantiomer of the *trans* epoxide **W6** (stereoselectivity) but of course a mixture of diastereoisomers as the

Workbook for Organic Synthesis: Strategy and Control Paul Wyatt and Stuart Warren
© 2008 John Wiley & Sons, Ltd

bromine centre is racemic. Slow acid-catalysed cyclisation with catalytic camphor sulfonic acid gives an efficient kinetic resolution (chapter 28). Only the cyclisation that puts the bromine equatorial **W7** in the THP occurs (50% yield) leaving 50% of the other diastereoisomer of the bromohydrin. The regioselectivity of the cyclisation is as expected: the tertiary centre is more electrophilic in acid solution.

Mechanism of the Reductive Demercuration

We showed in the textbook how reductive oxymercuration could be used to hydrate alkenes **8** to give the more substituted alcohol **62** regioselectively and we promised the mechanism of the last step here in the workbook.

Borohydride reduces the HgOAc group to the unstable HgH group **W8** and a radical R• abstracts the H atom as the Hg–H bond is very weak. Mercury is lost **W9** as mercury metal that separates as a grey suspension and then collects as an easily separated silver globule at the bottom of the flask. The identity of 'R' is now revealed as the resulting primary radical **W10**.

Applications of Hydroboration

We discussed the synthesis of **95** in the textbook and said that 'simple reactions led to ibogamine' **W12**. **Problem 17.1** Suggest in outline how this might be done.

Answer 17.1: The starting materials show us where the disconnections must be drawn **W12a** and it is clear that C-N bonds must be made between the primary amine (deprotected) and both the ester and the secondary alcohol. The obvious solution is to use acylation with the ester and reductive amination after oxidation to the ketone **W13**. All the chiral centres are correct except at the ester.[5]

W12a; ibogamine $\Rightarrow$ **W13**

Oxidation of **95** to the ketone **W16** was achieved with the hypervalent iodine species **W14** and the Cbz group removed by transfer hydrogenation with cyclohexene. Reductive amination also occurs under these conditions, as does epimerisation at the ester in 75% yield but in a 30:70 mixture with the minor diastereoisomer being the right one.

W14   **95** → **W15; 75% yield**   Pd/C →   **W16; 75% yield**

Cyclisation of the mixture gives cyclised major **W17** and cyclised minor diastereoisomer **W18** which are easily separated, as they are no longer stereoisomers, and the major isomer epimerised for further cyclisation. Reduction of **W18** gives ibogamine **W12**.

W16 →[220 °C] **W17** + **W18** →[LiAlH₄] **W12**

The Synthesis of Coriolin

The preparation of the starting material **77** illustrates other aspects of selectivity and control in electrophilic attack on alkenes. Using the method described in the textbook for the mono-epoxidation of conjugated dienes, cyclo-octadiene **W19** gave an excellent yield of the epoxide **W20**. Cyclisation in strong base (compare the similar step in imidomyremcin synthesis, see compound **127** in the chapter) gave the 5/5 fused system **W21** as a single diastereoisomer. The idea[6] was to make a bromohydrin from this compound but mixtures were obtained and it was better to protect the alcohol as the benzyl ether **W22**.

W19 →[MeCO₃H / Na₂CO₃] **W20; 78% yield** →[Et₂NLi / Et₂O] **W21; 71% yield** →[NaH / PhCH₂Br / DMF] **W22; 91% yield**

The cyclisation of **W20** deserves some comment. The intermediate is a carbene **W24** or carbenoid **W23** formed by removal of the epoxide proton on the allylic side. Protons on three-membered rings are more acidic than expected as you can see by the use of an unstrained ether solvent. The carbene inserts into the 'inside' C–H bond across the ring to give **W26**. The stereochemistry is explained by a twisted conformation **W25** showing just how close the carben(oid) carbon can get to the inside H. The two five-membered rings must be *cis*-fused: it is only the third centre (OH) that is in question.[7]

Bromination of the benzyl ether **W22** in aqueous DMSO gave one bromohydrin **W28**. The bromonium ion is formed on the *exo*-face of the folded bicyclic molecule **W27** and the water must now attack from inside the fold. This is much easier away from the hinge. Removal of the benzyl protecting group (what kind of selectivity is shown here?) gives the crystalline diol **W29** in good yield.[7]

Only one alcohol is oxidised with PCC, presumably because of the bromine atom adjacent to the other, and the ketone **W30** is protected and eliminated to give the enone **77** in good yield.

Problem 17.2: Suggest how **W31** might be converted into **77**. Warning: compare the oxygenation patterns of **W31** and **77** before you start!

Answer 17.2: There are obviously various possible ways. The chemists did it by: 1. Conjugate addition of Me$_2$CuLi (95%), 2. equilibrium di-methylation with *t*-BuOK/MeI (77%), 3. reduction, 4. THP protection, 5. removal of TBDMS group, and 6. oxidation.

Hydrosilylation

Hydrosilylation of alkynes is treated in chapter 16 of the textbook. Hydrosilylation of alkenes is quite like hydroboration and enol ethers are particularly interesting in the intramolecular silylation transfer of oxygen as they give one regioisomer very stereoselectively.[8]

W32 → W33 → W34 (R = n-C6H12, 60% yield)

The regio- and stereoselectivity of hydrosilylation of this type of enol ether goes in what Tamao describes as a 5-*endo* fashion.[9] We can say that platinum inserts into the Si–H bond **W35** and then adds to the alkene to give **W36**. Coupling of the Si and C ligands on the palladium gives the first isolable intermediate **W37**. As we know the stereochemistry of the product **W34**, the stereochemistry of the intermediate must be as shown in **W36**. **Problem 17.3**: Suggest a mechanism for the conversion of **W37** into **W34**.

Answer 17.3: The hydroperoxide anion must add to the silicon and the carbon atom must migrate from silicon to oxygen, again rather like hydroboration.[10] The fluoride adds a ligand to silicon before the migration step **W38** and the silyl ether **W39** is hydrolysed to **W34**.

An Asymmetric Synthesis of Fortamine

In the textbook we described a synthesis of the aglycone **44** of the antibiotic fortamine. As you can see from the first few steps (**32** to **34**) this synthesis achieved excellent diastereoselectivity but it is of course racemic as the starting materials and reagents are all achiral. The same authors later published an asymmetric synthesis in which they modified these steps slightly by the attachment of an easily removable chiral auxiliary. **Problem 17.4**: Suggest a suitable chiral auxiliary and where it might be attached.

32 → 33; 84% yield → 34; 99% yield → 44; 99% yield

Answer 17.4: The first chiral intermediate is **33** so ideally we should like to attach the auxiliary before that. This is not possible so the nitrogen atom becomes the next best choice. In the published paper[11] they opened the original racemic epoxide **32** with (*S*)-methylbenzylamine **W40** and separated the carbamates by chromatography before getting back on track with an iodolactonisation. There are of course many other solutions and this one has the disadvantage that there is no asymmetric induction in the formation of **W41** so it becomes a resolution rather than an asymmetric synthesis. Intermediate **W45** is almost identical to the bromolactone **36** used in the racemic synthesis.[11]

Vernolepin Synthesis

The intermediate **169** described in the chapter was reacted with glycol to give one of the two possible *ortho*-esters **W46** as a crystalline compound in reasonable yield. Danishefsky used this 'welcome stroke of good luck' to make vernolepin **154**. **Problem 17.5**: What remains to be done? State clearly which kinds of reactions are needed and what selectivity problems must be overcome. Make some suggestions for possible solutions.

Answer 17.5: The bridging lactone needs to be reduced to an aldehyde and then converted into a vinyl group with something like a Wittig reaction. An exo-methylene group needs to be added to the left hand ring in something like an aldol reaction and three extra carbon atoms added regio- and stereoselectively to the epoxide to give another exomethylene lactone. This looks like enolate addition followed by an aldol. It is a pity that both 'aldols' need formaldehyde as the electrophile.

In outline this is what Danishefsky and his group did. Reduction with DIBAL gave the required aldehyde **W47** and a Wittig reaction gave the vinyl compound **W48**. This was more difficult than it appears as protected derivatives of **W47** failed to react.

W46 →(*i*-Bu₂AlH / DME toluene)→ W47; 94% yield →(Ph₃P=CH₂)→ W48; 87% yield

Now came the first critical step. Would the lithium enolate of some derivative of acetic acid react at the right end of the epoxide? In fact the dilithium derivative of acetic acid itself did so and gave, after acidification and esterification, one compound **W49**. It appears that **W48** reacts in conformation **W48a** and the usual *trans*-di-axial stereochemistry is observed.

W48 →(1. [OLi/OLi]; 2. H⁺; 3. CH₂N₂)→ W49; 56% yield   W48a; reacting conformation

Treatment of **W49** with TsOH in benzene gave both lactones: chemoselectivity was not good. Attempts at protection of the OH group in **W48** before epoxide opening failed as the extra bulk of the protecting group prevented attack of the enolate on the inside of the folded conformation **48a**. Fortunately the required lactone **W50** was the major component–there was only 25% of **W51**. The double methylenation was best achieved with Eschenmoser's salt in a Mannich sequence. This is a great classic of organic synthesis dating from 1977 and we have given it the full treatment when the textbook and workbook are read together.[12]

W49 →(TsOH / benzene)→ W50; 57% yield →(1. excess LDA; 2. H₂C=NMe₂ ⊕; 3. MeI; 4. NaHCO₃)→ 154   W51

The first asymmetric vernolepin synthesis used the same key intermediate **155** as Danishefsky but prepared it by an asymmetric Heck reaction. In the prochiral starting material **W52** the two alkenes in the ring are enantiotopic and the chiral catalyst (BINAP/Pd, see chapter 26) directs the vinyl triflate to just one of them. You are not expected to explain this symmetry breaking but there are other selectivities that you can explain. **Problem 17.6**: Suggest a mechanism for the Heck reaction that explains the relative stereochemistry of the product and the regiochemistry of the β-elimination.

W52; prochiral dienone →(9 mol% Pd₂(dba)₃.CHCl₃, 11 mol% (*R*)-BINAP, *t*-BuOH, 60 °C, ClCH₂CH₂Cl)→ W53; 76% yield, 86% ee → (–)-155

Answer 17.6: Oxidative insertion into the C–OTf bond gives the Pd(II) complex **W54**, with of course BINAP as an asymmetric ligand.[13] The Heck reaction adds Pd and the vinyl group *syn* to the top face of the alkene, directed by the tether, with retention of configuration at the vinyl group to give **W55**. Elimination of H-Pd-OTf must occur *syn* so only the proton on the OH carbon is available and the product is the enol **W56** of **W53**.

Selenenyl Lactonisation

Problem 17.7: Draw a mechanism for the reactions involved in Curran's hirsutene synthesis.

Answer 17.7: Selenenyl lactonisation gives the five-membered ring lactone preferentially as it is difficult for the ester carbonyl oxygen atom to reach the far end of the selenonium ion in **W60**. The only twist is the removal of the TBDMS group from the intermediate by chloride ion **W61**. The selenide is oxidised to the selenoxide and elimination occurs in the usual *syn* fashion.[14]

The product **181** of one sequence shown in the chapter was used in a process called by the author 'carbolactonisation'. The selenium compound was treated with Bu₃SnH with catalytic AIBN as initiator in toluene at 110 °C in the presence of methyl acrylate to give a reasonable yield of the lactone with a new C–C bond. **Problem 17.8**: What is the mechanism of the reaction of the radical, formed from **181** with Bu₃SnH, with methyl acrylate and why does the reaction give only one stereoisomer?

Answer 17.8: Selenium is one of those elements, like Br and I, from further down the periodic table, that sacrifices itself **W64** to form a bond to tin and release a carbon radical, here **W62**. Radicals are excellent at conjugate addition **W65** as they work on bond strength and C–C is much weaker than C–O. The product radical–secondary and stabilised by the carbonyl group–captures a hydrogen atom from Bu₃SnH **W66** and completes the cycle. The radical **W62** is folded with an *exo* or convex and an *endo* or concave face and the addition **W65** occurs on the *exo* face.[15] The final radical **W66** is more stable than the first radical **W62** and so there is more of it and it is preferentially reduced by Bu₃SnH.

The Prins Reaction

Formation of 4-Chloro-Tetrahydropyrans

A modern version of this reaction (see compound **193** in the textbook chapter) uses allyl stannanes instead of propene and indium chloride as a Lewis acid. Other aldehydes, particularly aromatic ones, can be used and high yields of single diastereoisomers of 4-Cl-THPs **W67** result. **Problem 17.9**: Draw a mechanism for the reaction and explain the stereochemistry. If the tin/silicon compound **W68** is used, dihydropyrans **W69** are formed instead. Why is this?

Answer 17.9: The tin or the silicon directs the position of the cation in the intermediate and also sacrifices itself to fix the new alkene. So some initial interaction such as **W70** must lead to a tin-stabilised cation such as **W71** and hence a new alkene **W72** and, by repetition of the Prins reaction,[16] a new cation **W73**.

It is clear from the conformation of the final product that thermodynamic control is in action as all the substituents are equatorial. Presumably the various diastereoisomers of **W67** equilibrate through cations such as **W74** and **W75**.

W74   ⇌   W67   ⇌   W75   W67 conformation

Hydroboration

Another dialkyl borane that receives some applications is 'disiamyl borane'. This di-secondary amyl borane **W76** is prepared from 2-methylbut-2-ene and diborane and gives chemoselective hydroboration/oxidation of the triene **W77**. The primary alkyl group migrates preferentially from the trialkyl borane intermediate.[17]

W76   W77   W78; 88–91% yield

We met (mono-)siamyl borane in the textbook chapter 22[7]. It has been used in the synthesis of a steroid nucleus by a combination of hydroboration and carbonylation. Hydroboration of the diene **W79** with siamyl borane **227** gives a cyclic borane that gives the ketone **W80** on carbonylation and hence the steroid-like compound **W81** on treatment with acid. With two alkenes part of the same molecule, we can expect hydroboration of the less crowded alkene first and delivery of the primary alkyl borane stereoselectively to the more crowded alkene. **Problem 17.10**: With that clue, can you unravel the details of this process and reveal the structure of the cyclic borane?

W79   W80; 53% yield   W81

Answer 17.10: The first hydroboration occurs at the monosubstituted alkene with the expected regiochemistry but no stereochemistry. The second is intramolecular **W82** again with the expected regiochemistry. Now there is stereochemistry. The borane in **W82** is tethered to one, let us say bottom, face of the ring and so it delivers RB and H to that same face to give the isolable cyclic borane **W83**. Carbonylation follows the mechanism in the textbook chapter with the primary and then the secondary alkyl groups transferred from B to C while the branched secondary group (R) remains behind and is oxidised away with H_2O_2 in base.

W79   W82   W83   W80

Acid-catalysed cyclisation by electrophilic aromatic substitution **W84** now closes the remaining ring (B) of the steroid skeleton and the resulting alcohol **W85** dehydrates under the conditions of the cyclisation.[18] The regioselectivity of the dehydration is surprising.

Selectivity in Epoxidation

In the textbook we discussed the chemoselectivity evident in the epoxidation of *E,E*-22 to give one or other epoxide by choice of reagent but we did not discuss the mechanism of the less familiar formation of **21**. **Problem 17.11**: Which is the stronger acid, water or H_2O_2? Use this information to draw a mechanism for the epoxidation with alkaline H_2O_2 showing how the *anti*-epoxide **21** is formed.

Answer 17.11: Water is a weaker acid than H_2O_2 as the second OH group stabilises the HOO⁻ anion which adds in a conjugate fashion **W86** to the enone. The second step is an unusual S_N2 reaction at oxygen **W87** made possible by the weak O–O single bond. The reaction is stereoselective rather than specific as the stereochemistry of the alkene is lost in the intermediate enolate. However, if this intermediate is long lived the skeleton can rotate to give the more stable *anti*-epoxide **21**.

Epoxides of Conjugated and Symmetrical Dienes

Cyclo-octa-1,5-diene **26** is used to make 9-BBN, see chapter 16, and as a ligand for metals, see chapter 26. It is made by the Ni-catalysed dimerisation of butadiene. It gives the mono-epoxide **27** cleanly. **Problem 17.12**: Why are epoxides of dienes such as **25** so sensitive to acid? One of the products is cyclopentenone–how do you think that is formed?

Answer 17.12: The protonated epoxide can open **W88** to give a stable allylic cation that can lose a proton **W89** to give an extended enol **W90** (chapter 11) that could be protonated at C-α but protonation is reversible and γ-protonation gives the more stable conjugated ketone **W91**.

Conformationally Controlled Nucleophilic Attack on Epoxides

Problem 17.13: Convince yourself that the opening of the epoxides **37** and **41** (from the textbook) was indeed controlled by the formation of the *trans*-diaxial products.

Answer 17.13: The simplest way is to draw conformational diagrams of the products. After some trial and error, these seem to us to be the best.

Allyl Silane in the Preparation of an Alkene for Epoxidation

In the textbook we discussed how the exclusive *syn* relationship between the OH group and the epoxide in **51** showed that epoxidation was controlled by the free OH group in **50**. The starting material **50** was made by TiCl$_4$-catalysed addition of allyl silane to the sugar derivative **W92**. **Problem 17.14**: Explain this – you may find chapter 12 helpful.

Answer 17.14: The allylic acetate is removed by the Lewis acid to give a delocalised cation that is attacked by allyl silane at the end of the alkene remote from the silicon **W94** so that a stabilised β-silyl cation **W95** can be formed. Desilylation (with fluoride?) and ester hydrolysis gives **50**. The cation in **W94** is nearly flat and the allyl silane attacks the face not occupied by the acetate.[19]

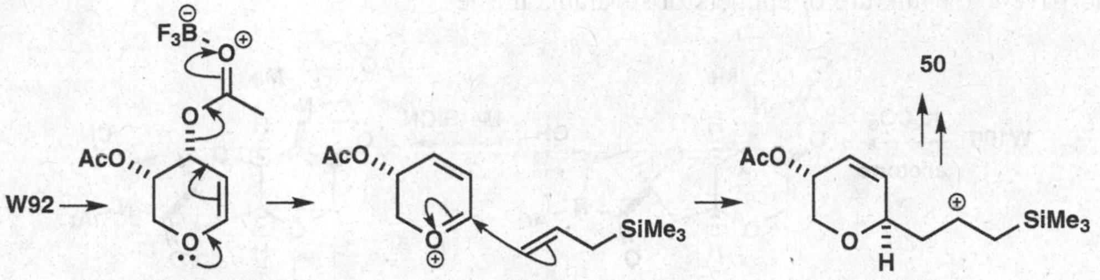

The Synthesis of Dysiherbaine

Problem 17.15: Suggest how the aldehyde **58** (described in the textbook) might be converted into (−)-dysiherbaine. This is quite difficult as the published synthesis is not too good.

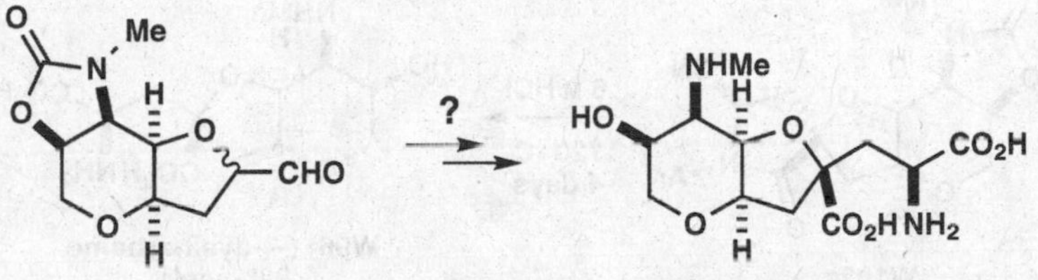

58

W96; (−)-dysiherbaine

Answer 17.15: There are many ways to proceed and it is simplest just to tell you what they did. First some extra carbon atoms were added by allylation of an enamine with moderate stereoselectivity at the new centre. Oxidation of the aldehyde allowed the epimers of the carboxylic acid to be separated by chromatography giving 69% of the wanted compound and 25% of its epimer.[20]

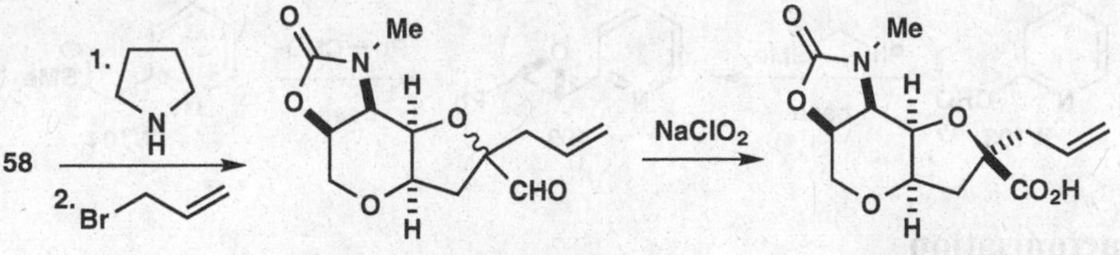

W97; 3:1 mixture of epimers   **W98; 69% yield**

Now peptide-style coupling of the acid to 2,4-dimethoxybenzylamine was followed by oxidative cleavage of the alkene in **W99** to give the new aldehyde **W100** in excellent yield as one diastereoisomer.

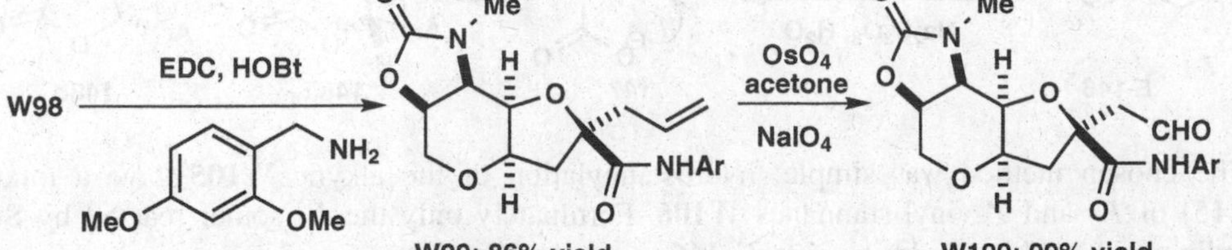

W99; 86% yield   **W100; 99% yield**

A Strecker reaction is needed to convert the aldehyde in **W100** into the amino acid in **W96**. Cyclisation in base gave a mixture of hemi-aminals and treatment with Me₃SiCN and a Lewis acid gave a 1:1 mixture of epimers of separable nitriles.

Finally the wanted nitrile **W102a** was hydrolysed under rather vigorous condition – 6 M HCl at 120 °C for 4 days – and all the protecting groups fell off to give disyherbaine in extraordinary yield considering the violence of the conditions.

Synthesis of Epoxides by the Sulfonium Ylid Method

We discussed the opening of the unsymmetrical diaryl epoxide **103** in the textbook and said that it was easily made by the sulfonium ylid method. This could be done in two ways – from **W103** or **W104**. In fact the first method[21] was chosen with NaOH as base to give 63% yield of **103**.

Halolactonisation

In the textbook we said that preparation of the starting materials **146** involves chemistry we met in the last chapter so it is reviewed in the workbook.[22]

The chosen method was simple: hydrostannylation of the alkyne **W105** gave a mixture (85:15) of *E*- and *Z*-vinyl stannanes **W106**. Fortunately only the *E*-isomer reacted by Stille coupling with aryl bromides to give *E*-**146** in reasonable yield.

W105, 2 x Bu₃SnH, AIBN, 100 °C, W106; 83% yield, 1. ArBr Pd(Ph₃P)₄ 2. HCl, E-146; 55–88% yield

Sulfenyl Lactonisation

We described the sulfenyl lactonisation of *E*- and *Z*-alkenes **182** in the textbook chapter as one of the most impressive demonstrations of the stereospecificity of the reaction. **Problem 17.16**: Choosing either example show how the stereochemistry of **183** arises.

E-182, R'SCl Et₃N, *anti*-183, *syn*-183, R'SCl Et₃N, Z-182

Answer 17.16: A top tip when working out the stereochemistry of a cyclisation is to draw the starting material in a conformation that makes cyclisation possible, thus *E*-**182a** gives the sulfonium salt of the carboxylate (Et₃N is a strong enough base to deprotonate a carboxylic acid) **W107** that cyclises by a 6-*endo-tet* mechanism **W107** with inversion giving *anti*-**183**. The compound that was wanted for leukotrienes was **W108**.

Problem 17.17: Given that the sulfenyl chloride R'SCl is made by the action of chlorine on a disulfide, what starting materials would be required?

E-182a, R'SCl Et₃N, W107, *anti*-183, W108

Answer 17.17: The symmetrical disulfide **W108** (actually made by oxidation of the thiol) gives the sulfenyl chloride **W110** and hence the required lactone[23] **W108**.

W109, Cl₂, −35 °C CH₂Cl₂, W110, n-C₁₄H₂₉ Et₃N, W108

The Prins Reaction

This section refers to the making of compounds such as **190** and **193** in the textbook chapter. With monoalkyl alkenes, two kinds of products can be formed depending on the conditions. With HCl the substituted 4-Cl-THPs **W111** are formed in good yield as mixtures of isomers. With H_2SO_4, products of both kinds **W112** and **W113** are formed in about equal quantities. It is notable that these reactions have been chiefly developed by chemists in industry and there is a strong flavour of practicality about the methods. With hex-1-ene a process has been developed to give about 45% yield of each product. **Problem 17.18**: Draw mechanisms for the formation of both products.

Answer 17.18: The dioxane is formed by the mechanism used in the chapter for **190** and the THPs from the same key intermediate[24] **W114** by the mechanism below where X = Cl or HO. If you proposed an oxo-ene mechanism (see **202** in the textbook) you may well be right, but read on.

The Oxo-Ene Mechanism for the Prins Reaction

The oxo-ene mechanism is exemplified by **209** from the textbook where we said: 'notice the excellent chemoselectivity in that the internal alkene is not attacked and the excellent regioselectivity in that hydrogen atoms at four other sites might have taken part in the reaction, but do not.' **Problem 17.19**: Why is the external alkene preferred, what are the four other sites from which H atoms might have been removed, and why are they not?

Answer 17.19: The original paper[25] is devoid of mechanisms or explanations as they are really interested in polymers, so we are on our own. The four alternative H atoms are marked on **W119** and the three on the six-membered ring are ruled out either because attack on the alkene would occur the 'wrong' way round (even in the oxo-ene mechanism we prefer to attack the less substituted end on the alkene with the formaldehyde). The most difficult question is: what is wrong with **W117**? There are two minor points: attack on the alkene is more hindered than **209** and the new alkene is *exo* to the six-membered ring and so less stable that the starting material. Do these factors add up to a convincing reason?

W117   W118   W119

Hydroboration

Reactions with α-Halo-Carbonyl Compounds

In the textbook we described the synthesis of **256** with the key step **254** and we said that the 9-BBN is removed from the product simply with ethanol. **Problem 17.20**: How does this work? Why does it not generally work with alkyl boranes?

Answer 17.20: The aryloxide is the hindered 2,6-di-*t*-butylphenoxide and this is too hindered to attack at boron. Ethanol is added at the end of the reaction and combines with **255** to give R$_2$B-OEt and the product **256**. Ethanol must attack the boron as a nucleophile and the enol(ate) of the product must be the leaving group. Without the ketone, the leaving group would be an alkyl anion and such reactions are unknown. An alternative mechanism would be a rearrangement of **255** to a boron enolate.[26]

W120   W121   W122

References

1. F. E. McDonald and X. Wei, *Org. Lett.*, 2002, **4**, 593.
2. Z.-X. Wang, Y. Tu, M. Frohn and Y. Shi, *J. Org. Chem.*, 1997, **62**, 2328.
3. M. Frohn, Z.-X. Wang and Y. Shi, *Org. Synth.*, 2003, **80**, 1.
4. D. J. Ager, K. Anderson, E. Oblinger, Y. Shi and J. VanderRoest, *Org. Process Res. Dev.*, 2007, **11**, 44.
5. K. J. Henry, P. A. Grieco and W. J. DuBay, *Tetrahedron Lett.*, 1996, **37**, 8289.
6. K. Iseki, M. Yamazaki, M. Shibasaki and S. Ikegami, *Tetrahedron*, 1981, **37**, 4411.
7. J. K. Crandall and L.-H. Chang, *J. Org. Chem.*, 1967, **32**, 532.
8. K. Tamao, Y. Nakagawa and Y. Ito, *Org. Synth.*, 1996, **73**, 94.
9. K. Tamao, T. Tanaka, T. Nakajima, R. Sumiya, H. Arai and Y. Itoso, *Tetrahedron Lett.*, 1986, **27**, 3377.
10. K. Tamao, N. Ishida Y. Ito, and M. Kumada, *Org. Synth.*, 1990, **69**, 96.
11. S. Knapp, M. J. Sebastian, H. Ramanathan, P. Bharadwaj and J. A. Potenza, *Tetrahedron*, 1986, **42**, 3405.
12. S. Danishefsky, P. F. Schuda, T. Kitahara and S. J. Etheredge, *J. Am. Chem. Soc.*, 1977, **99**, 6066.

13. K. Kondo, M. Sodeoka, M. Mori and M. Shibasaki, *Tetrahedron Lett.*, 1993, **34**, 4219.
14. D. P. Curran and D. M. Rakiewicz, *J. Am. Chem. Soc.*, 1985, **107**, 1448.
15. S. D. Burke, W. F. Fobare and D. M. Armistead, *J. Org. Chem.*, 1982, **47**, 3348.
16. G. S. Viswanathan, J. Yang and C.-J. Li, *Org. Lett.*, 1999, **1**, 993.
17. E. J. Leopold, *Org. Synth. Coll.*, 1990, **7**, 258.
18. T. A. Bryson and W. E. Pye, *J. Org. Chem.*, 1977, **42**, 3214.
19. S. Hosokawa, B. Kirschbaum and M. Isobe, *Tetrahedron Lett.*, 1998, **39**, 1917.
20. B. B. Snider and N. A. Hawryluk, *Org. Lett.*, 2000, **2**, 635.
21. A. Solladié-Cavallo, P. Lupatelli, C. Marsol, T. Isarno, C. Bonini, L. Caruso and A. Maiorella, *Eur. J. Org. Chem.*, 2002, 1439.
22. J. Thibonnet, M. Abarbri, J.-L. Parrain and A. Duchêne, *Tetrahedron Lett.*, 1996, **37**, 7507.
23. R. N. Young, W. Coombs, Y. Guindon and J. Rokach, *Tetrahedron Lett.*, 1981, **22**, 4933.
24. P. R. Stapp, *J. Org. Chem.*, 1970, **35**, 2419; 1971, **36**, 2505.
25. A. T. Blomquist and R. J. Himics, *J. Org. Chem.*, 1968, **33**, 1156.
26. H. C. Brown, H. Nambu and M. R. Rogic *J. Am. Chem. Soc.*, 1969, **91**, 6852, 6854, 6855.

18

Vinyl Cations

Problems and Further Examples from the Textbook

You will find it helpful for the first part of this chapter to have chapter 18 of the textbook open.

The Question of S_N2 on Vinyl Compounds

In the textbook we said that nucleophilic substitution of unactivated vinyl halides is a rare reaction. The S_N2 reaction at a trigonal sp^2 carbon atom **5** is generally thought to be unknown. It is puzzling that this innocent looking reaction should be impossible: could it be that the transition state, which would have to contain square co-planar carbon **6**, is too high in energy? In addition, the approach of the nucleophile whose lone pair electrons (HOMO) would have to overlap with σ^* of the C–X bond (LUMO) **6a** does look very sterically hindered by R^1. However this mechanism is possible.[1]

The S_N2 Reaction at sp^2 Carbon

Many mechanisms have been suggested[2] for the known examples of vinylic substitution. The reactions of *E*- and *Z*-iodonium salts with the weak nucleophile acetic acid are stereoselective as both isomers give the same product **W2**.

When R^1 is aryl, aryl participation **W4** for the *E*-iodonium salt would explain substitution with retention, though the intermediate cation in **W5** looks horribly strained. An S_N2 for the *Z*-iodonium salt that looks not too hindered would explain inversion.[3]

In other cases, α-elimination **W8** to give a vinylene carbene **W9** leads[4] to mixtures of *E*-
and *Z*-products **W10**. Though these mechanisms might be right for the exceptional leaving
group PhI, it is wiser to assume an addition-elimination mechanism in any unknown case.

The Benzyne Route to Meptazinol

We promised to reveal the benzyne route to meptazinol **33** in the workbook. It was already
known that treatment of *o*-chloro-anisole with strong bases (such as amide ions) **W11** gave
the only possible benzyne **W12** and that this reacted with nucleophiles (such as the base
used to prepare the benzyne) **W13** to give only the *m*-substituted product both because of
steric hindrance and because the anion (or lithium derivative) that is the immediate product
is stabilised by the OMe group (compare *ortho*-lithiation in chapter 7). This is a nucleophilic
substitution at an sp² carbon atom but the nucleophile does not end up at the same atom. It is
also a good entry into *meta* substituted compounds.

It seemed reasonable that, if LDA were used both to create the benzyne and the amide
enolate, coupling between these two **W16** would be observed. In practice, though the benzyne
was evidently formed, it captured LDA rather than the lithium enolate.[5]

A more hindered base was clearly needed and lithium tetramethyl piperidide (LiTMP) was
the answer. Initial experiments gave product **W18** but all refinements left the yield below 50%.

This is a warning that the product is somehow destroying one of the starting materials and labelling experiments showed that the lithium enolate **W19** of the product was being formed by the LiTMP. This was turned to advantage by using three molecules of LiTMP and adding ethyl iodide before quenching the reaction mixture with water. The overall yield for this very short process was an acceptable 40% but the cost of the LiTMP (three equivalents being necessary!) was too much and the process described in the chapter was preferred.[5]

Problem 18.1: When the route described in the chapter was initially tried, the same amide was used with the idea of quenching the resulting enolate with EtI. However this gave the wrong product **W22** so the ethyl group had to be in position at the beginning. What went wrong? What sort of enolate is it?

Answer 18.1: It is an extended enolate (see chapter 11) of an unusual kind having a carbonyl group at both ends. It reacts at the wrong place **W23** (steric and electronic?).

Conjugate Substitution

The ketone **81** was prepared in the textbook by conjugate substitution and used to make the hydroxy ester **W26**. **Problem 18.2**: Why is the Me₃Si group present in the ketone **81**?

Answer 18.2: To provide enantiomeric purity in all the compounds and to control the relative stereochemistry of **81**. More interestingly, it controls the regiochemistry of a Baeyer-Villiger reaction. The two groups that might migrate are both primary but an Me_3Si group stabilises a β-cation and the more stable cation migrates.[6]

Conjugate Substitution

It is even possible to use derivatives of 2-formyl ketones **W29** as vinyl cation equivalents, in this case in reaction with an unusual organo-lithium compound stabilised by sulfur and a cyclopropane ring **W30**. The product **W31** rearranges into a cyclobutane **W32**. **Problem 18.3**: Suggest detailed mechanisms for the reactions.

Answer 18.3: The Claisen ester condensation initially gives the formyl ketone that is deprotonated by ethoxide to give a stable enolate and hence the enols **W29** or **W31**. Conjugate addition of **W30** to **W29** or direct addition to the other tautomer **W35** gives **W31** on dehydration.[7] This is conjugate substitution with hydroxide as leaving group!

The rearrangement of **W31** is more interesting. Protonation gives cation **W36** that rearranges by ring expansion with the migration of the π-like σ-bonds of the cyclopropane. Tautomerism and hydration gives **W38** that loses PhSH to give **W32**. The PhS group stabilises the lithium derivative **W30** and controls the rearrangement **W36** besides being easy to remove.

This method was unsuitable on a larger scale partly because of the 'burning rubber' smell of PhSH and it proved better to convert the 2-formyl ketone **W29** into the pyrrolidine derivative **W39** (*E*- or *Z*-isomer does not matter). The reactions then went in good yield.

Pleraplysillin Synthesis

No doubt it is true, as we said in the textbook, that the cyclic enone for pleraplysillin-1 could be synthesised by a Robinson annelation. In fact it was not. The chemists noted the symmetry in the carbon skeleton and that it might be made by a Shapiro reaction (chapter 16). The starting material is dimedone **W40**, a very easily made cyclohexadione (*The Disconnection Approach* page 177). The tosylhydrazone **W41** does not need to be treated with the usual strong bases required for the Shapiro reaction – aqueous potassium carbonate will do. **Problem 18.4**: Suggest why this is so and draw a mechanism for the formation of the enone from the tosylhydrazone.

Answer 18.4: The first proton removed from **W41** will be from the sulfonamide to give **W43**. The removal of the second proton is usually the difficult step in a Shapiro reaction but here it is one of the acidic protons marked in **W42** as the dianion **W43** is an enolate.[8] Fragmentation gives **W44** as usual but no vinyl-lithium is formed as water will protonate the vinyl anion as it is released **W44**.

A Different Pleraplysillin Synthesis

An alternative synthesis of peraplysillin-1 **199** uses a rather different strategy.[9] The chemists decided to make the central *E*-alkene by a stereoselective Julia olefination (chapter 15) and to position the sulfone so as to be able to use alkylation of the keto-sulfone **W45** with the same available starting material (3-hydroxymethylfuran) used in the Stille synthesis in the main text.

The details of the reactions used in the synthesis are outlined below.[9] The second step is particularly interesting and you might like to try and write a mechanism for it. The regiochemistry of the alkene formation in this step deserves some attention. **Problem 18.5**: Explain all aspects of selectivity in this sequence and check that you understand all the reactions.

Answer 18.5: The epoxidation is straightforward – the only selectivity is that the epoxide **W50** has the same stereochemistry as the alkene **W49**. The elimination is more interesting: the epoxide opens to form the most stable (tertiary) cation and elimination **W55** presumably occurs that way so that the CMe_2 group can occupy genuine equatorial and axial positions away from the alkene. The tosylation is chemoselective in favour of the less hindered primary alcohol, MnO_2 is a selective oxidising agent for allylic alcohols and an unusual S_N2 reaction with the toluenesulfinate anion **W56** promoted by the ketone gives the starting material for the Julia olefination.[9]

W50 → W51 → W54

W55   W56

Conjugate Substitution

In the textbook we said that 2-methyl furan **45** is lithiated in the one remaining α-position **46** and it reacts with the enol ether **47** by conjugate substitution to give the enone **48** in quantitative yield.[10]

45   **46**   **47; enol ether**   **48; 100% yield**

We also described the reaction of three Prins products such as **53** with organo-lithium compounds: this particular example **53** was used to make bertyadionol.[11]

51   **51a**   **52**   **53; 72% yield**   **56; 94% yield**

Bertyadionol **W57** is a diterpene with interesting biological activity and no member of this lathyrane class of natural product had been made when Smith planned his synthesis. Removal of the methyl group and standard HWE (chapter 15) disconnection gave **W58** with one ketone masked with a dithian so that it could act as a nucleophile (chapter 14) in reaction with **53** representing the vinyl cation synthon **W59**.

W57; (−)bertyadionol   **W58 PO(OEt)₂**   **W59**

The synthesis started with enantiomerically pure *cis*-chrysanthemic acid **W60**, available by resolution. Chain extension with diazomethane gave **W61** and oxidation with SeO_2 (chapter 33) the aldehyde **W62** masked as a dithian **W63**.

W60; (−)-*cis*-chrysanthemic acid   W61   W62   W63

Acylation of the lithium derivative of $(EtO)_2PO.Et$ gave **W58** in 35% overall yield from **W60**. Now comes the vital step. Treatment with NaH and then BuLi produces a purple solution, presumably the lithium derivative of the dithian and the anion of the phosphonate, that combines successfully with **W63** to give, after oxidation, **W58** in good yield. The final coupling by HWE (chapter 15) reaction of the ketophosphonate with the aldehyde and stereoselective reduction of both ketones are described in detail in the paper.[11]

W64   W65; 35% yield from W70   W66; 58% yield   W58

Unsymmetrical Enol Ethers

In heterocyclic synthesis, one-carbon electrophiles are often needed and orthoformates can be used as in this example from allopurinol chemistry. Triethyl orthoformate condenses in acetic anhydride with the very 'enolisable' malononitrile **W67** to give the enol ether **W68**. Reaction with hydrazine occurs by conjugate substitution and addition to one nitrile to give **W69**. Cyclisation with formamide gives allopurinol **W70**, a treatment for gout.[12]

W67   W68   W69   W70

The same reagents are used to make the unsymmetrical enol ether **W72**. Conjugate substitution with one enantiomer of a hydroxyamine gives **W73** and hence ofloxacin **W75**. **Problem 18.6**: Explain the formation of **W74** and **W75**.

Answer 18.6: The nitrogen atom cyclises first as the anion **W76** formed with KF in the polar aprotic solvent DMF. This solvates cations but not anions and 'naked' fluoride is basic enough to form the anion. Nucleophilic aromatic substitution is promoted by the ketone and the four fluorine atoms. The *E/Z* mixture in **W73** or **W74** is unimportant as conjugation of the N lone pair with the carbonyl groups means there is little double bond character in the enamine and it can rotate.[13]

The second cyclisation has no conjugating groups and relies entirely on the three fluorine atoms that allow nucleophilic substitution. Fortunately this is a favourable cyclisation giving a six-membered ring. The final nucleophilic substitution with *N*-methyl piperazine is again promoted by the ketone.

Conjugate Substitution

In the textbook we revealed that lithium enolates react with the vinyl halide **73** by conjugate substitution. The lithium enolate of the enantiomerically pure heterocycle **74** gave one diastereoisomer of *E*-**75** in reasonable yield. **Problem 18.7**: Explain the two- and three-dimensional stereochemistry of **75**.

Answer 18.7: The lithium enolate adds **W79** to the unsaturated ester to give an enolate **W80** that can rotate to allow elimination to give the more stable *E*-product **75**. The lithium enolate is nearly flat but has a phenyl group on the top face so that any addition happens to the bottom face.[14] This is a Seebach style transfer of chirality to the benzylic centre and then back again (chapter 27).

Vinyl Sulfones as Vinyl Cation Equivalents

In the textbook we described briefly how the sulfone **98** allows a tandem (chapter 37) sequence of remarkable selectivity. The lithium derivative of the allylic phosphine oxide **96** (see chapter 12) adds to the enone **97** to give the lithium enolate **99** trapped by the sulfone **98** to give the complex product **100** in 96% yield: the only imperfection in stereochemical control is 3% of the *Z*-vinyl phosphine oxide.

The lithium derivative of **96** is an allyl anion equivalent that adds at the γ-position **W81** to the enone **97**. It is reasonable that the exceptionally stabilised allylic lithium derivative should prefer conjugate to direct addition (chapter 9). Conjugate substitution on *E*-**98** gave *E,E*-**100** almost exclusively while *Z*-**98** gave a 1:1 mixture of *E,E*-**100** and *E,Z*-**100** showing that the enolate intermediate has a reasonable lifetime and that the reaction is stereo-selective rather than specific.

Early experiments had shown that quenching enolates related to **99** from the simpler ketone **W83** gave good control over relative stereochemistry in **W84**. The proposed mechanism[15] **W85** involved a ten-membered cyclic transition state having a chair–chair conformation (like a trans-decalin with the middle bond missing) **W86**.

The product **100** was intended for conversion into **W87** with the idea that a steroid-like compound such as **W88** might be produced. **Problem 18.8**: Suggest how **100** might be converted into **W87** and how the functionality in **W87** would be useful for the synthesis of **W88**.

Answer 18.8: Hydrogenation of the two alkenes in **100** gives a saturated diketone that cyclises to **W87** in good yield.[16] The phosphine oxide is ideal for a Horner-Wittig olefination (chapter 15) and hydrolysis of the acetal, chain extension by aldol or Wittig, and cyclisation should give **W88**.

The Diels-Alder Reaction on β-Bromo and β-Sulfonyl Alkynes

In the textbook we reported that the Diels-Alder reaction that gives the required β-substituted alkenes **103** without conjugate addition and that the bromine atom can be displaced by carbon nucleophiles such as methyl copper.[17]

Though furans take part well in normal Diels-Alder reaction with electron-deficient dienophiles, simple pyrroles do not. The problem is that, say, *N*-methylpyrrole is too nucleophilic and alkynes such as **102** too electrophilic and a simple ionic addition **W90** occurs instead. Other epibatidine syntheses appear in chapter 35.

We also discussed the more elaborate diene **105** reacting with the β-tosyl alkyne acid **106** to give a good yield of one regio- and stereo-isomer of the lactone **107**. **Problem 18.9**: Give a detailed mechanism for the formation of **107**.

Answer 18.9: The stereochemistry of **107** suggests that the ester **110** is an intermediate.[18] The alkynyl sulfone is delivered by the ester tether to the lower face of the diene and the methyl group is pushed upwards. Later the ester **W92** was prepared and gave a good Diels-Alder reaction. The product was converted into forskolin **W93**.

Problem 18.10: What aspects of **W92** are better for this synthesis and what problems remain?

Answer 18.10: Diels-Alder reaction and conjugate substitution by MeCu give **W94**. The biggest improvement is the oxygen atom on the double bond as it proved very difficult to add this to **108**. Reduction of one alkene and dihydroxylation of the other look good stereochemically. Side chain addition requires the carbonyl in the lactone bridge but stereoselective oxidation between the two alkenes looks tricky (see chapter 33 for such reactions). See the original papers[18,19] for more details: there is also an asymmetric synthesis.[20]

The Heck Reaction

We devoted a large chunk of the textbook chapter to this important reaction so we shall just add a few examples here. A double version between two aromatic heterocycles **W95** and **W96** gives only moderate yield but the achievement is correspondingly great. The product **W97** gives a new heterocycle **W98** on heating. **Problem 18.11**: Suggest a mechanism for this reaction.

Answer 18.11: A Diels-Alder reaction **W99** is followed by a reverse Diels-Alder reaction **W100** and the loss of nitrogen gas.[21] The stereochemistry of the nitrogen bridge in **W100** does not matter.

A Synthesis of Strychnine

We discussed a synthesis of strychnine in the textbook chapter that included this step. **Problem 18.12**: Justify the regio- and stereochemistry of this step.

Answer 18.12: The vinyl iodide forms the Pd(II) intermediate **W101** with retention and closes the ring also with retention to give **W102**. The Pd-vinyl has been delivered to the top of the alkene as it is tethered to the top face through the nitrogen atom. Elimination cannot return the alkene to its original position as that hydrogen is on the bottom face. So elimination to the only H atom on the top face has to occur.[22]

After a simple-looking reduction, the final ring is closed without palladium. **Problem 18.13**: Explain what is happening in these last two steps.

Answer 18.13: One of the alkenes in the conjugated system might be reduced by LiAlH$_4$ and it does not much matter which as deconjugation occurs in acid solution by kinetic protonation of the extended enol (chapter 11). Treatment with base isomerises back to the conjugated enone so that conjugate addition of the alcohol can occur to give strychnine. The stereochemistry of **163** is determined by the more favourable *cis* ring junction and that of the final product **164** by the tether delivering the oxygen to the top face of the enone.

Recent Developments in the Heck Reaction

Bromoalkenes for the Heck reaction can be made by the Corey-Fuchs olefination that was used in the textbook for the synthesis of **169**. **Problem 18.14**: Suggest a mechanism. (Warning: no reactions are known in which a nucleophile attacks the *carbon* atom of CBr$_4$. One of the bromine atoms must be attacked instead.)

Answer 18.14: An S$_N$2 reaction at bromine **W103** displaces Br$_3$C$^-$ which immediately returns to displace bromide from phosphorus **W104**. Only then can an S$_N$2 reaction occur at bromine **W105** to create the ylid **W106** needed for the reaction.

Transfer of Groups from Tin to Carbon in the Stille Coupling

When we want to transfer a group from tin we do not want to waste three of them. The rule is that tin transfers the best *anion* to palladium (strengthening our view that the organotin compound is the nucleophile). Alkynyl groups are transferred best and the order for other groups is roughly as below. Note that this is the opposite order from that of copper (chapter 9) which prefers to *keep* the most stable anion.

Starting Materials for Two Syntheses of Crocacins

In the first crocacin synthesis in the textbook, the vinyl iodide **W107** was coupled with the vinyl stannane **188** using $Pd_2(dba)_3$ as the palladium source to give crocacin C **210** in 69% yield. We described the synthesis of **188** in the textbook.[23]

The synthesis of the iodide **W107** uses an asymmetric aldol reaction and was left to the workbook. Disconnection by some kind of olefination gives **W108** made from an epoxide **W109** and hence from an alkene made from the aldol product **W111**.

The first step was an asymmetric aldol using an Evans chiral auxiliary, as described in chapter 27, giving **W112** after protection. The next two steps were straightforward, an HWE reaction (chapter 15) with $(EtO)_2PO.CH_2CO_2Me$ being preferred to the conventional Wittig. Epoxidation was highly stereoselective with *m*CPBA (92:8 in favour of **W114**) directed by the OH group to ensure regioselectivity. The stereoselectivity was, as they say, 'remarkable'. Presumably in the Houk conformation the side chain with the OTBDMS group is larger than the methyl group.

Ph⟍═⟍ Me CHO OR **W112; R=SiMe₂t-Bu**   1. Swern 2. HWE 3. DIBAL   Ph⟍═⟍ Me ⟍═⟍ OH OR **W113; 82% yield**   *m*CPBA   Ph⟍═⟍ Me ⟍═⟍ O OH OR **W114; 94% yield**

The higher order cuprate $Me_2CuCNLi_2$ was used to get regioselectivity in the epoxide opening and the various protecting groups were removed and the OMe groups installed. Finally two steps we have barely mentioned: oxidation with a hypervalent iodine compound, the Dess-Martin periodinane, and a Takai olefination using a Cr(II) species. You can find references to these methods in the paper.[23]

W114 1. $Me_2CuCNLi_2$ 2. Exchange of protecting groups   Ph⟍═⟍ Me Me ⟍═⟍ OH OMe OMe OH **W115**   1. Dess-Martin 2. $CrCl_2$, CHI_3 **W107** 67% yield >95:5 *E:Z*

The same starting material **W115** for the other synthesis was also made by an asymmetric aldol reaction, but using chiral pool strategy (chapter 23), and the vinyl stannane was introduced by another Takai olefination with a modified Cr(II) reagent.[24]

Ph⟍═⟍ Me Me OH OMe OMe **W115**   1. DMP 2. $CrCl_2$ Bu_3SnCHI_2 Ph⟍═⟍ Me Me ⟍═⟍ SnBu₃ OMe OMe **215; 88% yield**

The Synthesis of β-Carotene

The starting materials **223** and **224** for the synthesis of β-carotene described in the textbook are made by methods relevant to this chapter. The vinyl iodide starts with a Heck coupling followed by carbometallation of an alkyne.

W116 $Pd(PPh_3)_4$ **W117**   →   **W118**   1. Me_3Al Cp_2ZrCl_2 2. I_2, THF   **223**

The *bis*-stannane **224** is made from simpler stannanes by the Kocienski version of the Julia coupling (chapter 15). Both halves – the sulfone and the aldehyde – are made from the same starting material **W119**. A Moffat-style oxidation gave the aldehyde while a Mitsunobu coupling followed by oxidation with a high valent Mo species with H_2O_2 gave the benzthiazole[25] **139**.

Development of GR151004

Glaxo's (now GSK's) selective 5HT agonist for the treatment of migraine **W122** has two different aromatic rings joined together by a single bond. Disconnection here suggests the type of reaction in this chapter. The compound came from the research labs as the product of a Stille coupling with the pyridine intended as the nucleophile and the benzene as an electrophilic triflate. However, unwillingness to use large amounts of toxic tin compounds in production led to development chemists at Glaxo proposing a Suzuki coupling instead.

W122; GR151004 – new anti-migraine drug   W123; M = Bu₃Sn or (RO)₂B   W124

This is one of the very best methods of joining two aromatic rings together. Indeed before these methods became available there were no really efficient procedures for this sort of reaction. Treatment of **W123**; M = (HO)$_2$B and **W124**; X = OTf with catalytic (Ph$_3$P)$_2$Pd(OAc)$_2$ in the presence of NaHCO$_3$ in aqueous isopropanol gave a good yield of the drug. Sadly, we believe the drug is not being developed as difficulties appeared during late trials.[26]

Acrivastine

Alternative syntheses of an intermediate **W127** in the production of Acrivastine, Borroughs Welcome's treatment of allergic rhinitis show the value of Pd-catalysed coupling.[27] Both start from the ketone **W126** easily made by lithiation of 2,6-dibromopyridine **W125** and reaction with *p*-tolyl nitrile.

One method uses protection, formylation, HWE, and deprotection to make the unsaturated ester **W127**. An electrophilic formyl group is added to the pyridine rather than a vinyl cation equivalent.

The other goes straight there in one step with a Heck reaction.

A Synthesis of Ebelactone

In the textbook we gave the Suzuki coupling that made **287** at the heart of Mandal's ebelactone **W131** synthesis and we promised some more discussion here.[28]

Fragment A came from a *syn* selective asymmetric Evans aldol reaction (chapter 27) on the boron enolate of the oxazolidinone **W132**. Conversion to a Weinreb amide **W134** and reaction with a vinyl Grignard reagent gave the enone **W135**. The Evans 1,3-diastereoselective reduction with acetoxy-borohydride (chapter 21) gave the *syn*-alcohol **W136** which was protected as **286**. Yields and diastereoselectivities are uniformly excellent.

The other half of the Suzuki reaction came from the aldehyde **W137** (chiral pool) and the crotyl boronate incorporating di-isopropyl tartrate (double asymmetric induction). The product **W138** is formed in excellent yield as a 97:3 mixture of diastereoisomers. Protection and reduction gave **W139**.

Removal of the TBDMS group and oxidation gave the aldehyde **W140** needed for the reaction with the crotyl stannane to put in the next two stereocentres. The result was a high yield of **W141** in a 96:4 diastereomeric ratio. You will notice that both crotyl boranes and crotyl stannanes react at the remote end of their allylic systems (chapter 12).

Protection and oxidative cleavage give the aldehyde **W142** ready for a Corey-Fuchs reaction (see above) to prepare the alkyne **284** ready for conversion to a vinyl borane. The coupling of **286** and **284** is described in the main text.

You will have noticed and admired the efficiency of this chemistry: the lowest yield was 83% and the poorest diastereoselectivity was 96:4.

References

1. M. Ochiai, S. Yamamoto and K. Sato, *Chem. Comm.*, 1999, 1363.
2. Z. Rappoport, *Acc. Chem. Res.*, 1981, **14**, 7.
3. T. Okuyama and M. Ochiai, *J. Am. Chem. Soc.*, 1997, **119**, 4785; *J. Am. Chem. Soc.*, 2001, **123**, 8760.
4. T. Okuyama and M. Ochiai, *J. Org. Chem*, 1999, **64**, 8563.
5. G. Bradley, J. F. Cavalla, T. Edington, R. G. Shepherd, A. C. White, B. J. Bushell, J. R. Johnson and G. O. Weston, *Eur. J. Med. Chem.*, 1980, **15**, 375.
6. M. Asaoka, K. Takenouchi and H. Takei, *Tetrahedron Lett.*, 1988, **29**, 325.
7. J. A. Kaydos, J. H. Byers and T. A. Spencer, *J. Org. Chem.*, 1989, **54**, 4698.
8. G. A. Hiegel and P. Burk, *J. Org. Chem.*, 1973, **38**, 3637.
9. Y. Masaki, K. Hashimoto, Y. Serizawa and K. Kaji., *Chem. Lett.*, 1982, 1879.
10. F. J. Moreno-Dorado, F. M. Guerra, F. J. Aladro, J. M. Bustamante, Z. D. Jorge and G. M. Massanet, *Tetrahedron*, 1999, **55**, 6997.
11. A. B. Smith, B. D. Dorsey, M. Visnick, T. Maeda and M. S. Malamas, *J. Am. Chem. Soc.*, 1986, **108**, 3110.
12. R. K. Robins, *J. Am. Chem Soc.*, 1956, **78**, 784.

13. H. Egawa, T. Miyamoto and J. Matsumoto, *Chem. Pharm. Bull.*, 1986, **34**, 4098.
14. D. Ma, Z. Ma, J. Jiang, Z. Yang and C. Zheng, *Tetrahedron: Asymmetry*, 1997, **8**, 889.
15. M. R. Binns, R. K. Haynes, A. G. Katsifis, P. A. Schober and S. C. Vonwiller, *Tetrahedron Lett.*, 1985, **26**, 1565; *J. Am. Chem. Soc.*, 1988, **110**, 5411; M. R. Binns, O. L. Chai, R. K. Haynes, A. G. Katsifis, P. A. Schober and S. C. Vonwiller, *Tetrahedron Lett.*, 1985, **26**, 1569.
16. R. K. Haynes, S. C. Vonwiller and T. W. Hambley, *J. Org. Chem.*, 1989, **54**, 5162.
17. C. Zhang and M. L. Trudell, *Tetrahedron*, 1998, **54**, 8349.
18. E. J. Corey, P. D. S. Jardine and J. C. Rohloff, *J. Am. Chem. Soc.*, 1988, **110**, 3672.
19. E. J. Corey and P. D. S. Jardine, *Tetrahedron Lett.*, 1989, **30**, 7297.
20. E. J. Corey, P. D. S. Jardine and T. Mohri, *Tetrahedron Lett.*, 1988, **29**, 6409.
21. G. J. Bodwell and J. Li, *Org. Lett.*, 2002, **4**, 127.
22. M. Nakanishi and M. Mori, *Angew. Chem., Int. Ed.*, 2002, **41**, 1934.
23. L. C. Dias and L. G. de Oliveira, *Org. Lett.*, 2001, **3**, 3951.
24. J. T. Feutrill, M. J. Lilly and M. Rizzacasa, *Org. Lett.*, 2002, **4**, 525.
25. B. Vaz, R. Alvarez and A. R. de Lera, *J. Org. Chem.*, 2002, **67**, 5040.
26. GlaxoSmithKline, personal communication.
27. A. Kleeman and J. Engel, *Pharmaceuticals*, p. 37.
28. A. K. Mandal, *Org. Lett.*, 2002, **4**, 2043.

19

Allyl Alcohols: Allyl Cation Equivalents (and More)

Allylation of Enolates

A good example[1] of the lack of regioselectivity in normal allylation of enolates arises when the PhS group is used to stabilise the enolate **W2** from **W1**. Sodium hydride cleanly removes the rather acidic proton from between the PhS and carbonyl groups but reaction with an unsymmetrical allylic halide **W3** gives a nearly 50:50 mixture of regioisomers **W4** and **W5**.

This problem can be overcome in what seems to be a bizarre manner. Simply use a weaker base! Reaction of **W1** with the same allylic halide **W3** using sodium carbonate in water gives only the 'rearranged' products **W5** in 94% yield. The mechanism must be different as this base (Na_2CO_3) is not strong enough to make the enolate in the first place. What really happens is allylation without rearrangement at sulfur followed by proton removal from the very much more acidic sulfonium salt **W6** and [2,3]-sigmatropic rearrangement of the ylid **W7**.

The new double bond will be *E* (if it has any geometry) because the five-membered transition state of the [2,3]-sigmatropic rearrangement (*remember that the sum of the numbers in the square bracket give the ring size in the transition state*) has a 'half chair' transition state with pseudoequatorial substituents. This is explained in chapter 15. **Problem 19.1**: What would be the product from the alkylation of **W8** with **W9** under the same conditions?

Answer 19.1: Alkylation at sulfur by the less hindered end of the allylic bromide gives **W10** and hence the [2,3]-sigmatropic rearrangement of the ylid **W11** to give a new quaternary carbon atom in the product[1] **W12**.

Problems and Further Examples from Chapter

The next section relates directly to compounds featured in chapter 19 in the textbook and it will help if you have that open while you work on it.

The Synthesis of Shikimic Acid

Problem 19.2: Suggest mechanisms for and explain the stereochemistry of the first step in the completion of the shikimic acid synthesis from **127**. How would you do the final stages?

Answer 19.2: Hydrolysis of the esters gives the carboxylate anion that cyclises[2] **W14** onto the epoxide to give the *trans*-di-axial product **W15**. Acetylation gives **W13**.

Hydrolysis of all the esters presents no problem but the elimination step works better with acetate as the leaving group. The answer is to open the lactone bridge and acetylate everything giving **W16** then eliminate with DBU to give **W17** (E1cB) and hydrolyse all the esters.

Controlling the Stereochemistry of Epoxidation

Problem 19.3: These two reactions both give the same diastereoisomer of an epoxide **W19**. Explain the mechanism and stereochemistry.

Answer 19.3: Conjugate addition by hydroperoxide anion must occur from the same face of the alkene as the OH group. This looks like guidance by H-bonding **W21**. However, the methyl group adds from the opposite face to the epoxide oxygen.[3] This is presumably because the oxygen blocks the top face and activates the bottom face in Felkin style (chapter 21).

Problem 19.4: Suggest a synthesis of the Z-vinyl lactone **283** used in allyl palladium cation complex chemistry.

Answer 19.4: The synthesis is outlined[4] in Trosts's footnote 5. Here are the disconnections–the synthesis uses the Grignard reagent **W26**; M = MgBr added to the ester of the aldehyde-acid **W27**.

In the textbook we described allyl cation complexes made from compounds with electron-withdrawing groups on the allyl system such as the nitrile **269**. Here we are concerned with the synthesis of the starting materials. Those with an ester group at one end come from γ-acetoxy enoate esters and they are best made by acetate attack catalysed by palladium in this style. We said: 'Superficially this might look rather like a Heck reaction but the nucleophile is added in a different position. These complexes are reagents for the a^4 synthon **262** having *umpolung* while the Heck reaction gives products from natural conjugate addition (a^3 **263**).' If an ester group (CO_2Me) is the activating group, only the *E*-isomer **W29** is formed. **Problem 19.5**: Explain why Pd(II) is needed for this reaction instead of the usual Pd(0). What is the role of the alkyl nitrite RONO?

Answer 19.5: The conversion of **W28** to **W29** is an oxidation so Pd(II) is needed.[5] The palladium comes out of the reaction as Pd(0) **W31** and the nitrite re-oxidises it to Pd(II).

These γ-acetoxy enoate esters **W29** combine with malonates under Pd(II) catalysis to give *E*-**W33**. **Problem 19.6**: Suggest a mechanism and explain the regioselectivity.

Answer 19.6: Normal Pd(II) attack on the allylic acetate in **W29** gives the expected allyl cation complex—indeed the same complex shown in **W31**. The malonate anion attacks the end remote from the ester group, just as the acetate does in **W31**. In many palladium-catalysed reactions, the natural regioselectivity of the system is enhanced: here attack at the γ-position gives the thermodynamically preferred conjugated product[6] **W33**.

Preparation of Allylic Alcohols

In the textbook we reported that reduction of the enone **14** gave the allylic alcohol **15** and said that **14** was 'prepared by a Robinson annelation.' In detail, this is not obvious. The enone **14** is in fact made by cyclisation of the dienone **W34** with EtAlCl₂. **Problem 19.7**: Suggest a mechanism for this reaction and a synthesis of the starting material **W34**.

Answer 19.7: Obviously the alkene attacks the Lewis acid complex of the enone **W35** but the problem is what to do next. The simplest solution is a cascade of H migrations **W36** that neatly explains the stereochemistry of **14**. This will work only if each H is *anti* to the other. The ring junction must be *cis* as the alkene is tethered. The stereochemistry of the *i*-Pr side chain is reasonable as approach to the bottom face of the enone as drawn will push the H atom down. This centre is inverted in the H migrations. This route gives 90% yield of just **14** but a thermal ene reaction (heat at 240 °C in a sealed tube) gives 87% of a 53:47 mixture of diastereoisomers.[7]

The synthesis of **W34** finally brings in the Robinson annelation. Conjugate addition of butenone to the enamine of aldehyde **W37** gives the keto-aldehyde **W38** that cyclises to the only six-membered ring possible.[8] The starting material **W37** can be made by alkylation of an aldehyde.

Claisen, Claisen-Cope and Claisen-Ireland Stereochemistry

The formation of any allyl vinyl ether such as **152** or **155** involves no change in the stereochemistry of the allyl alcohol **75**–this is acetal exchange at the vinyl ether or acetal centre (this also applies to **174** to **176**). No bond is broken to the allyl alcohol and the double bond is not involved at all.[8] **Problem 19.8**: Draw a mechanism to convince yourself that this is right.

Answer 19.8: In the synthesis of **152** repeated substitutions at the acetal centre by the S_N1 mechanism are followed by an E1 elimination. All the intermediates are oxygen-stabilised cations. There is no reaction and therefore no change in stereochemistry at the chiral centre in the allylic alcohol.

Problem 19.9: Using the method shown in the textbook, explain the stereochemistry of this Claisen-Ireland rearrangement.

Answer 19.9: The formation of the enolate uses HMPA so it will give the Z-enolate **W41** and the resulting alkene **W43** must also be Z as it is in a six-membered ring. The boat-like transition state develops from delivery of the silyl enol ether to the bottom face of the THP ring **W42** and the two marked Hs are *anti* in the product.[9]

The next stage transforms **W40** into a new allylic alcohol **W46** as it is to be followed by a Sharpless asymmetric epoxidation (chapter 25). **Problem 19.10**: Interpret these reactions, suggest a reagent for the last step, and justify the stereochemistry of the new alkene.[10]

Answer 19.10: The first step is a Swern oxidation giving an aldehyde–the key step is an elimination of a sulfonium ylid **W47**. It is used as the centre next to the new CHO group can easily be epimerised in acid or strong base. The olefination is the Still-Gennari method (chapter 15) for making Z-alkenes of conjugated carbonyl compounds and DIBAL is the best reagent for the final reduction, though there are other possibilities.

W47 W44 W45

Vinyl Nucleophiles: Revision of Chapter 16

The vinyl stannane **304** used in the synthesis of the fungal metabolite guanacastepene,[11] was made by a palladium-catalysed coupling of a vinyl iodide **W49** with hexamethylditin. **Problem 19.11**: Suggest mechanisms for the reactions.

W48 → I_2, pyridine / CH_2Cl_2 → W49; 77% yield → $Me_3Sn-SnMe_3$ / $(Ph_3P)_4Pd$ / 80 °C → 304; 90% yield

Answer 19.11: There are two reasonable mechanisms for the first step: either electrophilic iodine addition to the alkene **W50** followed by a pyridine-catalysed elimination or a Baylis-Hillman style (chapter 10) addition, iodination elimination sequence. The problem with the first mechanism is that the alkene is electron-deficient and will not add iodine very well.

W50 I_2 → W51 → W52 ⇌ W53 → W49

The Baylis-Hillman route is more convincing and is preferred by Johnson's group.[12] Conjugate addition by pyridine **W54** gives a nucleophilic enolate that will iodinate **W55** well. The rest is the standard E1cB elimination.

W54 → W55 → W56 ⇌ W57 → W49

Stille coupling of **W49** had already been studied by Johnson[13] and the next step amounts to a Stille coupling to make a reagent for Stille coupling. The usual oxidative addition gives a vinyl-Pd(II) complex **W58** and transmetallation with loss of Me_3SnI allows the other Me_3Sn group to join the vinyl group on the Pd atom **W59** ready for reductive elimination of Pd(0).

The next steps in the guanacastepene synthesis are a photochemical reaction of **306** and a fragmentation. **Problem 19.12**: Suggest how these reactions might proceed. This problem revises photochemistry and radical reactions. (*Hint*: Sm(II) often attacks O atoms to form a radical.)

Answer 19.12: The first step is a photochemical 2 + 2 cycloaddition: the enone is excited and adds to the alkene. Normally 'reverse' regioselectivity is expected (the nucleophilic end of the alkene adds to the less electrophilic end of the enone) but intramolecular reactions tend to give the more stable ring. The stereochemistry of **W60** shows that **306** must fold as **306a** to give the more stable *cis* 5/6 ring fusion and keep the two methyl groups away from each other. The yield is an impressive 76%. The radical fragmentation **W62** breaks the weakest bond (most strained) even though it produces a secondary rather than the alternative tertiary radical. The immediate product **W63** collects a hydrogen atom and the samarium enolate reacts with PhSeBr in the usual way.[11]

Pd-Catalysed Reactions of Monoepoxy-Dienes with Heteroatom Nucleophiles

Though less dramatic, reactions of the mono-epoxides of dienes such as **W64** with heteroatom nucleophiles show interesting regioselectivities and can be very useful. In preparing starting materials for the Claisen rearrangement, Kirschleger[14] reacted phenols with **W64** using Pd(PPh$_3$)$_4$ as catalyst. **Problem 19.13**: Suggest an explanation for the regioselectivity of the palladium-catalysed reaction and a mechanism for the second step.

Answer 19.13: The usual Pd allyl cation complex **W68** is formed and evidently reacts with the nucleophile at the more substituted position. There must be an electronic explanation for this and S_N1 reactions are usually preferred at a tertiary rather than a primary centre. The details are in the patent literature–see reference 8 in Kirschleger's paper.[14]

Another surprise is the Claisen rearrangement **W65a** catalysed by HCl. Acids are often avoided in such reactions as a dihydrofuran can be formed by cyclisation of the OH group back onto the tertiary centre. The *E*-alkene in **W66** comes from the equatorial position of the larger substituent (CH_2OAc) in the chair-like transition state for the [3,3]-sigmatropic rearrangement (chapter 15).

The second example of such reactions involves a kinetic resolution. Trost[15] made macrocyclic bis-indoles from the simple bromoalcohol **W74** which was in turn made asymmetrically by nucleophilic attack of bromoethanol on a simple vinyl epoxide **W73** catalysed by a chiral palladium complex of **W75**. **Problem 19.14**: How is it possible for both enantiomers of the epoxide to give a single enantiomer of the product?

Answer 19.14: One seductively attractive answer is to say that the palladium allyl cation complex is flat and has no stereochemistry. This is not true because the Pd attacks the allylic epoxide from the opposite face to the leaving group (here epoxide O). So one enantiomer of **W73** gives **W76** while the other gives **W77**. The key observation is that these two complexes are in equilibration faster than they react with the alcohol. Under the influence of the asymmetric catalyst (*L) 77% yield of **W74** in 92% ee results.

[2,3]-Sigmatropic Rearrangements of Allylic Sulfur Compounds

We described [2,3]-sigmatropic rearrangements such as **41a** in the textbook and explained the role of the thiophile, here $(MeO)_3P$. **Problem 19.15**: The other products are PhSMe and the phosphate ester $(MeO)_3P = O$. Suggest a mechanism.

Answer 19.15: The first step is nucleophilic attack on sulfur by the phosphite **43** which gives a salt that is attacked at P by methoxide (this may be an addition-elimination mechanism rather than an S_N2) **W78** to give a new salt that can be attacked at C by PhS$^-$ (definitely S_N2) to give the final products.

Simmons-Smith Cyclopropanation

In the textbook we described the synthesis of **117** by the Simmons-Smith reaction and outlined its transformation into **119**. We said: 'The solution to the regioselectivity problem of cyclopropanation with three OH groups and three alkenes embedded in the skeleton of **119** is elegant.' **Problem 19.16** Which is the compound 'with three OH groups and three alkenes' that might have been cyclopropanated to give a precursor to **118**? What could have gone wrong?

Answer 19.16: The mystery compound is **W80** found simply by opening the lactone and removing protecting groups and the three-membered ring. Two of the alkenes (C–9 and C–11) are also allylic alcohols and could have been cyclopropanated by the Simmons-Smith method so the cyclopropanation had to be conducted while two OHs were protected and the third alkene was not yet in place.[16]

The Synthesis of Shikimic Acid

The starting material for Koreeda's shikimic acid synthesis described in the textbook was the diene **123** prepared as a 4:1 E:Z mixture from allyl silane **W81**. **Problem 19.17**: Explain this reaction and why it is not worthwhile separating the unwanted Z-isomer before the Diels-Alder reaction.

Answer 19.17: Allyl silanes form lithium derivatives (chapter 12) that react at the γ-position with electrophiles such as DMF **W82**. Acetylation of the enol of the product **W83** gives **123**. The 4:1 ratio of **123** presumably reflects the stereochemistry of the enol. The E,Z-diene cannot easily adopt the required s-*cis* conformation and so only E,E-**123** reacts.[2]

The last intermediate described in the textbook was **127** which was converted into a derivative of shikimic acid **W85**. **Problem 19.18**: Explain what is happening in these reactions: treatment with Ac_2O and pyridine merely acetylates alcohols and needs no explanation.

Answer 19.18: LiOH hydrolyses the esters in **127** and the carboxylate anion cyclises **W84a** onto the near end of the epoxide to give the *trans*-di-axial product **W84a**. Treatment with acidic MeOH opens the lactone to give **W87** after acetylation: this sequence has inverted one OH group. Elimination of one acetate occurs by the E1cB mechanism via the enolate of the ester.

The Synthesis of 11-Deoxytetrodotoxin

We described one step in the synthesis of **145** in the textbook. This compound was used in turn to make **146**. **Problem 19.19**: Can you trace the skeleton of **146** in the simpler epoxide **145**?

Answer 19.19: The easiest way is to number the carbon atoms in one compound and try to see where each comes from in the other. We have chosen to number **W88**−tetrodotoxin without the guanidine−and compare it with **W89**−**145** without the acetal. The easiest atoms to identify are the carbon with the methyl group C-6 and the *t*-alkyl amine C-3.

We can then see that two atoms are between C-6 and C-3 whichever way we count but C-8 has the side chain attached (C-9) and, if we reveal the aldehyde from which C-9 must be made, we can fill in the other numbers. This approach also reveals what remains to be done, for example C-1 and C-2 need to be oxidised and C-10 to be removed altogether. If you consult the paper you will see that this analysis is not quite right as, for example, C-1 is removed and a new side chain added.[17]

[3,3]-Sigmatropic Shifts: The Cope Rearrangement

In the textbook we described the Cope rearrangement **147** used by Corey in one of his syntheses of gibberellic acid. The starting material was in fact **147a**. **Problem 19.20**: Suggest how this compound might be made.

Answer 19.20: Some Diels-Alder product is obviously involved and, if we remove the silyl enol ether, it looks like **W92**. Reversing the Diels-Alder gives us a simple dienophile **W94** but an odd-looking diene **W93**.

In fact the diene **W93** can be made by alkylation of the anion **W95** of cyclopentadienone. The initial product is **W97** but this equilibrates by 1,5-H shifts to the more stable **W93**. Diels-Alder reaction with **W94** is very selective for the right diastereomer of **W92** if BF_3 catalysis is used. This binds to the ketone and increases *endo*-selectivity.[18]

The Claisen Rearrangement in a Synthesis of *cis*-Chrysanthemic Acid

We described Funk's synthesis of chrysanthemic acid **197** from the seven-membered lactone **195** in the textbook. **Problem 19.21**: Suggest a synthesis for the seven-membered lactone **195**.

195   196; 97% yield   197; 94% yield

Answer 19.21: Disconnection of the lactone and transformation of the Z-alkene into a triple bond reveals a simple route from **W100**. In the synthesis the dianion (2.1 equivalents. BuLi) of **W100** reacts with acetone to give **W99** in 65% yield. Hydrogenation with a poisoned Pd catalyst gave **196** directly as lactonisation was spontaneous.[19]

195a   W98   W99   W100

Palladium Allyl Cation Complexes

We described the reaction of the lactone **243** in the textbook with malonate esters to give one regio- and stereoisomer of the triester product **244**. We now describe the synthesis of **243** from cyclopentadiene by ketene cycloaddition and Baeyer-Villiger rearrangement.[20]

243   244; 84% yield

Cycloaddition of dichloroketene **W103**, generated *in situ* from dichloroacetyl chloride **W101** and Et₃N, with cyclopentadiene gives the bicyclic ketone **W104**. Removal of the chlorines with zinc gives the important ketone **W105**. The 'cycloaddition' may be an ionic reaction. The regioselectivity is as expected from either as both use the LUMO of the ketene (larger coefficient on the carbonyl) and the HOMO of the diene (largest coefficient at the ends).

W101   W102   W103   W104   W105

Baeyer-Villiger rearrangement of **W105** occurs easily as the ring is strained so competes well with epoxidation (chemoselectivity). Only the vinyl group migrates (regioselectivity) as it helps to support the positive charge in the transition state. The product is **243**.

In the textbook we reported the synthesis of the unsaturated hydroxy diester **255** by palladium allyl cation chemistry. **Problem 19.22**: How might **255** be converted into *trans*-chrysanthemic acid **W106**?

Answer 19.22: We need the anion of the malonate to cyclise onto the nearer end of the allylic alcohol with a leaving group instead of OH. In fact the chemists decarboxylated **255** by the Krapcho method and benzoylated the alcohol to give **W107** as a *syn:anti* mixture in a 3:2 ratio.[21] Equilibration of this mixture by refluxing with NaOEt in ethanol gave 85% of the *trans* ester and hydrolysis to the pure *trans* acid **W106** was quantitative.[22]

We showed in the textbook that addition of the nitroalkane **300** to the allyl esters **298** or **299** required base (Cs₂CO₃) with **298** but not with **299**. **Problem 19.23**: Why is base required for the reaction of **298** but not for that of **299**?

Answer 19.23. Acetate, the leaving group from **298**, is not a strong enough base to deprotonate the nitroalkane but the carbonate from **299** decomposes to CO_2 and *i*-PrO⁻ which is strong enough.

Problem 19.24: Suggest a mechanism for the conversion of **301** into **302** when Cs₂CO₃ is present.

Answer 19.24: The carbonate removes the second proton from the nitroalkane and the anion adds Michael-fashion to the unsaturated ester **301a**. It appears that **301** folds **W109** to put both large groups (aryl and ester) equatorial in the inevitably chair-like transition state.[23]

References

1. K. Ogura, S. Furukawa, and G.-I. Tsuchihashi, *J. Am. Chem. Soc.*, 1980, **102**, 2125.
2. M. Koreeda and M. A. Ciufolini, *J. Am. Chem. Soc.*, 1982, **104**, 2308.
3. J. Carnduff, M. Hafiz, R. Hendrie and F. Monaghan, *Tetrahedron Lett.*, 1984, **25**, 6033.
4. B. M. Trost and T. P. Kuhn, *J. Am. Chem. Soc.*, 1979, **101**, 6756.
5. J. Tsuji, H. Ueno, Y. Kobayashi and H. Okumoto, *Tetrahedron Lett.*, 1981, **22**, 131.
6. J. Tsuji, H. Kataoka and Y. Kobayashi, *Tetrahedron Lett.*, 1981, **22**, 2573.
7. B. B. Snider, D. J. Rodini and J. van Straten, *J. Am. Chem. Soc.*, 1980, **102**, 5872.
8. R. E. Ireland, T. C. McKenzie and R. I. Trust, *J. Org. Chem.*, 1975, **40**, 1007.
9. M. A. M. Fuhry, A. B. Holmes and D. R. Marshall, *J. Chem. Soc., Perkin Trans. 1*, 1993, 2743.
10. R. E. Ireland, J. D. Armstrong, J. Lebreton, R. S. Meissner and M. A. Rizzacasa, *J. Am. Chem. Soc.*, 1993, **115**, 7152.
11. W. D. Shipe and E. J. Sorensen, *Org. Lett.*, 2002, **4**, 2063.
12. C. R. Johnson, J. P. Adams, M. P. Braun, C. B. W. Senanayake, P. M. Wovkulich and M. R. Uskokovic, *Tetrahedron Lett.*, 1992, **33**, 917.
13. C. R. Johnson, J. P. Adams, M. P. Braun and C. B. W. Senanayake, *Tetrahedron Lett.*, 1992, **33**, 919.
14. J.-Y. Goujon, A. Duvai and B. Kirschleger, *J. Chem. Soc., Perkin Trans. 1*, 2002, 496.
15. B. M. Trost and W. Tang, *Org. Lett.*, 2001, **3**, 3409.
16. Y. Baba, G. Saha, S. Nakao, C. Iwata, T. Tanaka, T. Ibuka, H. Ohishi and Y. Takemoto, *J. Org. Chem.*, 2001, **66**, 81.
17. T. Nishikawa, M. Asai and M. Isobe, *J. Am. Chem. Soc.*, 2002, **124**, 7847.
18. E. J. Corey and J. E. Munroe, *J. Am. Chem. Soc.*, 1982, **104**, 6129.
19. R. L. Funk and J. D. Munger, *J. Org. Chem.*, 1985, **50**, 707.
20. B. M. Trost and T. R. Verhoeven, *J. Am. Chem. Soc.*, 1980, **102**, 4730.
21. J. P. Genêt, F. Piau and J. Ficini, *Tetrahedron Lett.*, 1980, **21**, 3183.
22. S. Julia, M. Julia and G. Linstrumelle, *Bull. Soc. Chim. Fr.*, 1966, 3499.
23. C. Jousse-Karinthi, C. Riche, A. Chiarioni and D. Desmaële, *Eur. J. Org. Chem.*, 2001, 3631.

D
Stereochemistry

20

Control of Stereochemistry–Introduction

In most chapters in the workbook we extend the chemistry in the textbook chapters and introduce other examples from the literature. This chapter is rather different. It is about the understanding and manipulation of stereochemistry–one of the most taxing skills the organic chemist requires. This chapter therefore consists of problems to allow you to practise just those skills. The terms GSR (Group Swap by Rotation) and CBP (Change of Bonds in the Plane) are defined in chapter 20 of the textbook.

Chirality

The simplest concept! Yet it catches so many people out. **Problem 20.1**: Which of these compounds are chiral? (*Hint*: You are advised NOT to use criteria such as 'Is there a chiral centre?' but look for symmetry elements. If a molecule has a plane or a centre of symmetry it cannot be chiral. A C_2 axis (or C_3 etc) is the only symmetry compatible with chirality.)

(a)

W1

W2

W3

(b)

W4

W5

W6

(c)

W7

W8

W9

Workbook for Organic Synthesis: Strategy and Control Paul Wyatt and Stuart Warren
© 2008 John Wiley & Sons, Ltd

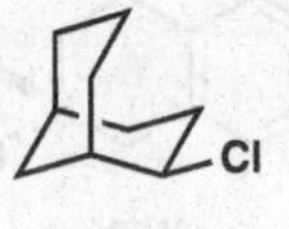

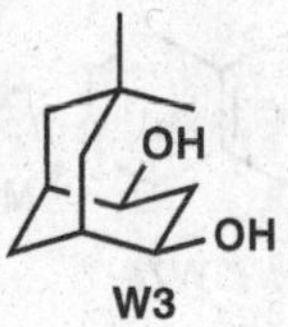

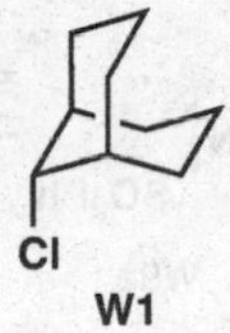

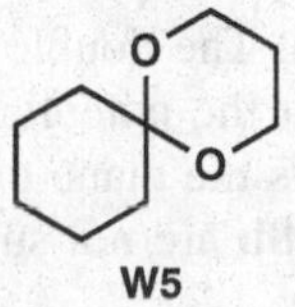

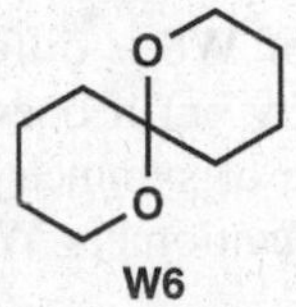

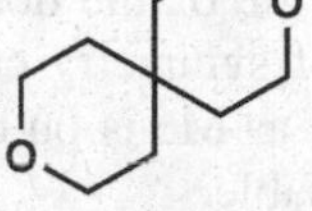

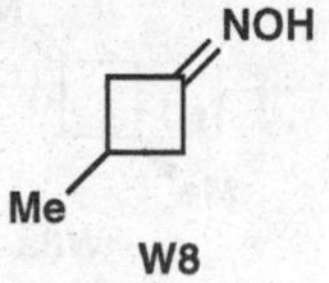

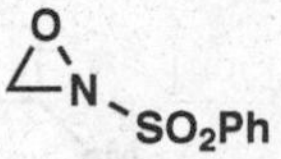

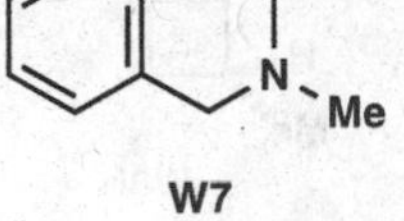

Answer 20.1: Series (a) is simple–only the conformational drawing might deceive you. Two compounds **W1** and **W3** have planes of symmetry shown by the dashed rectangle and must be achiral. The other **W2** has no plane or centre of symmetry and is chiral.

(a)

Series (b) is more complicated. The three compounds are isomers but in **W4** and **W5** each ring is a plane of symmetry for the other ring. This is more obvious with planar structures than with conformational diagrams but neither compound is chiral. However, the acetal **W6**, a pheromone of the olive fly is chiral as neither ring is a plane of symmetry for the other. Its enantiomer, *ent*-**W6** cannot be superimposed on **W6**.

(b)

Series (c) concerns nitrogen. The simple amine does have three different substituents and N is pyramidal. Rapid pyramidal inversion between **W7a** and **W7b** means that such nitrogen atoms are not chiral as the molecule has, on average, a plane of symmetry. The three-membered ring **W9** is different. The transition state for pyramidal inversion at N has a trigonal N atom with bond angles of $120°$. This is fine for **W7** but not for **W9**. The pyramidal N is strained with bond angles of $109°$ but trigonal N is worse so pyramidal inversion between **W9a** and **W9b** is slow, and the molecule is chiral.

(c)

The case of **W8** is different again. The double bond of the oxime does not strictly have *E* or *Z* geometry but it does mean that the potential plane of symmetry cutting the ring in half is not a plane of symmetry and nor is the plane of the ring as Me is on one side and H on the other. The enantiomers **W8a** and **W8b** are not superimposable.

How Many Stereoisomers Are There?

Problem 20.2: We advise you *not* to use the traditional mnemonic that a compound with *n* chiral centres has 2^n 'stereoisomers'. One problem is that it makes no distinction between diastereoisomers and enantiomers and another is that it often does not work! Try it out on these two compounds: one isomer of **W11** can be drawn **W11a** to resemble **W10**. Then try the much more helpful question: how many diastereoisomers are there, and which are chiral?

W10   **W11**   **W11a**

Answer 20.2: Both compounds have four chiral centres (ringed). So by the mnemonic, both should have $2^4 = 16$ 'stereoisomers'. In fact **W10** has only one diastereoisomer: it is impossible to invert any centre in this tricyclic cage structure. It is chiral as there are no elements of symmetry and so there are two enantiomers of **W10** and that is that. However **W11** has the full 16 'stereoisomers', that is eight diastereoisomers as all are chiral. We could put all the substituents on one face **W11b** or one substituent on the other face **W11c–f** or two on one and two on the other such as **W11a** and if you complete this catalogue you will have eight diastereoisomers. The enantiomer of each is easily drawn by inverting every centre.

W10   **W11**   **W11a**   **W11b**

W11c   **W11d**   **W11e**   **W11f**

Problem 20.3: Now apply this method to rather simpler examples: how many diastereoisomers of the alcohols **W12 and W13** are there, and which is chiral?

W12   **W13**

Answer 20.3: The diol **W12** is simple: either both OH groups are on the same face **W12a** or there is one on each **W12b**. The *syn*-diol **W12a** has a plane of symmetry and is achiral (you may have called this a *meso* compound). However the *anti* diol **W12b** has C_2 symmetry and is chiral.

plane of symmetry C_2 axis

W12a: achiral   **W12b: chiral**   ***ent*-W12b**

The phenyl-substituted diol **W13** has three diastereoisomers **W13a–c**. The two *syn*-diols **W13a,b** have a plane of symmetry and are achiral. The *anti*-diol **W13c** is chiral (no plane of symmetry) and has an enantiomer *ent*-**W13c**. It may puzzle you that the phenyl group has no stereochemistry in **W13c**. To convince you, try giving it stereochemistry and then rotating it though 180° around a horizontal axis (as drawn) and you will 'invert' the phenyl centre.

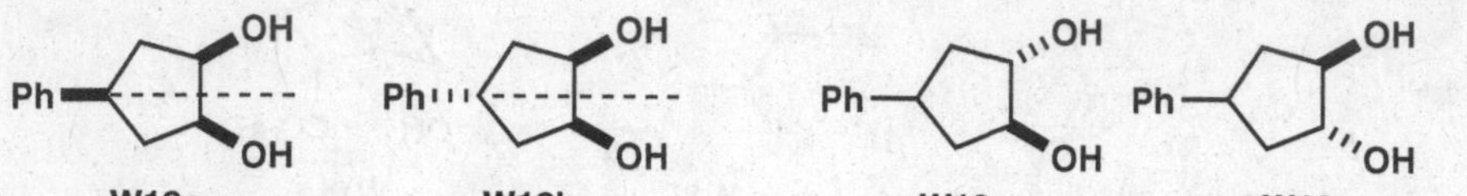

W13a   W13b   W13c   *ent*-W13c

Drawing Stereochemical Structures

This is no trivial matter. Experienced chemists spend time working out how to draw the stereochemistry of their compounds in the most helpful way. Even then they may get it wrong.

The top line of scheme 2 in a recently published paper on the synthesis of indolizidine alkaloids appears below (compound numbers from the paper and not from the textbook chapter). **Problem 20.4**: What is wrong with the presentation of the stereochemistry in these diagrams? Suggest an improved version.

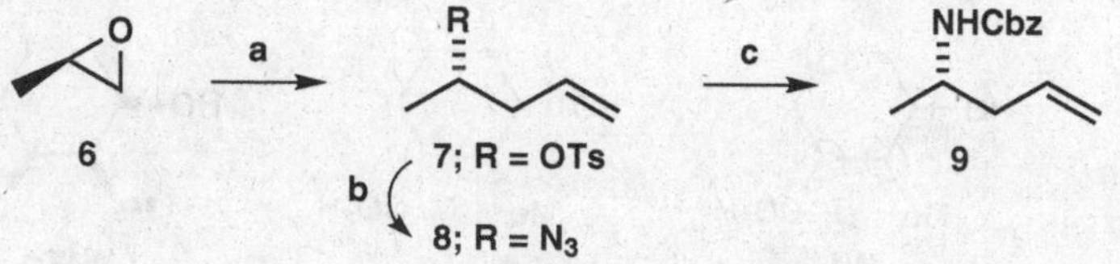

6   7; R = OTs   9

                  b
                  8; R = N₃

Reagents and conditions: (a) (i) C_2H_3MgBr, Cu(COD)Cl (10 mol %), THF, −78 °C to rt, 12 h, (ii) TsCl, DMAP (10 mol %), CH_2Cl_2, 36 h (65%); (b) NaN_3, DMF, 40 °C, 12 h (92%); (c) (i) $LiAlH_4$, Et_2O, 0 °C to rt, 2 h, (ii) Cbz-Cl, K_2CO_3, THF, 12 h (89%).

Answer 20.4: Since the displacement of OTs by azide occurs with inversion, the same diagram cannot be used for compounds **7** and **8**. One possible improvement is shown below. Note that the stereochemistry of R in **7** is related to that of the methyl group in **6** by GSP.

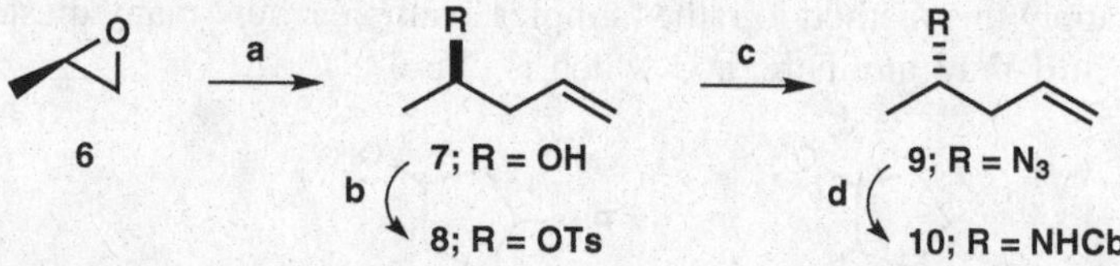

6   7; R = OH   9; R = N₃

                  b   d
                  8; R = OTs   10; R = NHCbz

Reagents and conditions: (a) C_2H_3MgBr, Cu(COD)Cl (10 mol %), THF, −78 °C to rt, 12 h, (b) TsCl, DMAP (10 mol %), CH_2Cl_2, 36 h (65%); (c) NaN_3, DMF, 40 °C, 12 h (92%); (d) (i) $LiAlH_4$, Et_2O, 0 °C to rt, 2 h, (ii) Cbz-Cl, K_2CO_3, THF, 12 h (89%).

Problem 20.5: A single enantiomer of hydrobenzoin **W14** has been drawn in two slightly different ways with the stereochemistry omitted. Draw in the stereochemistry.

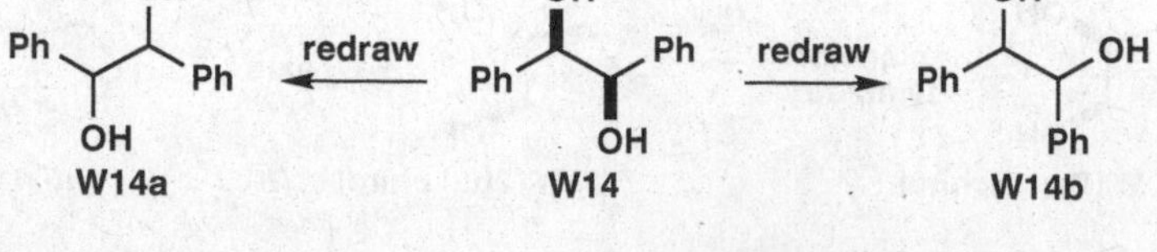

W14a   ←redraw   W14   redraw→   W14b

Answer 20.5: With **W14b**, one side has been left alone. On the other side, two of the groups have swapped position by rotation (GSR) and so we need to draw one of them downwards. There are two possibilities: **W14b(ii)** is more usual as the main chain is in the plane of the paper.

W14 W14b(i) W14b(ii)

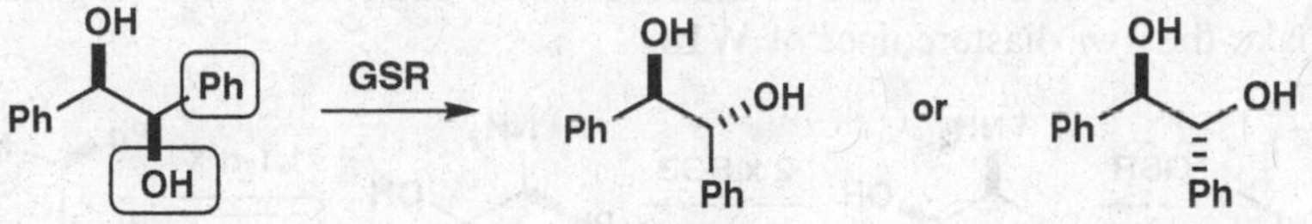

With **W14a**, the whole molecule has been rotated about an axis. Both the hydroxyl groups that were coming forwards will swing to the back **W14a(i)**. Another way to look at this is to rotate the molecule anticlockwise by about 60° to get the carbon–carbon bond in the correct orientation **W14a(ii)** and then do two GSRs. The final structure is **W14a(i)** by either method.

W14a(i) W14a(ii) W14 W14a(i)

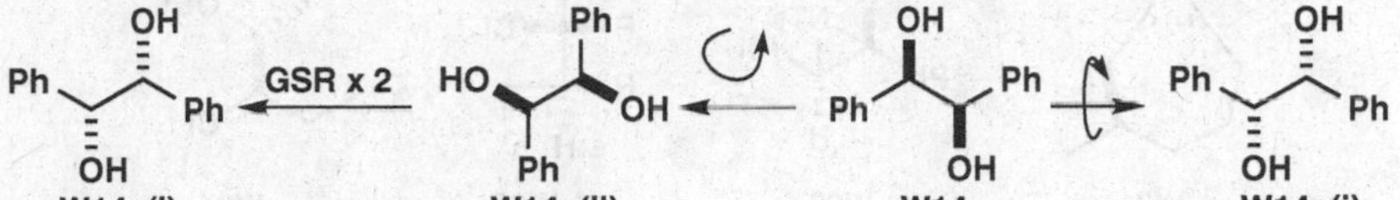

Problem 20.6: The following scheme has been drawn by a stereochemical denialist. Reagents have also been omitted. Draw in the stereochemistry and suggest how the transformations may be achieved.

W15 W16 W17 W18

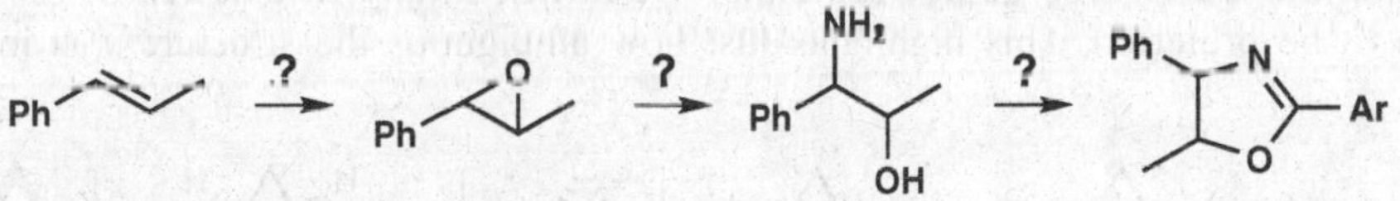

Answer 20.6: The oxygen atom must make both bonds on the same side of **W15** so the epoxide is either **W16a** or **W16b**. We would expect a reagent like mCPBA to do this.

W15 W16a W16b W19 W20

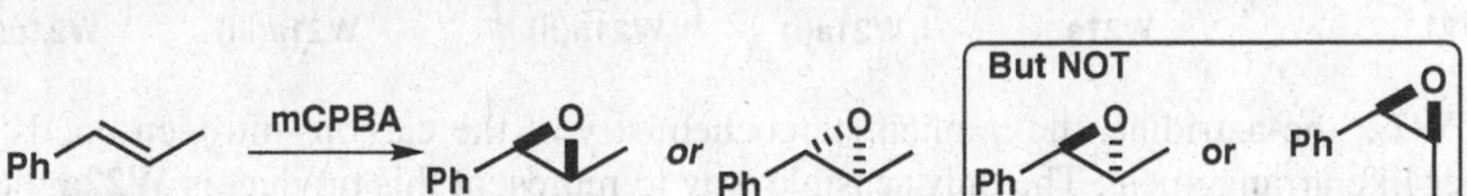

The regioselectivity of the ring opening of the epoxide **W16** (with, say, ammonia) might be an issue but assuming complete S_N2 character in reaction at the benzylic position, then we will have one inversion. Note that although the liberated hydroxyl in **W17b** is drawn pointing downwards, it has not changed sides of the molecule, there has been no inversion, we can just draw it normally now that the strain of the three-membered ring has gone. Note that **W17a** and **W17b** are enantiomers of the same diastereomer.

W16a W17a W17b W16b

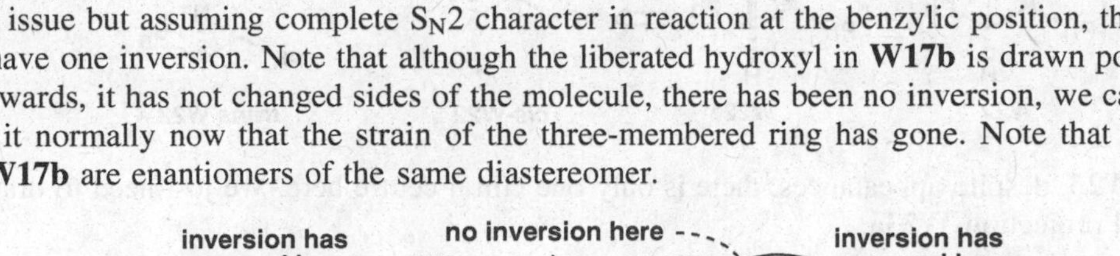

The last step merely involves the formation of a ring with the two heteroatoms. There are no inversions to worry about and all we need to do is to draw the amine and hydroxyl groups on the same side of the molecule. A GSR brings the OH onto the correct side. We would like both functional groups to be in the plane and we can push them into the plane with two CBP manoeuvres. A 1,1-diX disconnection on the ring would suggest to us a reagent like benzoyl chloride to make the *syn* diastereomer of **W18**.

Problem 20.7: A student comes to you for help on **W21–24**. How should they be redrawn properly?

Answer 20.7: For **W21** the exocyclic methyl group is easy and a CBP sorts it out. For the stereochemistry around the cyclopropane we need to ask ourselves where the hydrogen atoms are. Although we might initially imagine them to be coming forwards as in **W21a(i)**, it is not unreasonable for them to be going back either **W21a(iii)**. In any case **W21a(ii)** or **W21a(iv)** are much to be preferred. This highlights just how ambiguous the structure was in the first place!

With **W22**, the aziridine, the implicit stereochemistry of the carbon–nitrogen bonds reveals an Escher like arrangement. The only sensible way to represent this product is **W22a**, however, this is *probably not* what the drawer meant! They probably meant the *trans* compound.

In **W23**, despite appearances, there is only one chiral centre here. We just need to draw the normal projection **W23a**.

The real problem with **W24** is that a stereochemical bond has been used to connect two chiral centres. Additionally, when two stereochemical bonds are connected to the same carbon atom then the angle between them is best kept quite shallow (as we have done with the dihalide **W23a**). If this is *not* done (and instead a 120° angle is used), then *three* stereochemical bonds are needed to represent the tetrahedron on a page–and this can get messy. Let us consider each chiral centre in turn. *The lower chiral centre:* The thick bond is probably there for the benefit of the lower chiral centre. The hydrogen atom would be pointing backwards and using this method will dispense with the thick bond.

The upper chiral centre: The final hydrogen is probably also coming forwards. This would automatically be the case if the thick bond was not there at all. So we just dispense with the thick bond. We can now put the chiral centres back together with a normal (non stereochemical) bond.

The Words We Use

Problem 20.8: Are the CH_2 groups in **W26–28** homotopic, enantiotopic or diastereotopic?

Answer 20.8: Ways to make this decision are described in the textbook chapter. You can look for symmetry elements or you can replace one of the Hs in the CH_2 group with X and see if you get no stereochemistry (homotopic), an enantiomer (enantiotopic) or a diastereomer (diastereotopic). In **W25**, the molecule is planar so the plane dividing the two Hs is a plane of symmetry but there is no other symmetry. If you replace one H by X, you have an enantiomer **W25a**. The Hs are enantiotopic. In **W26** the plane dividing the two Hs is not a plane of symmetry but the plane passing through the two Hs is. If you replace one H by X you get a diastereoisomer **W26a**. The Hs are diastereotopic.

The conformational drawing of **W27** made it look as though there were no planes of symmetry but **W27** is in rapid equilibrium with **W27a** having the same energy but the 'up' Hs, equatorial in **W27**, have become axial in **W27a**. So a flat diagram **W27b** is better for our purpose. It shows that both the plane of the molecule and the plane at right angles to it are planes of symmetry. If we replace the 'down' Hs with X we get no stereochemistry. The Hs are homotopic.

The spiro compound **W28** has three CH_2s. Now the conformation is vital as there is no 'ring flipping' equilibrium. If we look at the CH_2s on the right hand ring **W28a** we see that the other ring makes the plane of the RH ring a plane of symmetry passing between the two Hs. If we replace one with X **W28b** we have an enantiomer. These Hs and the Hs of the other CH_2 group in the same ring **W28c** are enantiotopic.

Looking at the left-hand ring **W28d** we see that the right-hand ring does not make the plane of the left-hand ring a plane of symmetry. The front Hs are on the same side of the ring as O and the back Hs on the same side as CMe_2. They are diastereotopic **W28e**.

Now some chemistry as a relief! **Problem 20.9**: Suggest reagents for the conversion of **W29** into **W30–33** and describe what the reaction might achieve in stereochemical terms using words like 'diastereoselective'. Would the reagents you have suggested actually achieve these results? If not, suggest what reagents might do so.

Answer 20.9: The starting material **W29** is achiral. The achiral alcohol **W30** could be formed diastereoselectively with reagents such as NaBH$_4$ (*syn* alcohol major product). The unsaturated ester could be made by a Wittig-style reaction using a phosphonium ylid **W34** or the anion of a phosphonate ester **W35**. The new alkene has no geometry but the molecule is now chiral and could be formed enantioselectively but not with the achiral reagents **W34** or **W35**. Phosphonium ylids with three different substituents on phosphorus have been used successfully. The acetal **W32** requires the chiral diol **W36** and if one enantiomer is used the reaction will be enantiospecific as no change occurs at the chiral centres. The lactone **W33** is the result of a Baeyer-Villiger reaction and the product is chiral. Normal reagents like mCPBA would give racemic **W33** but asymmetric reagents such as **W37** might do the job. The best way is probably to use an enzyme (chapter 29).

Asymmetric or Not?

Problem 20.10: Are the products of these reactions single enantiomers or racemic?

Answer 20.10: Bu$_3$SnH removes the Br in a radical reaction but does not affect the other chiral centre. The lactone **W39** is a single enantiomer. Reduction of the ketone **W40** does not affect the chiral centre already present but we are not told whether **W40** is a single enantiomer or not. If **W40** is a single enantiomer, so is **W41** but if not, **W41** is racemic. The epoxy-diketone **W42** is achiral (horizontal plane of symmetry) so the product cannot be a single enantiomer as the reagent is achiral too. The reaction is diastereoselective. The decision in the formation of **W45** requires us to know the mechanism of the reaction, just as it did in the others, but this is less familiar. It is a sequence of two electrocyclic reactions (the first con- and the second dis-rotatory but this has no relevance here) and the intermediate **W47** is planar and therefore achiral. The product must be racemic.

Dangers of Loss of Stereochemical Integrity

Problem 20.11: Each starting material is a single stereoisomer. Is there a danger of loss of stereochemical integrity in the reaction?

Answer 20.11: Enolate alkylation of **W48** will lose all stereochemistry as the enolate is planar and achiral. Elimination (E1) on **W50** will of course lose the stereochemistry at the tertiary alcohol **W56** but the other two centres are not involved and will be unchanged. The Heck reaction on **W52** goes through an intermediate **W57** (chapters 8 and 18) that has lost the double bond. It is reformed by reductive elimination in the last step but may be reformed with *E* or *Z* geometry. Cyclisation of **W54** goes through the planar tertiary benzylic cation and the product must be racemic.

21

Diastereoselectivity

Problems and Further Examples Relating Directly to the Text

You will find it helpful to have chapter 21 from the textbook open as you look at this first section. **Problem 21.1**: This problem concerns the reactions of exocyclic enolates with electrophiles. Methylation of the enolate of the aldehyde **W1** gives only one product **W2** which can be isolated in 72% yield. However, the equivalent methylation of the unsaturated aldehyde **W3** gives a 50:50 mixture of two diastereoisomers of **W4**. Explain.

Answer 21.1: This example comes from work by Ireland and Mander.[1] If you draw the conformation **W1a**, you will see that methylation must occur equatorially to give **2**. This is presumably because of the unfavourable 1,3-diaxial interaction with the ringed hydrogen if the electrophile attacked axially. The conformation of the extended enolate (chapter 11) **W3a** shows that there are no 1,3-diaxial interactions on either face and so little discrimination is expected.

Problem 21.2: Corey's synthesis[2] of erythronolide B included the bromolactonisation of **122** (the numbers are those of the chapter). The synthesis continued with the aqueous alkaline treatment of **123** to give **W5**. Draw mechanisms for both these steps and explain the stereochemistry.

Answer 21.2: Reversible bromination of the alkene gives a bromonium ion that can cyclise with inversion **W6** to give **123**. Hydrolysis and a second inversion **W7** gives the epoxide **W5**.

Problem 21.3: Bromolactonisation of the other alkene gives **W8** and debromination with Bu₃Sn radicals gives a mixture of isomers, **W9** predominating. Comment on the stereoselectivity of this last step.

Answer 21.3: Radical removal of Br **W10** gives a planar, tertiary, carbonyl-stabilised radical that can pick up a hydrogen atom from either face **W11**. The major product **W9** has an equatorial methyl group.

Problem 21.4: Direct epoxidation of **108** gives the epoxide **109** with the epoxide on the lower face of the molecule. Draw a conformational diagram of **108** to show why this is so. However, if the iodolactone **104** is treated with basic methanol, an epoxide is formed on the top face of the molecule. Explain the reaction and the stereochemistry.

108   109   104   110

Answer 21.4: The lactone ring forms a diaxial bridge across the top face of the alkene **108a** and prevents attack from the top. Methanolysis of **104** gives a perfect arrangement for cyclisation **W12** to the epoxide **110**.

108a   104a   W12   110

Problem 21.5: Draw a mechanism for the bromoetherification in Kishi's synthesis of monensin. The stereochemistry is more difficult to explain but you could have a go, if you like!

124   125

Answer 21.5: Bromination of the more substituted and therefore more nucleophilic alkene is expected and the nucleophile prefers the nearer end of the bromonium ion **W13** so the reaction is definitely *exo* in Baldwin's terms. The relative stereochemistry (*anti*) of the addition is not in dispute but the tricky bit is why does the bromine attack the top face of the alkene? One explanation would be thermodynamic as the product has a 2,5-*trans*-disubstituted THF. Another would be a Houk-style kinetic explanation with the allylic chiral centre (ringed in **W14**) having H eclipsing the alkene, Me above and the rest of the molecule below.[3]

124 →   W13   → 125   W14

Predicting Stereochemistry

Reaction between **W15** and pentadien-4-one gave one stereoisomer of one adduct **W16** in high yield. **Problem 21.6**: Predict the structure and stereochemistry of **W16** using the evidence that, on treatment with a Lewis acid, it gives the diketone **W17**.

W15

25 °C, 4 hours → **W16** one adduct in 94% yield → $TiCl_4$ / CH_2Cl_2 → **W17**

Answer 21.6: This is a Diels-Alder reaction with the expected regioselectivity between the HOMO of the cross-conjugated silyl enol ether **W15** (chapter 11) and the enone.[4] You could get the structure of **W16** by drawing the mechanism **W18** or the *endo* conformation for the reaction **W19** or by working backwards from **W17** realising that the key step is a conjugate addition of the silyl enol ether onto the remaining enone **W20**.

W18   **W19**   **W16**   **W20**

Stereochemical Control in Six-Membered Rings

One class of indole alkaloids, the *Iboga* alkaloids such as **W21**, might be made from a simple cyclohexene oxide **W23** as the disconnections show. **Problem 21.7**: What would be the regioselectivity of attack by an amine on epoxide **W23**?

W21   **W22** **X**   **W23**   **W23a**

Answer 21.7: This will be a conformationally controlled reaction and the reaction will give the *trans*-di-axial product **W24** (R = 3-indolyl) as required for the synthesis.[5] On heating **W24** cyclised to the amide **W25** and tosylation gave **W26** (= **W22**; X = OTs) which cyclised to **W21** with $AlCl_3$.

W23   **W24**   **W25**   **W26**

Reduction of the Robinson annelation product **W27** with $NaBH_4$ gives the alcohol **W28** whose tosylation and hydrogenation gives a further chiral centre with control. **Problem 21.8**: Explain the origin of the stereoselectivity in these reactions.

W27 → W28 → W29

Answer 21.8: We expect reduction of a cyclohexanone with a small reagent to give the more stable equatorial alcohol **W28**. Catalytic hydrogenation occurs on the outside (*exo* or convex) of the folded molecule **W30** as that face of the alkene can sit flat on the catalyst surface.

W28 conformation → W30 conformation → W29 conformation

The conformation of the *cis*-decalin **W29** allows a remarkable cyclisation.[6] Treatment with the anion of DMSO creates an enolate that can cyclise **W31** to give a new four-membered ring **W32**.

W29 conformation → W31 → W32

Thermodynamic control can also lead to stereoselective formation of new chiral centres around a six-membered ring. The attempted aldol condensation between pentan-2-one and a simple, aromatic aldehyde creates a new six-membered ring with four centres. **Problem 21.9**: How is this control achieved?

W33 → W34   Ar = *p*-chlorophenyl

Answer 21.9: Presumably a double aldol reaction gives the dienone **W35** and reversible conjugate addition of water makes the ring with all the substituents equatorial. The coupling constant between the axial hydrogens is 10.4 Hz confirming the conformation and the structure.[7]

W33 → W35 ⇌ W34 conformation

Kinetic and Thermodynamic Control

The bicyclic alkene **W36** gives only the epoxide **W37** with peracids. This epoxide rearranges with the weak Lewis acid LiI into the aldehyde **W38** but this slowly converts into the aldehyde **W39** on standing. **Problem 21.10**: Explain the rearrangement and the stereoselectivities.

Answer 21.10: This problem concerns folded molecules with *endo* and *exo* faces. Epoxidation occurs only on the *exo* face. The aldehyde **W38** with the CHO group on the *exo* face must be the kinetic product of the rearrangement as the *exo* aldehyde **W39** is more stable.[8] Presumably **W38** gives **W39** by enolisation. The mechanism is most likely to be opening of the strained epoxide with iodide and rearrangement with inversion **W40**. Both steps are stereospecific. The alternative formation **W41** of a cyclobutyl cation **W42** looks less likely to lead to a stereospecific reaction.

Problem 21.11: Explain the stereochemistry of these reactions.

Answer 21.11: Acylation of the enolate gives the stable enolate of **W44** which equilibrates to give the more stable *exo* ester–thermodynamic control. Sodium borohydride is prepared to approach cyclohexanones axially to make the more stable equatorial alcohol but approach from the inside (*endo* face) of **W44** is too much to ask so kinetic control with *exo* approach of the reagent gives *endo*-alcohol **W45**. It may be significant that one oxygen atom of the acetal must be right inside the folded molecule. This chemistry was an early stage in Nicolaou's prostacyclin synthesis.[9]

enolate of **W44**

conformation of **W44**
ester group omitted for clarity

A more subtle example is the cleavage of 'isosorbide' (a product derived from glucose) with Me$_3$SiI and acetone to give the iodo-alcohol **W47** that can be converted into the useful epoxide **W48**. **Problem 21.12**: Identify and explain the selectivity in the formation of **W47**.

Answer 21.12: Isosorbide **W46** is not quite symmetrical and the two OHs are different. You might expect preferential attack on the less hindered *exo* alcohol but this does not lead to the product and is reversible. Instead acetone, activated by Me$_3$SiI, must attack the *endo* alcohol **W49** as it can then react intramolecularly with a THF oxygen atom **W51**. Now *endo* functionality is an advantage. The surprising conclusion here is that the only stereoselectivity becomes chemoselectivity in the choice between the two OH groups. No stereochemistry changes in the reaction.[10]

Drawing Stereochemistry

Problem 21.13: Explain the stereochemistry of **W54** and the formation of **W55**. You may notice that four centres in **W54** are undefined. Why is this? Suggest what they might be.

Answer 21.13: The reaction is a double Diels-Alder and evidently the two cyclopentadienes add to opposite faces of the quinone **W53**. Assuming *endo* addition **W56** it is clear that the ring junction hydrogens are on the same face of the first adduct **W57** as the CH_2 bridge. The stereochemistry of the second addition is the same but to the opposite face.[11] No stereochemistry was shown in **W54** as just one bromination allows elimination to the diphenol as the double enol **W58** makes clear.

Creating Stereochemistry by Cyclisation

Problem 21.14: Explain the stereochemistry of these products. The hydroxyacid **W60** is 80:20 *trans:cis* but the hydroxyester **W61** is 100% *trans*. Describe the relationship between the two primary alcohols in **W59**.

Answer 21.14: Ozonolysis with oxidative work-up gives the diacid **W62** and spontaneous cyclisation gives five-membered lactones *trans*-**W60** and *cis*-**W60**. The primary alcohols in **W59** and, more significantly, in **W62** are diastereotopic. One cyclises to give *trans*-**W60** and the other to *cis*-**W60**. Cyclisation to *trans*-**W60** is preferred by 4:1 but during acid-catalysed esterification *cis*-**W61** isomerises to *trans*-**W61** by enolisation.[12]

Remote Control in the Aldol Reaction

In the textbook chapter we described how boron enolates of ketones gave good 1,2- and 1,3-control in aldol reactions. They are not the only specific enol equivalents to do this. As part of a synthesis of peloruside A, Pagenkopf[13] used the silyl enol ether **W64** in a Mukaiyama aldol with the aldehyde **W65**. **Problem 21.15**: Define and comment on the stereoselectivity of the aldol reaction.

Answer 21.15: Both products are *anti*-aldols (OH and OBn are *anti*) so 1,2-control is perfect. As might be expected, it is the 1,3-control that is not so good, but it was the best they could achieve with a variety of methods. They comment:[13] 'This modest level of stereoselectivity is none the less noteworthy given the lack of an α stereocentre and the complexity of the aldehyde.' If there were an α stereocentre, it would be 1,2-control rather than 1,3-. The complexity of the aldehyde is significant as good 1,3-control depends on centres other than the nearest being ignored. You might compare this result with compounds **58** and **60** in the textbook chapter.

Felkin-Anh Selectivity

Enantiomerically pure aldehyde is easily made from phenylalanine.[14] **Problem 21.16**: Explain the stereoselectivity in the formation of the products **W71** and **W72**.

Answer 21.16: The products are rather unhelpfully drawn as we need to know which face of the aldehyde is attacked by, for example, the lithium enolate of ethyl acetate. Redrawing **W71** in a more normal way by two GSRs **W71a** makes it clear that a Felkin-Anh selectivity is responsible. The largest and most electronegative group is Bn_2N and attack occurs opposite that group and alongside H **W74**. This gives **W75** redrawn **W75a** in the same way as **W71a**.

Reaction with the Danishefsky diene **W72** is by oxo-Diels-Alder **W76** catalysed by BF$_3$ but with the same Felkin-Anh stereoselectivity for the same reasons. The initial product is the cycloadduct **W77** which hydrolyses and eliminates methanol on work-up to give **W73**.

W76   W77   W73

An Example from the Recent Literature

Diastereoselectivity is very much alive and well in 2007. The Novartis drug fluvastatin[15] (Lescol) **W78** reduces cholesterol and is needed on a >50 tonnes *per annum* scale. The racemic drug is used so the major stereochemical issue is ensuring 1,3-control in the side chain diol. This can be achieved by stereoselective reduction of the hydroxy-ketone **W79**. **Problem 21.17**: What reagent(s) did they have in mind for the reduction?

W78; fluvastatin   W79

Answer 21.17: The standard route is the Prasad method that complexes OH and carbonyl in a six-membered ring **W80** and delivers an axial hydrogen from NaBH$_4$. There is a full description in the textbook.

W79   W80   W81

The starting material **W79** was actually used as the *t*-Bu ester **W82** rather than the sodium salt and can be made by an aldol reaction from the enal **W83** and the enolate synthon **W84**.

W82   W83   W84

Double deprotonation of the ketoester **W85** via the stable enolate **W86** gives **W87**, the reagent for synthon **W84** and this gives 80% yield of **W82** on reaction with the aldehyde **W83**.

A Revision Session

The starting material for all this chemistry is the indole **W83**. **Problem 21.18**: For a concentrated session of revision, suggest first of all general strategies by which this compound might be made.

Answer 21.18: The most obvious disconnections are of the enal **W83b**, with an aldol or Wittig style reaction in mind, or of the whole side chain **W83a** with a Heck reaction in mind.

The two starting materials **W84** and **W85** would be derived from the indole **W86** by bromination or perhaps a Vilsmeier acylation. **Problem 21.19**: Suggest a synthesis of **W86**.

Answer 21.19: The *N*-i-Pr group could be added to the anion of the complete indole so any indole synthesis will do. Indoles with a free C-2 are usually made indirectly by a Fischer indole using an ester blocking group **W87**. This has the advantage that the anion is more easily made. In fact the Fischer indole route was followed in a slightly unusual way.[16]

The basic indole **W85** could be made by a different synthesis and the formyl group put in by a Vilsmeier acylation. The indole synthesis is very simple: just a Lewis acid catalysed cyclisation of the amino ketone **W89**.

The complete propenal side chain was eventually inserted by a new method invented at the company.[17] A kind of conjugate Vilsmeier approach uses an aminoenal and $POCl_3$ on the indole **W86** to give the enal **W83** in one step. The full story of this inventive approach to laboratory and production methods is told in two papers.[15,16]

Houk Selectivity

Problem 21.20: Dihydroxylation of the *E*-allyl silane gives one diastereoisomer of a diol (92:8 selectivity) that gives the *Z*-allylic alcohol on treatment with NaH. Deduce and explain the stereochemistry of the diol.

Answer 21.20: The alkene is formed by the base-catalysed Peterson reaction—a *cis* elimination (chapter 15). If we draw the diol **W91a** in the shape of the allylic alcohol, it must have the $PhMe_2Si$ and nearer OH groups *cis* to each other. The right hand half of the molecule is already in the conformation of **W90**, so the two OH groups must also be *cis* as that is the stereospecificity of dihydroxylation. The intermediate is **W91a** or, if we draw it in the shape of **W90**, with the silyl group *anti* to the two OH groups.

W92 $\Longrightarrow$ [structure W93: PhMe₂Si, Ph, OH, OH] $\Longrightarrow$ [structure W91a: PhMe₂Si, Ph, OH, OH] = [structure W91: PhMe₂Si, Ph, OH, OH]

This is an impressive example of Houk control in open chain compounds. The H atom on the chiral centre eclipses the alkene and attack by OsO_4 occurs *anti* to the large and electropositive silyl group.[18]

[structure W90: PhMe₂Si, Ph]   [Houk conformation of W90: Ph, PhMe₂Si, H] $\xrightarrow[\text{pyridine}]{OsO_4}$ [= W91: Ph, HO, OH, PhMe₂Si, H]

W90   Houk conformation of W90   = W91

We say 'impressive' because in related reactions, such as the epoxidation of *E*- and *Z*-**W94**, high Houk selectivity is observed only for the *Z* isomer. **Problem 21.21**: Explain.

[structure E-W04: PhMe₂Si, Me, Me] $\xrightarrow[\text{NaH}_2\text{PO}_4]{m\text{CPBA}}$ [structure W95: PhMe₂Si, Me, O, Me] [structure Z-W94: PhMe₂Si, Me, Me] $\xrightarrow[\text{NaH}_2\text{PO}_4]{m\text{CPBA}}$ [structure W96: PhMe₂Si, Me, Me, O]

E-W04   W95; 61:39 *anti:syn*   *Z*-W94   W96; >95:5

Answer 21.21: Drawing the same Houk conformation as for *E*-**W91**, the eclipsing H comes up against the H atom at the far end of the alkene. This is greatly preferred to eclipsing Ph or silyl groups. However, with **W94** the methyl group on the chiral centre can eclipse the H atom in *E*-**W94** without much trouble. So the 61% comes from *E*-**W94a** and the 39% from *E*-**W94b**. With *Z*-**W94** things are very different. Whatever eclipses the alkene clashes with the other Me group rather than an H atom. Only *Z*-**W94a** is favoured. The fullest account of this chemistry comes in a remarkable summary paper.[19]

[structure E-W94a: Me, 61%, PhMe₂Si, H]   [structure E-W94b: H, 39%, PhMe₂Si, Me]   [structure Z-W94a: Me, >95%, PhMe₂Si, H Me, H]

E-W94a   *E*-W94b   *Z*-W94a

References

1. R. E. Ireland and L. N. Mander, *J. Org. Chem.*, 1967, **32**, 689.
2. For a full discussion of Corey's synthesis of Erythronolide B, see Nicolaou and Sorenson, p. 167.
3. T. Fukuyama, C. L. J. Wang and Y. Kishi, *J. Am. Chem. Soc.*, 1979, **101**, 260; For a full discussion of Kishi's synthesis of Monensin see Nicolaou and Sorenson, p. 185.
4. M. C. Jung and Y. G. Pan, *Tetrahedron Lett.*, 1980, **21**, 3127.
5. J. W. Huffman, C. B. S. Rao and T. Kamiya, *J. Org. Chem.*, 1967, **32**, 697.
6. C. H. Heathcock, *Tetrahedron Lett.*, 1966, 2043.
7. T. P. Clausen, B. Johnson and J. Wood, *J. Chem. Ed.*, 1996, **73**, 266.
8. D. L. Garin, *J. Org. Chem.*, 1971, **36**, 1697.
9. K. C. Nicolaou, W. J. Sipio, R. L. Magdola, S. Seitz and W. E. Barnette, *J. Chem. Soc., Chem. Commun.*, 1978, 1067.

10. S. Ejjiyar, C. Saluzzo and R. Amouroux, *Org. Synth.*, 2000, **77**, 91.

11. R. Rathore, C. L. Burns and M. I. Deselnicu, *Org. Synth.*, 2005, **82**, 1.

12. F. M. Hauser and R. C. Huffmann, *Tetrahedron Lett.*, 1974, 905.

13. D. W. Engers, M. J. Bassindale and B. L. Pagenkopf, *Org. Lett.*, 2004, **6**, 663.

14. M. T. Reetz, M. W. Drewes and R. Schwickardi, *Org. Synth.*, 1999, **76**, 110.

15. P. C. Fünfschilling, P. Höhn and J. P. Mutz, *Org. Process Res. Dev.*, 2007, **11**, 13.

16. O. Repic, K. Prasad and G. T. Lee, *Org. Process Res. Dev.*, 2001, **5**, 519.

17. G. T. Lee, J. C. Amedio, R. Underwood, K. Prasad and O. Repic, *J. Org. Chem.*, 1992, **57**, 3250.

18. I. Fleming, N. J. Lawrence, A. K. Sarkar and A. P. Thomas, *J. Chem. Soc., Perkin Trans. 1*, 1992, 3303.

19. I. Fleming, *J. Chem. Soc., Perkin Trans. 1*, 1992, 3363.

22

Resolution

Problems and Further Examples Relating Directly to the Text

You will find it helpful to have chapter 22 from the textbook open as you look at this first section. **Problem 22.1**: Suggest a mechanism for the synthesis of **11** by this reaction. One of the functional groups now should be protected so that the other can be used for the resolution and the amine was blocked with a Boc group to give **14**.

Answer 22.1: Ammonia must have added in a conjugate fashion so the intermediate must be the unsaturated acid **W2** formed by an aldol reaction and decarboxylation **W1**. This is a standard Knoevenagel-style condensation avoiding strong acid or base.[1] We are sure you can draw the mechanism for the addition of ammonia to **W2** without our help.

The reason for adding the epoxide to (R,S)-**30** was to make the amide **31** easier to hydrolyse after separation of the diastereoisomers. There is in fact an intermediate in the hydrolysis. **Problem 22.2**: Suggest a mechanism for the formation of the intermediate, reasons why it should form, and how it makes the hydrolysis easier.

(R,S)-30 → 31 → (R)-$(-)$-27 + 32

Answer 22.2: Intramolecular nucleophilic attack by the OH group on the amide **W3** is faster than attack by external water. The tetrahedral intermediate **W4** could cleave back to **31** or forward to **W5**. In basic solution, the more stable amide **32** would be the product, but in acid solution the ester is more stable because the amine in **W5** is protonated and does not act as a nucleophile. Now hydrolysis of the ester in **W5** is much easier than hydrolysis of an amide.[2]

W3 ⇌ W4 ⇌ W5 → W6 → (R)-$(-)$-27

The Synthesis of (+)-*syn*-Sertraline

In the textbook we said. 'There is no point in resolving any earlier compound in the synthesis (of (+)-*syn*-sertraline) as even more material would be wasted in the reductive amination step.'
Problem 22.3: Why would it be particularly foolish to resolve the lactone **43**?

Ar = 3,4-dichlorophenyl   **43** → **44** → **45**   (+)-*syn*-**45**; sertraline

Answer 22.3: Because in the next step, the Friedel-Crafts alkylation to give the acid **W8**, the chiral centre is lost in the cationic intermediate **W7** so racemic **W8** would inevitably result.[3] Cyclisation by Friedel-Crafts acylation would not restore it!

43 → **W6** → **W7** → **W8**

The reductive amination on racemic **44** gives 70% racemic *syn*-**45**. **Problem 22.4**: Supposing you resolved **44**. What would be the product from the reductive amination of enantiomerically pure **44**?

Answer 22.4: Each enantiomer of **44** must give exactly the same diastereoisomer of **45**. The product would be enantiomerically pure **45** in the same (70:30) ratio of *syn*:*anti* diastereoisomers.

The Synthesis of L-Methyl DOPA

The synthesis of methyl DOPA **80** by the Strecker reaction[4] was straightforward and of course produced racemic material. Fortunately, resolution[5] gave pure L-methyl DOPA. **Problem 22.5**: Suggest a detailed mechanism for each step.

77; 57% yield   **78; 88% yield**   **(±)-79, 100% yield**   **(±)-80, 84% yield**

Answer 22.5: The Strecker reaction often gives amino nitriles **W10** as products of cyanide addition to the imine **W9**. Here an extra molecule of CO_2 (from the carbonate) has been incorporated to make the stable heterocycle (a hydantoin) **78** in the Bücherer modification.[6] Amino nitriles react with ammonium carbonate to give hydantoins so a likely mechanism is outlined here.

As hydantoins are also formed from cyanohydrins **W13** with ammonium carbonate, that provides an alternative route. One suggestion for the rearrangement of **W12** into **78** is via **W14** and **W15**.

The Synthesis of DARVON

We explained in the textbook that the enantiomerically pure ketone $(+)$-(R)-**102** can be converted to DARVON **104** by a chelation controlled diastereoselective addition of benzyl Grignard and acylation. **Problem 22.6**: How will the diastereoselectivity change if the other enantiomer of the ketone **97** were used?

(+)-(*R*)-102   103   104

Answer 22.6: Magnesium must chelate the amine and the carbonyl group **W16** giving two possible conformations **W17** and **W18** for the reaction.[7] The nucleophile can approach alongside H in **W18** so this will be the more reactive conformation and it gives **103**.

W16   W17   W18   103

The Synthesis of Vincamine

In the textbook we made **109**, a precursor for vincamine, by combining the imine **107** and the silyl enol ether **108**. **Problem 22.7**: How would you make **108** and what is the mechanism of the transformation of **107** and **108** into **109**?

107   108   racemic (±)-109
                                                             mixture of diastereoisomers

Answer 22.7: The silyl enol ether **108** is obviously made by silylation of the aldehyde **W19** and the bromine best disconnected back to the alcohol **W20**. This is a 1,5-diO compound normally made by conjugate addition.

108 $\Longrightarrow$ $\xrightarrow{\text{FGI}}$ $\Longrightarrow$ conjugate addition?
        W19   W20

However, Oppolzer and his group[8] preferred to use the alkene **W21** easily made from a specific enol equivalent of butanal, say an enamine, with allyl bromide. Protection of the aldehyde, hydroboration and bromination gave **W23**, easily hydrolysed to **W19** in 68% yield from **W22**. The silyl enol ether was made in 85% yield with Me_3SiCl and *i*-Pr_2NEt.

In the mechanism, the imine N in **107** must displace Br from **108** allowing cyclisation of the silyl enol ether onto the iminium salt **W24** in a Mannich reaction.

Problem 22.8: Outline how vincamine **105** might be made from (+)-*syn*-**109**. You might like to be reminded that vincamine is in equilibrium with the keto-ester **106**.

Answer 22.8: If we make **106**, the remaining chiral centre can take care of itself. The aldehyde **109** needs to be extended by two functionalised carbon atoms and there are many ways to do this. Oppolzer and his group chose to extend the chain with a HWE reaction (chapter 15) and make the cyclic amide **W27** so that the extra carbonyl group could be introduced by nitrosation.[8] Sadly the hydrolysis of the oxime **W28** was a poor step, even with formaldehyde to scavenge the hydroxylamine by-product. Cleavage of the keto-amide **W29** with methoxide gives **106**.

The Synthesis of Fusilade

Problem 22.9: Discuss the selectivity of this synthesis. Why does the first reaction occur on one of the two OH groups and not on the other? Why are two molecules of NaOH required for the second reaction? Draw the mechanisms for the two nucleophilic substitutions. Why might the reaction on the chloroacid have gone with retention?

Answer 22.9: Monoalkylation **W30** of the anion of **116** occurs at least twice as fast as dialkylation. Two molecules of NaOH are needed for the second step as one will remove the carboxyl proton and the second remove the phenol proton. The dianion **W31** is more nucleophilic at the phenolate anion for nucleophilic aromatic substitution activated by both the pyridine nitrogen and the CF_3 group.[9]

Application of the Product of an Enzymic Resolution

We revealed in the textbook that resolution of the simple furan alcohol by a lipase from a *Candida* species gave the alcohol (−)-**124** while the other enantiomer reacted to give an ester. **Problem 22.10**: This enantiomer of **131** was needed for conversion into the pyranone **W35** via **W34**. Suggest mechanisms for these steps and explain the stereochemistry of the product. Why is one centre of **W35** not defined?

Answer 22.10: The reaction with bromine in methanol is a famous furan reaction. Bromine adds to one α-position **W36** and MeOH can now add to the other **W37**. Now loss of the bromine **W38** gives another cation and addition of a second molecule of methanol gives **W34**. The absence of stereochemical control in this reaction is irrelevant as both methoxy groups are lost in the next step.[10]

The key step in the second reaction is participation by the ring oxygen atom in the loss of the alcohol with inversion **W40**. The rest is straightforward hydrolysis of acetals by addition of water to **W42** and **W45**. The undefined centre in **W35** is left that way deliberately as it is a hemiacetal epimerising continuously by ring opening.

Improvement in Enantiomeric Excess by the Enzyme Used for Kinetic Resolution

The amide (*R*)-**162** was formed in 45% yield and 98.5% ee by kinetic resolution during acylation. Once it has been separated from free (*S*)-**2**, it can be hydrolysed to free (*R*)-**2** with the same enzyme! This automatically perfects the ee. **Problem 22.11**: How much (*S*)-**162** is there in the product from the acylation? Explain what happens to it.

Answer 22.11: One way to work this out is to put a '1' in front of the ee giving 198.5 and divide by 2 giving 99.25% of (*R*)-**162** and the rest (0.75%) (*S*)-**162**. Another is to divide the difference between 100 and 98.5 (= 1.5) by 2 (= 0.75) and add it to 98.5. What happens? The enzyme is selective for the (*R*) compounds in both directions so it will hydrolyse (*R*)-**162** much faster than (*S*)-**162**. So (*S*)-**162** remains as the amide.

(R)-162; amide of (R)-2   (R)-2   (S)-162; amide of (S)-2

The Asymmetric Synthesis of Fortamine by Resolution

In the synthesis of fortamine described in chapter 17 the product **44** is racemic (compound numbers from chapter 17). The same synthesis has in fact been used to make optically active fortamine. **Problem 22.12**: Which compound would you resolve, how, and why?

32   33; 84% yield   34; 99% yield   35; 88% yield

36; 86% yield   37; 89% yield   38

42   43; 92% yield   44; 99% yield

Answer 22.12: Questions you should ask yourself are: (i) Which is the first chiral intermediate? (ii) Is it suitable for resolution? (iii) What happens afterwards to the chiral centres in the molecule resolved? (iv) Which intermediates could be conveniently resolved? Your aim should be to resolve as early as possible.[11]

The answers are: (i) The epoxide **33** is the first chiral intermediate. (ii) It is not suitable for resolution as it has no functional group 'handle'. (iii) Irrelevant. (iv) The amino alcohol **34** is ideal: it is an early intermediate, it has two suitable functional groups, especially the amine, and nothing happens to its two chiral centres thereafter. The chemists decided not to add a resolving agent to **34** but to add the simple amine (S)-**W46** instead of MeNH$_2$. The 1:1 mixture of the two *anti*-diastereomers of **W47** could not be separated but the carbamates **W48** could and the yield of the enantiomerically pure **W48** was excellent (48% out of 50%). It merely remained to get back on track with the synthesis and **W51** reacted with NaN$_3$ as well as **38**.

The Asymmetric Synthesis of Tazadolene by Resolution

Problem 22.13: In a similar vein, here is a synthesis of racemic tazadolene **W56**. Where would you resolve to make this an asymmetric synthesis? Evaluate resolution as a strategy in this case.

Answer 22.13: The first chiral intermediate **W53** should not be resolved as the mechanism of the next step destroys the only chiral centre. Compound **W54** has plenty of functional groups ideal for resolution but it is evidently formed as a mixture of diastereoisomers that we do not want to separate as they all give *E*-**W55** on elimination (E1cB presumably). Conversion of **W55** into **W56** does not affect the chiral centre so either **W55** or **W56** could be resolved using the amino or hydroxyl group, but have not as far as we know. Resolution is only moderate as a strategy since it must be done at a late stage.[12]

The Asymmetric Synthesis of Praziquantel by Resolution

Problem 22.14: In the same way, suggest how resolution could be incorporated into this synthesis of praziquantel **W61** and evaluate resolution as a strategy.

Answer 22.14: The starting material **W57** is the first chiral intermediate but the primary amine **W58** is much more attractive for resolution. There is a trap here: the mechanism of the final step needs careful study as all chiral centres are lost in the electrocyclic opening of the cyclobutene in **W60a** giving, after loss of acetate to form an imine, the achiral intermediate **W62** for cycloaddition.

Only praziquantel itself could profitably be resolved so resolution is not a good strategy.[13] Of course, while these drugs are being made, both enantiomers are needed for biological evaluation so resolution is a good strategy at that stage. However, once the active enantiomer is known, a more advanced strategy would be better.

References

1. H. Boesch, S. Cesco-Cancian, L. R. Hecker, W. J. Hoekstra, M. Justus, C A. Maryanoff, L. Scott, R. D. Shah, G. Solms, K. L. Sorgi, S. M. Stefanick, U. Thurnheer, F. J. Villani and D. G. Walker, *Org. Process Res. Dev.*, 2001, **5**, 23.
2. P. E. Sonnet, P. L. Guss, J. H. Tumlinson, T. P. McGovern and R. T. Cunningham, ACS Symposium 355 *Synthesis and Chemistry of Agrochemicals*, eds D. R. Baker, J. G. Fenyes, W. K. Moberg and B. Cross, American Chemical Society, New York, 1987, p. 388.
3. M. Williams and G. Quallich, *Chem. and Ind. (London)*, 1990, 315; M. Lautens and T. Rovis, *J. Org. Chem.*, 1997, **62**, 5246; E. J. Corey and T. G. Gant, *Tetrahedron Lett.*, 1994, **35**, 5373.
4. G. A. Stein, H. A. Bronner and K. Pfister, *J. Am. Chem. Soc.*, 1955, **77**, 700.
5. S. Terashima, K. Achiwa and S. Yamada, *Chem. Pharm. Bull.*, 1965, **13**, 1399.
6. E. C. Wagner and M. Baizer, *Org. Synth. Coll.*, 1955, **3**, 323.

7. E. A. Pohland, L. R. Peters and H. R. Sullivan, *J. Org. Chem.,* 1963, **28**, 2483; E. A. Pohland and H. R. Sullivan, *J. Am. Chem. Soc.*, 1953, **75**, 4458.

8. W. Oppolzer, H. Hauth, P. Pfäffli and R. Wenger, *Helv. Chim. Acta*, 1977, **60**, 1801.

9. 'Fluazifop-butyl' Technical Data Sheet, Zeneca Plant Protection, Fernhurst, 1981; D. W. Bewick, Eur. Pat., 133,033, *Chem. Abstr.*, 1985, **102**, 165–249.

10. D. G. Drueckhammer, C. F. Barbas, K. Nozaki, C.-H. Wong, C. Y. Wood and M. A. Ciufolini, *J. Org. Chem.*, 1988, **53**, 1607.

11. S. Knapp, M. J. Sebastian and H. Ramanathan, *J. Org. Chem.*, 1983, **48**, 4786; S. Knapp, M. J. Sebastian, H. Ramanathan, P. Bharadwaj and J. A. Potenza, *Tetrahedron*, 1986, **42**, 3405; S. Knapp, *Chem. Soc. Rev.*, 1999, **28**, 61.

12. J. Szmuszkowicz, *Chem. Abstr.*, 1983, **100**, 6311.

13. W. F. Berkowitz and T. V. John, *J. Org. Chem.*, 1984, **49**, 5269

23

The Chiral Pool: Asymmetric Synthesis with Natural Products as Starting Materials

Problems and Further Examples Relating Directly to the Text

You will find it helpful to have chapter 23 from the textbook open as you look at this first section.

Compounds Derived from Amino Acids

The important chiral auxiliary **W2** is made from proline **W1** and used to create new chiral centres as in the aminal **W4** formed with phenylglyoxal **W3**. **Problem 23.1**: Suggest how to make **W2** from proline and explain any selectivity in the formation of **W4**.

Answer 23.1: The carbonyl group must be reduced, a molecule of aniline added, and the integrity of the chiral centre ensured throughout. These steps could be done in various orders and by various means. Mukaiyama's synthesis[1] involved Cbz protection at nitrogen **W6**, activation of the acid by a mixed anhydride formed and reacted immediately as this is the danger moment for racemisation, deprotection and reduction of the amide **W7**.

The stereochemistry of **W4** was not entirely defined but you should have seen that there was no change in the chiral centre from proline. The benzoyl group is on the same side of the bicyclic molecule as the ring junction H **W4a**. Chemoselectivity comes from the aldehyde in **W3** being more electrophilic than the ketone.[2] It does not matter which amine initially forms an imine, say **W8** as aminal formation, like acetal formation, is under thermodynamic control and the detailed mechanism does not matter. The folded molecule **W4a** must have a *cis* ring junction between two five-membered rings so the ring nitrogen atom is also a chiral centre and the larger substituent prefers to be on the outside/*exo*/convex face.

Hydroxy Acids

Problem 23.2: Comment on this extract from Aldrich catalogue: L-(+)-lactic acid: $[\alpha]_D - 13.5°$, $c = 2.5$ in 1.5 N NaOH solution.

Answer 23.2: How can a (+) compound have a negative rotation? The solution is the key: in NaOH we do not have lactic acid but the sodium salt and Aldrich do say 'Most salts of L-(+)-lactic acid are laevorotatory, hence the negative rotation in NaOH solution.' There is a warning here to be wary about relying on rotations for hard information. We might also wonder what 'L' lactic acid is. In fact it is (S) lactic acid. Neither (+) nor 'L' tells us what we want to know. In fact most (S)-lactic acid derivatives are laevorotatory. These data are also from the Aldrich catalogue which is one of the few reliable ways to find out what the absolute stereochemistry of a (+) or (−)compound might be. Why is there no data on lactic acid itself in water? Again we quote: 'Lactic acid solution: contains variable amounts of intermolecular esterification products.'

L-(+)-(S)-lactic acid   L-(−)-(S)-sodium lactate   L-(−)-(S)-lactamide   L-(−)-(S)-ethyl lactate

The Synthesis of Bestatin 161 from Malic Acid 34

In the textbook we said: First the benzyl group must be added. The di-lithium derivative of diethyl malate **164** is alkylated on the opposite side to the OLi group (chapter 30). Now the two ester groups must be distinguished. Reaction of the free diacid with trifluoroacetic anhydride and then with ethanol gives **165**. Presumably the two acids form a cyclic anhydride and the free OH forms an ester. Reaction with ethanol occurs at the more reactive (next to $OCOCF_3$) and less hindered (not next to benzyl) carbonyl group of the anhydride and the trifluoroacetyl groups fall off during work-up. **Problem 23.3**: Draw out the intermediates to see what we mean.

164; diethyl malate → **165** → **166**

Answer 23.3: The dilithiated intermediate is chelated so that the benzyl group adds from the face opposite the CO_2Et group[3] **W9**. The cyclic anhydride is attacked by EtOH at the more electrophilic carbonyl group **W10**.

W9 · · · **W10**

The Synthesis of Streptazolin

There have been several syntheses of streptazolin and two are described in part in the textbook. We now fill in some of the gaps.

The Synthesis of Streptazolin from Tartaric Acid

The details of the conversion of **190** to **189** are that a sequence of reactions gives a compound **W11** that gives the imide **189** in 100% yield when refluxed with acetyl chloride. **Problem 23.4**: What is the structure of the intermediate and why is acetyl chloride needed?

190 → **W11?** → **189**

Answer 23.4: The first step formes the diacid that is not isolated but converted to the cyclic anhydride **W12** with AcCl. Ammonia now gives the monoamide **W11** (it does not matter which carbonyl group is attacked) and cyclisation to the imide now requires the activation of the remaining acid as the acid chloride.[4]

190 → **diacid** → **W12; 62% yield** → **W11; 91% yield** → **189**

Problem 23.5: Suggest a synthesis for the Z-vinyl silane **188** and a mechanism for the cyclisation to **186**.

Answer 23.5: There are many syntheses of **188** perhaps the simplest is stereoselective reduction of an alkyne **W14** assembled in the usual way.

The cyclisation to **186** is a typical vinyl silane reaction (chapter 16). The Lewis acid (BF$_3$) removes the acetate to make an iminium ion **W16** attacked by the vinyl silane at the *ipso* carbon (the one attached to silicon) to give a β-silyl cation. Retention of configuration at the alkene is expected and here obligatory. The stereochemistry of **W16** is Felkin-like as the vinyl silane attacks the top face of the iminium salt opposite the large electronegative OBn group, pushing the H atom down.

Streptazolin: Conclusion of the First Synthesis

Now to the final steps including the second cyclisation. Reduction with the 'ate' complex from DIBAL and BuLi cleaves the amide in **186** releasing the aldehyde and the free NH is protected with a carbamate **W17**. Next a Corey-Fuchs reaction gives the dibromoalkene **W18** and elimination with two molecules of BuLi gives the lithium acetylide, methylated as it is formed to give the alkyne **W19**.

The palladium-catalysed cyclisation probably starts with the formation of the double π-complex **W21**. The PdH group hydrometallates the alkyne to give the Z-geometry of the product **W22**. Carbometallation of the alkene closes the five-membered ring **W23** in a *syn* process that puts the palladium on the wrong side for elimination with the ring-junction hydrogen. Loss of Pd with a *syn*-H from the other side of the six-membered ring with 'movement' of the alkene forms **W24** in 84% yield from which streptazolin **184** can be made.

The conversion of **W24** into streptazolin requires moving the alkene back to where it came from and the Fe(CO)$_3$ diene complex **W25** does that.[4] The later Trost synthesis avoids this problem.[5]

Palladium-Catalysed Alkyne Coupling

The Pd(0)-catalysed cyclisation of **198** gives both alkenes in **205** in the right place with the right geometry.[5] The mechanism of the cyclisation was already established by Trost and we now discuss it. Cyclisation and deprotection gave streptazolin **190** in only 11 steps from the chiral pool.[6]

The first two steps give the differentially protected diol **W26** and hence a π-complex **W27** with a species corresponding to PdH$_2$ (+ ligands). Double hydropalladation gives the *bis*-vinyl-Pd(II) σ-complex **W28** (the transferred H atoms are marked) with geometry dictated by the intramolecular transfer of H to one end of each alkyne and Pd to the other. Reductive elimination of Pd couples the two alkenes with retention and eliminates Pd(0) for the next cycle.

W26; R = TBDMS   **W27; R = TBDMS**   **W28; R = TBDMS**

The Synthesis of the Alkaloid Luciduline from a Terpene

Luciduline **W29** is a tricyclic alkaloid having a tertiary amine and a ketone and, more to our point, five chiral centres. Three are contiguous but two are separated from any other. We shall describe Oppolzer's interesting asymmetric synthesis[7] from the chiral pool with a terpene as starting material.

W29
(+)-luciduline

five stereogenic centres marked with circles

Our only clue is the 1,3-relationship between the two functional groups. This could suggest a Mannich reaction but Oppolzer saw the possibility of considerable simplification if he used a 1,3-dipolar cycloaddition **W31** as that reveals a much simpler decalin **W33** with all the chiral centres around six-membered rings. It also reveals a cyclohexene so a Diels-Alder disconnection **W34** is very attractive. The chiral starting material is the cyclohexenone **W35** having seven carbon atoms so three would have to be removed from a terpene.

W30   **W31**   **W32**

W33   **W34**   **W35**

The monoterpene (+)-pulegone **W36** is readily available and a reverse aldol reaction removes the extra three carbon atoms as acetone. The alkene is introduced by sulfenylation, oxidation to the sulfoxide, and thermal elimination to give our starting material **W35**.

The Diels-Alder reaction works well with Lewis acid catalysis, putting in two more chiral centres, and the reductive amination via the oxime **W39** puts in a fourth. The final one comes from the 1,3-dipolar cycloaddition which must give the regio- and stereoselectivity of **W31** because the dipole is tethered to the alkene **W40**. The reduction cleaves the weak N–O bond and oxidation to the ketone gives luciduline in excellent yield. **Problem 23.6**: Explain in more detail how the stereochemistry of **W34** and **W31** is controlled.

Answer 23.6: The *endo* arrangement **W27** for the Diels-Alder reaction has the diene on top of the enone to avoid the Me group and the two marked Hs are then forced down. The conformation of the oxime **W33a** with the Me equatorial shows that this folded molecule has an *exo* and *endo* face and that the reagent will approach from the *exo* face to give the axial hydroxylamine **W33a** needed for the 1,3-dipolar cycloaddition.

The Synthesis of Camphor-10-Sulfonic Acid, an Important resolving Agent

In the textbook we showed two important chiral auxiliaries **95** and **96** derived from camphor sulfonic acid **94** and we said that sulfonation on camphor **93** occurs at the bridgehead methyl group by a series of rearrangements described in the workbook. The preparation is best done with sulfuric acid and acetic anhydride[8] and gives only a moderate (44–47%) yield.

93; (+)-camphor   **94; (+)-camphor-   95; Oppolzer's   96; Davis's
                        10-sulfonic acid**   **chiral sultam**   **asymmetric oxidant**

A well known terpene rearrangement on the protonated ketone **W43** gives a tertiary carbocation that can lose a proton **W44** to put an alkene just where we need it for sulfonation by some suitable electrophilic sulfur cation **W45**. Then the rearrangement sequence reverses **W46** to give **94**. There are other products, hence the low yield.

The Synthesis of Captopril from Proline

In the textbook we described a synthesis of captopril from proline in which proline was used as a combined chiral pool starting material and resolving agent. The mixture of diastereomers of **131** was separated and the crystalline one **W47** treated with ammonia in methanol to release enantiomerically pure captopril. The rotations of the two isomers are: **W47**, – 164.4 and **W48**, +17.8. **Problem 23.7**: Comment on these figures.

mixture of diastereomers of 131   **W47; crystallises   W48; in mother
                                        out [α]$_D$ –164.4   liquor [α]$_D$ +17.8**

Answer 23.7: There are (at least!) two things to say. First these compounds are not enantiomers but diastereomers so there is no reason why they should even have rotations of opposite sign, let alone of equal value.[9] Secondly, the compound that crystallises out is likely to have a higher ee than the compound in solution as we pointed out in chapter 22.

Hydroxy Acids: The Synthesis of Omapatrilat

The losses in yield when these semi-resolutions occur are not ideal and Bristol-Meyers Squibb now have a new ACE inhibitor omapatrilat **W49**. This is clearly also derived from amino acids but two amide disconnections reveal that they are joined by a sulfide **W52**.

If the 1,1-diX relationship in all these molecules is disconnected before the second amide **W51a** an imine **W53** is revealed that comes from a simpler dipeptide **W54**. Neither component amino acid is found in proteins but both can probably be made from other amino acids.

In the published synthesis neither part of **W54** was made from the chiral pool. The left hand half was made from the available racemic thiolactone **W55** by resolution of the racemic acid **W57** with the simple amine **W58** (chapter 22).

The other half **W61** was made by standard amino acid chemistry from acylaminomalonate **W59** and asymmetry introduced by an enzymatic method (chapter 29). Stereochemistry appears during the decarboxylation of the acid formed by the hydrolysis of **W60**.

Protected forms of the two amino acids **W57** and **W62** were joined by peptide coupling to give **W63**. Removal of the SAc group and oxidation to an aldehyde **W64** allows imine formation and cyclisation in one step. The new chiral centre between N and S is thermodynamically controlled **W65**.

The only chiral pool contribution to omapatrilat is the thiol acid **W50**. Thanks to an earlier careful study of reaction of hydroxy acids, it was known the phenyl-lactic acid **W66** (both enantiomers are available) could be converted into **W68** providing that the S_N2 displacement was carried out in ethanol. In DMF the product was racemic. Deprotection of **W50** and coupling to **W65** gives omapatrilat.[10]

Novartis E-selectin inhibitor

Sialyl Lewis X is an important tetrasaccharide involved in the inflammation of tissues. Novartis has developed[11] a much simpler compound, their E-selectin inhibitor **W70**. One obvious C–O disconnection separates the trisaccharide component, obviously made by sugar coupling, from the simple (*R*)-cyclohexyl-lactic acid **W71**.

W70; Novartis E-selectin inhibitor

W71; (*R*)-phenyl-lactic acid

We are going to consider just that compound. There is an obvious chiral pool approach from phenylalanine. Diazotisation and reaction with sulfuric acid gives the aryl version of the compound **W73** with some loss of ee but hydrogenation was efficient and went with stereochemical integrity. However, the starting material is unnatural and expensive D-phenylalanine **W72**.

W72; (*R*)-(D)-Phe

1. NaNO$_2$
H$_2$SO$_4$
2. crystallise from hexane

W73; 78% yield 94.8% ee

5% Rh/C
15 bar H$_2$
EtOH, 50 °C

W71; 97% yield 95.0% ee

The alternative order of events (hydrogenation first, diazotisation second) was worse. The hydrogenation was excellent but the diazotisation of the saturated compound **W74** gave a very poor yield and there was no point in determining the ee. You are referred to the paper for the discussion of the alternatives. In the end an asymmetric dihydroxylation method (chapter 25) was preferred. Even when there is a natural compound available, the chiral pool approach may not be best.

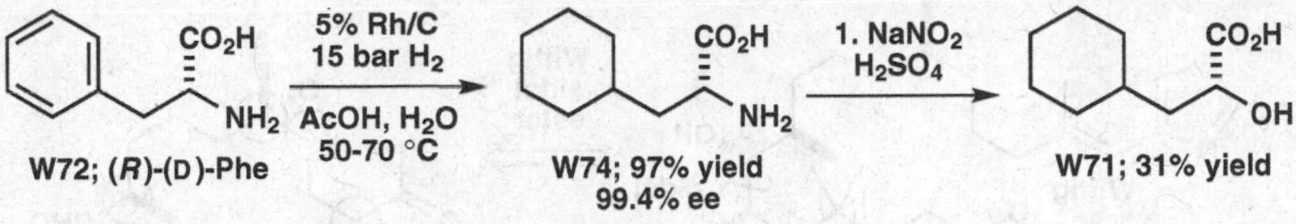

Glyceraldehyde, Glycerol and Epichlorohydrin

Glyceraldehyde occurs naturally but is unstable and its acetal **W77**, easily made from C_2-symmetric mannitol **W75** by oxidative cleavage, is the point of entry for a whole family of C_3 chiral compounds with functionality on each carbon.

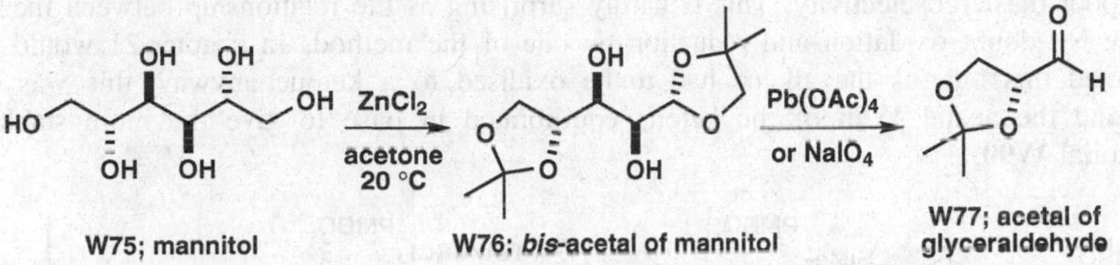

These compounds are obviously closely related and the diagram shows that all can be made from **W77**. Once the oxidation level of the three carbon atoms is the same (OH, acetal, Cl, OTs, OMs) it is essential to keep the two ends different. Hydrolysis of many of the intermediates in the chart leads to achiral glycerol. Beware!

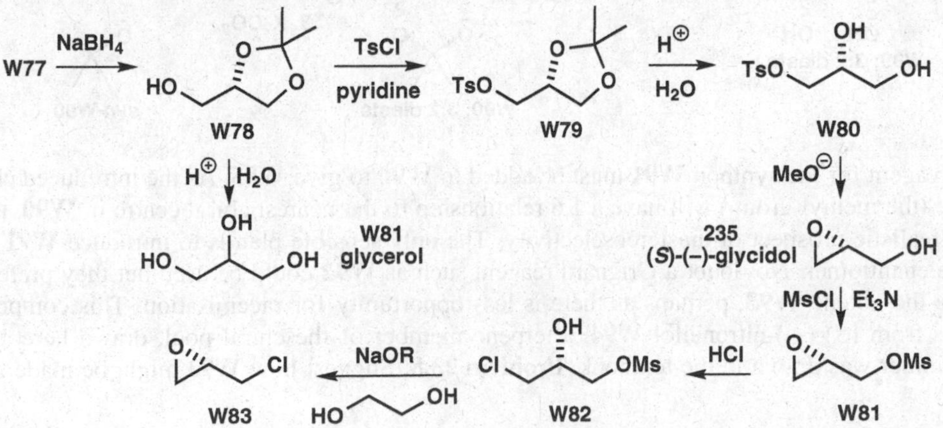

Chiral Pool Synthesis of Epothilone B

The anti-tumour compound epothilone B **W84** has been synthesised by a chiral pool strategy from one of these glyceraldehyde-derived compounds in a remarkable sequence of reactions in which the original chiral centre disappears. We are concerned with the northern section of the macrocyclic ring **W85** revealed by Wittig, ester, and aldol disconnections.[12]

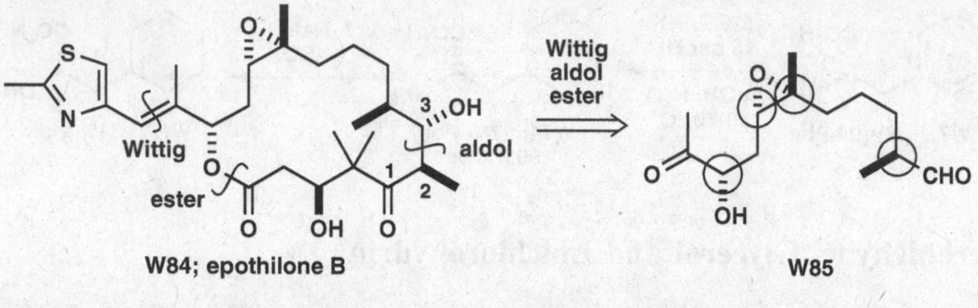

W84; epothilone B   W85

The starting material was the PMB protected hydroxy aldehyde derived from lactic acid.[13] Addition of the allyl silane gave the adduct **W87** the product of chelation control, which was protected and ozonised to the aldehyde **W88**. Addition of propenyl Grignard gave **W89** with poor diastereoselectivity. This is hardly surprising as the relationship between the diols is 1,3. No doubt oxidation and reduction by one of the methods in chapter 21 would have corrected this but, as the alkene had to be oxidised to a ketone anyway, this was done first and the acetal **W90** of the ketone equilibrated in base to give the more stable di-equatorial **W90**.

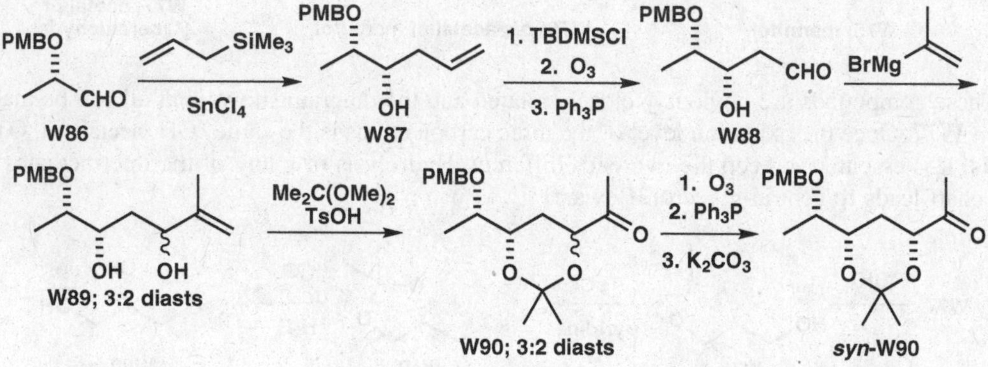

W86   W87   W88

W89; 3:2 diasts   W90; 3:2 diasts   syn-W90

A reagent for the synthon **W91** must be added to **W90** to give **W85**. As the introduced chiral centre (the methyl group) will have a 1,6 relationship to the nearest chiral centre in **W90**, there is no realistic prospect of diastereoselectivity. The only sensible plan is to introduce **W91** as a single enantiomer. No doubt a Grignard reagent such as **W92** could be used but they preferred to use the alkene **W93**, perhaps as there is less opportunity for racemisation. This compound comes from (*S*)-(−)-citronellol **W94**, a terpene member of the chiral pool, drawn here as its enantiomer was drawn in the textbook. **Problem 23.8**: Suggest how **W93** might be made from **W94**.

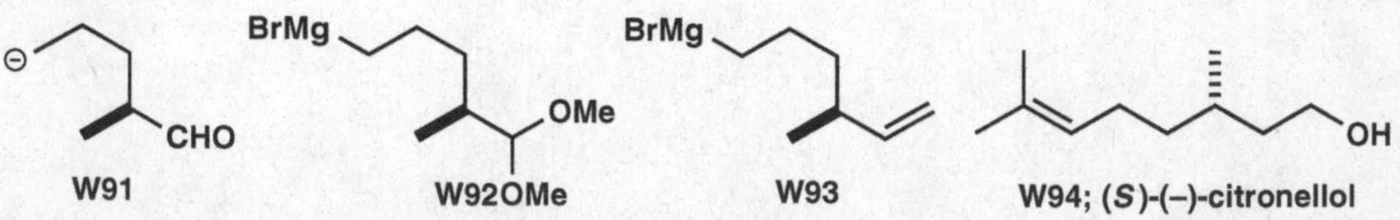

W91   W92 OMe   W93   W94; (*S*)-(−)-citronellol

Answer 23.8: The Grignard reagent **W93** can be made from the alcohol **W95** and redrawing **W94** to resemble **W95** more closely, we see we need ozonise the alkene and reduce the aldehyde to the alcohol and eliminate the alcohol to give the alkene. It is a close decision as to which to do first as obvious selectivity problems arise.

In fact it was already known[14] that dehydration of **W94** to the diene allowed a surprisingly selective ozonolysis of the trisubstituted alkene in the presence of the new monosubstituted alkene. Ozone is an electrophilic reagent.

Addition of the Grignard **W93** to *syn*-**W90** gave the correct diastereomer of **W96** providing that they added extra MgBr$_2$ to the reaction. This is clearly a chelation-controlled addition with Mg bridging two oxygen atoms. The chemists decided to do the Wittig reaction next to avoid selectivity problems if they ozonised first. The secondary alcohol was deprotected and oxidised to give the required ketone **W97**. **Problem 23.9**: Suggest how **W97** might be converted into **W85**.

Answer 23.9: You were told that the Wittig reaction was done first though you might not have used the phosphonium salt **W98**–you cannot predict exact reagents or conditions. This is even more true in the next step where to make the epoxide in **W85** it is obviously necessary to protect some of the alcohols in the triol **W100** selectively. They say that **W100** 'much to our surprise underwent completely regioselective *O*-silylation at the 15-position even when 3 mol equivalents of TESCl had to be applied for complete conversion.' Given the result we can suppose that steric hindrance allows the selection between the two secondary alcohols.

The rest is somewhat easier. The C-13 alcohol needs to be converted into a leaving group—they chose mesylate **W102**—and the elimination to give the epoxide carried out in base. Retention at the tertiary alcohol and inversion at the secondary makes this step stereospecific. All that remains is the oxidative cleavage of the alkene.

The remote chiral centre in the lower half of epothilone came from an asymmetric allylation catalysed by a TADDOL-Ti complex, a catalyst also derived from the chiral pool. This is an interesting synthesis worth further study.

References

1. M. Asami, H, Ohno, S. Kobayashi and T. Mukaiyama, *Bull. Chem. Soc. Japan*, 1978, **51**, 1869.
2. T. Mukaiyama, *Tetrahedron*, 1981, **37**, 4111.
3. B. H. Norman and M. L. Morris, *Tetrahedron Lett.*, 1992, **33**, 6803.
4. J. Ohwada, Y. Inoue, M. Kimura and H. Kitasawa, *Bull. Chem. Soc. Jpn*, 1990, **63**, 287; H. Yamada, S. Aoyagi and C. Kibayashi, *J. Am. Chem. Soc.*, 1996, **118**, 1054.
5. B. M. Trost, C. H. Chung and A. B. Pinkerton, *Angew. Chem., Int. Ed.*, 2004, **43**, 4327.
6. B. M. Trost and D. C. Lee, *J. Am. Chem. Soc.*, 1988, **110**, 7255; B. M. Trost, F. J. Fleitz and W. J. Watkins, *J. Am. Chem. Soc.*, 1996, **118**, 5146.
7. W. Oppolzer and M. Petrzilka, *J. Am. Chem. Soc.*, 1976, **98**, 6722; *Helv. Chim. Acta*, 1978, **61**, 2755.
8. P. D. Bartlett and L. H. Knox, *Org. Synth.*, 1965, **45**, 12.
9. M. A. Ondetti, B. Rubin and D. W. Cushman, *Science*, 1977, **196**, 441; J. Saunders, *Top Drugs*, chapter 1.
10. J. A. Robl, C.-Q. Sun, J. Stevenson, D. E. Ryono, L. M. Simpkins, M. P. Cimarusti, T. Dejneka, W. A. Slusarchyk, S. Chao, L. Stratton, R. N. Misra, M. S. Bednarz, M. M. Asaad, H. S. Cheung, B. E. Abboa-Offei, P. L. Smith, P. D. Mathers, M. Fox, T. R. Schaeffer, A. A. Seymour and N. C. Trippodo, *J. Med. Chem.*, 1997, **40**, 1570.
11. T. Storz and P. Dittmar, *Org. Process Res. Dev.*, 2003, **7**, 559.
12. H. J. Martin, P. Pojarliev, H. Kählig and J. Mulzer, *Chem. Eur. J.*, 2001, **7**, 2261.
13. C. H. Heathcock, S. Kiyooka and T. A. Blumenkopf, *J. Org. Chem.*, 1984, **49**, 4214; K. Gerlach, M. Quitschalle and M. Kalesse, *Tetrahedron Lett.*, 1999, **40**, 3553.
14. D. R. Williams, B. A. Barner, K. Nishitani and J. G. Phillips, *J. Am. Chem. Soc.*, 1982, **104**, 4708.

24

Asymmetric Induction I: Reagent-Based Strategy

Problems and Further Examples Relating Directly to the Text

You will find it helpful to have chapter 24 from the textbook open as you look at this first section.

Asymmetric Reduction with Ipc$_2$BCl

This borane, Ipc$_2$BCl, has been used in the preparation of enantiomerically pure Prozac, (S)-fluoxetine **38**. The simple ketone **36** is reduced to give the alcohol **37**, easily transformed[1] into Prozac **38**. **Problem 24.1**: What is the transition state for this reduction, i.e. which group goes 'outside'? Suggest how **37** might be converted into Prozac.

36 **37** **38; (S)-Fluoxetine (Prozac)**

Answer 24.1: Using the absolute stereochemistry of **37** as our guide, the amine-containing side chain must be on the outside **W1** suggesting some bonding interaction between the amine and the rather electron-deficient boron atom.

35; R = Ipc **W1** **37**

Chlorine must be replaced by the amine and the trifluorophenyl group added at oxygen. There are various ways to do this but the reported way simply uses base to make the oxyanion as a better nucleophile than the neutral amine and chooses F as the leaving group to accelerate the addition step.

The Alcon Synthesis of Azopt

The available starting material **W3** in the Alcon synthesis of Azopt **39** was converted into the ketone **42** before asymmetric reduction. **Problem 24.2**: Explain the selectivity in the formation of **42**.

Answer 24.2: Nucleophilic addition to the electron-rich thiophene ring can occur only if the intermediate anion is stabilised by conjugation with the ketone. This would not be possible in an attempted displacement of the other chlorine.

Problem 24.3: Later in the synthesis **46** was converted into **47**. Explain the role of each reagent in this sequence.

Answer 24.3: Normally BuLi does not exchange with chlorine to give an aryl-lithium (chapter 7). Here it obviously does and we can offer the presence of sulfur as an incentive. Addition to SO$_2$ **W6** gives the sulfinic acid **W7** and oxidative substitution with hydroxylamine-*O*-sulfonic acid gives **42** perhaps via **W8** and **W9** though there are other ways to draw these intermediates.[2]

Asymmetric Hydroboration

In the textbook we said: 'the Me_2S-BH_3 complex reacts rapidly once with α-pinene, nearly as rapidly a second time to give good yields of Ipc_2BH **51**, but fails to react a third time.'
Problem 24.4: As α-pinene is rarely optically pure, mixtures of stereoisomers of Ipc_2BH **51** are formed in this reaction. Ipc_2BH **51** is crystalline, and the hydroboration reaction is reversible so that Ipc_2BH and $IpcBH_2$ **W10** are in equilibrium. Recrystallisation improves the optical purity of **51** so that α-pinene of 95% ee gives Ipc_2BH of 98% ee after recrystallisation. Explain.

Answer 24.4: Other diastereoisomers of **51** will normally contain one enantiomer of the isomer drawn and one different isomer—the most likely combination being *meso*-**51**. We know **51** is crystalline and is in excess so it will crystallise out of the mixture preferentially and more of *meso*-**51** will equilibrate to **51**, that will crystallise and so on.[3] This is a separation of diastereoisomers and the *meso* compound should remain in solution and contain most of the wrong enantiomer. See the discussion of the Horeau principle in chapter 20 of the textbook.

More Hindered Pinene Derivatives for Hydroboration

Brown has improved on most of the ees by using modified 'pinenes' with larger bridgehead substituents[4] but these 'pinenes' have to be prepared and the Ipc reagents retain their popularity. Thus the ethyl equivalent **W12** could easily be made from nopol **W11** and hydroborated in the usual way. This borane **W13** (actually in equilibrium with the dialkylborane) gives better results[5] with many alkenes than does Ipc_2BH.

Problem 24.5: Suggest a synthesis of the phenyl derivative **W14** from either α- or β-pinene.

Answer 24.5: β-Pinene **W15** is better both because it is available in high ee and because oxidative cleavage of the alkene gives a ketone **W16** ideal for the synthesis. Note that β-pinene **W15** belongs to the other enantiomeric series to α-pinene **22**. However, the boranes derived from **W14** give low ees with most olefins.[6]

Asymmetric Addition of Nucleophiles to Carbonyl Groups

In the textbook we said: 'Merck's HIV-1 reverse transcriptase inhibitor Efavirenz **83** is one of the simplest anti-HIV drugs yet produced. The most straightforward synthesis is based on closure of the amino alcohol **84** with some phosgene equivalent and the preparation of **84** by asymmetric addition of cyclopropylethynyl-lithium **86** to the ketone **85**.' **Problem 24.6**: Suggest a synthesis of the ketone **85**.

Answer 24.6: The obvious route is acylation of a lithium derivative **W18** made by adding an *ortho*-director (chapter 7) to *p*-chloroaniline **W20**. There is a wide choice of options.

The *t*-butyl amide **W21** was actually used.[7] Treatment with two molecules of BuLi removed the amide proton and the *ortho* proton (to give **W22**) and acylation with ethyl trifluoroacetate gave the product as the diol **W23**.

W20 → **W21; 97% yield** → **W22**

W23; 87% yield → **85; 99% yield**

Asymmetric Epoxidation with Chiral Sulfur Ylids

The starting material for Metzner's sulfur ylids **95** is made by the baker's yeast reduction of pentan-2,4-dione.[8] **Problem 24.7**: How does this work?

95 → **96; 95% yield** → **97; 92% yield 88% ee**

Answer 24.7: Baker's yeast controls the reduction according to a mnemonic given in chapter 29 of the textbook.[9] Here reduction of the first ketone **W24** has a Me group as the small R and the rest of the molecule as the large R. Rotation of **W25** reveals that the second reduction follows the same enantioselectivity of the first – only the large group is slightly different. Evidently the first asymmetric unit in **W26** is too distant to affect the second induction.

Mnemonic for baker's yeast reduction

W24 → **W25** = **W26** → **95**

The alkylation of the sulfide **96**, the formation of the ylid **99**, and the reaction with the aldehyde are all carried out in one operation. The intermediates are **98, 99** and **100**. **Problem 24.8** What are the alternatives at each stage and why do they not happen?

Answer 24.8 The cyclic sulfide **95** is simply a nucleophile. It might attack the aldehyde **W27** but it will simply fall out again **W28**. In any case it is a soft nucleophile better at attacking alkyl halides. The sulfonium salt might dealkylate with bromide ion but that would occur at the benzyl group and give starting materials so we should not notice. The alternative ylid **W29** is much less stable than **99** as it has a tertiary carbanion and is not conjugated with anything. Another possibility would be hydroxide-initiated elimination **W30** but the proton to be removed is not very acidic.[8]

In the last step, the most obvious alternative is a Wittig-style elimination of the sulfoxide **W33** and formation of stilbene. (We should get Z-stilbene from **100**.) We already know that this alternative is available to all such adducts of sulfonium ylids and carbonyl compounds and they all prefer to make epoxides. Sulfonium ions are less easily attacked (at S) than are phosphonium salts (at P).

Davies Asymmetric Conjugate Addition of Amines

In the textbook we offered at least a partial explanation of this method. We said: 'The lithium derivative **122** forms a Li–O bond to the carbonyl group of the ester **123** in its best "s-Z" conformation and the nitrogen atom adds to the conjugate position "like a butterfly settling on a leaf" **126** to give the lithium enolate in the same conformation **124a**.'

The various alternatives include the conformation of the unsaturated ester being 's-*Z*' or 's-*E*' the s referring to the s-bond shown below. There is not much to choose between these but H probably prefers to eclipse the carbonyl oxygen atom rather than the *t*-BuO group. In any case, if Li is first bonded to the carbonyl oxygen atom, the reagent can be delivered only to the 's-*Z*' conformation. An alternative to **126** is **W34** and the two Ph groups on the reagent probably prefer to keep out of the way of the Ph on the ester.[10]

s-*Z*-123 s-*E*-123 126 W34

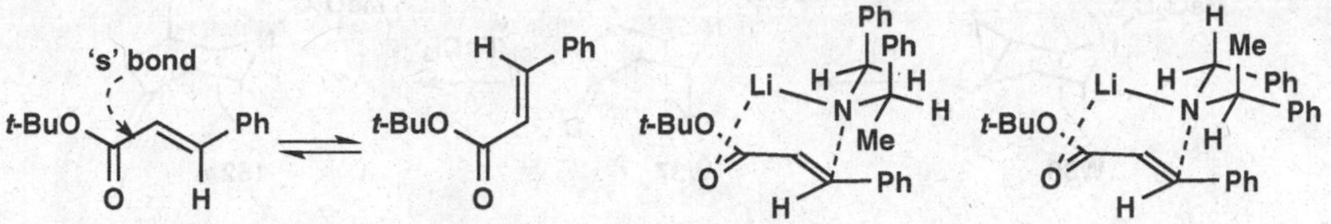

The Asymmetric Synthesis of Thienamycin

In the textbook we gave the aldol reaction used to set up two new chiral centres in Davies's synthesis of the antibiotic thienamycin.

131 132; '*E*' boron enolate 133; 75% yield

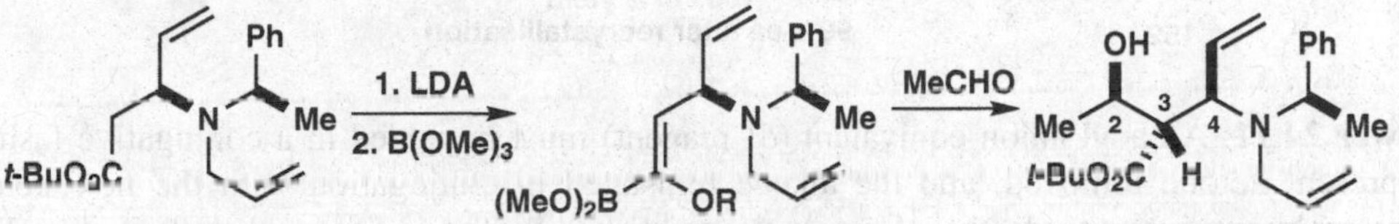

Problem 24.9: The aldol **133** is formed in a 91:9 ratio with another isomer **W35**. Explain.

W35

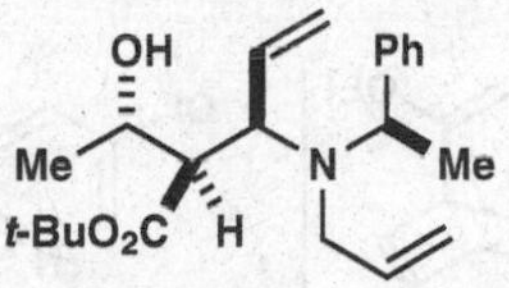

Answer 24.9: In both products the 2,3-stereochemistry is *anti* between ester and OH groups. So aldol selectivity is perfect. It is the transmission of the asymmetric information from C-4 to C-3 that is less than perfect. It is less likely that the original centre on the chiral auxiliary affects the result. However, it is important that the chiral auxiliary is still there as **133** and **W35** are diastereoisomers and were separated by flash chromatography.[11]

Enantioselective Deprotonation and the Synthesis of Anatoxin

In the textbook we gave an enantioselective synthesis of anatoxin including the key step **151** to **152**. **Problem 24.10**: Suggest a mechanism for the Fe(III) cleavage of the cyclopropane **151**.

151; 99% yield 152 153; 71% yield
99% ee after recrystallisation

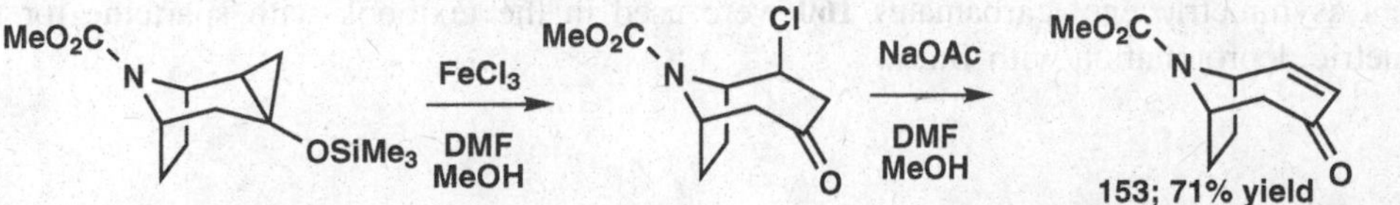

Answer 24.10: The stable oxidation states of iron are Fe(III) and Fe(II) so Fe(III) is a one-electron acceptor and a radical reaction is likely. Oxidation and removal of the silyl group (by chloride?) gives the oxygen centred radical **W36** Cleavage of the weakest bond in the cyclopropane and formation of a carbonyl group is thermodynamically very favourable.[12] The new radical **W37** accepts chlorine to give **152**.

Problem 24.11: suggest how the enone **153** might be converted into anatoxin **145**.

Answer 24.11: An acyl anion equivalent (d[1] reagent) must be added in a conjugative fashion, the present ketone removed, and the alkene reinstated in conjugation with the new ketone. There are many ways to do this. Here is the published method. The copper derivative **W38** adds in a conjugative fashion and the enolate **W39** is trapped by triflation to give **W41**. Transfer hydrogenation gets rid of the OTf group and the alkene is moved with RhCl3 catalysis to give **W43**, a protected form of the target molecule.[12]

Asymmetric Homoenolates (d³ reagents)

Hoppe's asymmetric enol carbamates **160** were used in the textbook with sparteine for the asymmetric deprotonation with BuLi.

159; Z:E 92:8   **160: Sp = sparteine**   **161; 69% yield, 97% ee**

Hoppe also tried a number of promising C_2 symmetric diamines that simply demonstrated the unique qualities of sparteine. They gave best results of 2%, 16%, 20%, and 22% ee respectively.[13]

W44   **W45**   **W46**   **W47**

Desymmetrisation by Asymmetric Deprotonation of Epoxides

Any reaction that desymmetrises a *meso* compound by deprotonation may become asymmetric if sparteine is used. Hodgson's desymmetrisation of cyclic epoxides is a good example.[14] The diastereoselectivity of the reaction was already known.[15] The *cis* epoxide **W48** gives the *endo*-bicyclic alcohol **W49** on treatment with BuLi, presumably *via* the lithium derivative **W50** and the carbene **W51** that inserts into the C–H bond shown.

W48   **W49**   **W50**   **W51**

This makes sense if the epoxide has a folded conformation **W48a** and the lithium derivative **W50a** is formed and reacts with retention of configuration. The carbene inserts across the inside of the folded eight-membered ring with retention at the only hydrogen it can reach. There is a certain amount of disagreement over the details of the mechanism.[16] Asymmetry appears in the lithium derivative **W50** and, when the same reaction is conducted with a slight excess of (–)-sparteine, the enantiomer shown **W49** is formed in up to 86% yield and 84% ee.

W48a   **W50a**   **W51**   **W52**   **W49**

The Asymmetric Synthesis of Indoles by Asymmetric Addition of ArLi to Alkenes

Two reports[17] of cyclisation of aryl-lithiums onto alkenes to make dihydro-indoles **W56** appeared back-to-back in 2000. Virtually the same compounds **W53** were used, the only difference being R (allyl, benzyl or methyl). Cyclisation of an aryl-lithium onto an unacti-vated alkene **W54** is unusual enough to be worth comment. Halogen-lithium exchange puts the Li atom on the benzene ring. Presumably carbometallation follows *via* a π-complex to give **W55** and hence **W56** on work-up. With 1.5 equivalents of (−)-sparteine, ees of around 90% were achieved.

Asymmetric Synthesis with Reagents

Problem 24.12 Suggest syntheses of alcohol **W57** (either enantiomer) by two different sorts of reagent.

Answer 24.12 The two most obvious answers are by asymmetric reduction of a ketone or by hydroboration of an alkene. The ketone is **W58** and you can take your choice among CBS, DARVON, BINAL or pinene-derived reducing agents. It is not possible to predict which will give the best result. Hydroboration of **W59** or **W60** could give **W57** but **W59** is much to be preferred as we can safely predict that the borane will add the right way round. Again you have a choice of the various, mostly pinene-derived, boranes.

Problem 24.13: Suggest how to make the amino-alcohol **W61** (either enantiomer) using an asymmetric reagent.

Answer 24.13: The relative stereochemistry (*anti*) suggests the epoxide **W62** derived from **W59** as an intermediate. There are asymmetric epoxidations of alkenes, mostly in chapter 25, but the only method in this chapter was the Metzner epoxidation with sulfur ylids from **96**. It is better to put the benzyl group on the ylid and use the aliphatic aldehyde.[8]

References

1. D. W. Robertson, J. H. Krushinski, R. W. Fuller and J. D. Leander, *J. Med. Chem.*, 1988, **31**, 1241.
2. R. E. Conrow, W. D. Dean, P. W. Zinke, M. E. Deason, S. J. Sproull, A. P. Dantanarayana and M. T. DuPriest, *Org. Process Res. Dev.*, 1999, **3**, 114.
3. H. C. Brown M. C. Desai and P. K. Jadhav, *J. Org. Chem.*, 1982, **47**, 5065.
4. H. C. Brown and U. P. Dhokte, *J. Org. Chem.*, 1994, **59**, 2365.
5. H. C. Brown, R. S. Randad, K. S. Bhat, M. Zaidlewicz, S. A. Weissman, P. K. Jadhav and P. T. Perumal, *J. Org. Chem.*, 1988, **53**, 5513.
6. H. C. Brown, S. A. Weissman, P. T. Perumal and U. P. Dhokte, *J. Org. Chem.*, 1990, **55**, 1217.
7. A. Thompson, E. G. Corley, M. F. Huntington, E. J. J. Grabowski, J. F. Remenar and D. B. Collum, *J. Am. Chem. Soc.*, 1998, **120**, 2028; M. E. Pierce, R. L. Parsons, L. A. Radesca, Y. S. Lo, S. Silverman, J. R. Moore, Q. Islam, A. Choudhury, J. M. D. Fortunak, D. Nguyen, C. Luo, S. J. Morgan, W. P. Davis, P. N. Confalone, C. Chen, R. D. Tillyer, L. Frey, L. Tan, F. Xu, D. Zhao, A. S. Thompson, E. G. Corley, E. J. J. Grabowski, R. Reamer and P. J. Reider, *J. Org. Chem.*, 1998, **63**, 8536.
8. K. Julienne, P. Metzner, V. Henryon and A. Greiner, *J. Org. Chem.*, 1998, **63**, 4532; K. Julienne, P. Metzner and V. Henryon, *J. Chem. Soc., Perkin Trans. 1*, 1999, 731; C. L. Winn, B. R. Bellenie and J. M. Goodman, *Tetrahedron Lett.*, 2002, **43**, 5427.
9. J. K. Lieser, *Synth. Comm.*, 1983, 765.
10. S. G. Davies, N. M. Garrido, O. Ichihara and I. A. S. Walters, *J. Chem. Soc., Chem. Commun.*, 1993, 1153; S. G. Davies and D. R. Fenwick, *J. Chem. Soc., Chem. Commun.*, 1995, 1109; S. G. Davies, C. J. R. Hedgecock and J. M. McKenna, *Tetrahedron: Asymmetry*, 1995, **6**, 827, 2507.
11. S. G. Davies and D. R. Fenwick, *Chem. Commun.*, 1997, 565.
12. N. J. Newcombe and N. S. Simpkins, *J. Chem. Soc., Chem. Commun.*, 1995, 831.
13. J. G. Peters, M. Seppi, R. Fröhlich and D. Hoppe, *Synthesis*, 2002, 381.
14. D. M. Hodgson and G. P. Lee, *Chem. Commun.*, 1996, 1015.
15. R. K. Boeckman, *Tetrahedron Lett.*, 1977, 4281.
16. J. K. Crandall and M. Apparu, *Org. React.*, 1983, **29**, 345, 354–359.
17. W. F. Bailey and M. J. Mealy, *J. Am. Chem. Soc.*, 2000, **122**, 6787; G. S. Gil and U. M. Groth, *Ibid.*, 2000, 122, 6789.

25

Asymmetric Induction II: Asymmetric Catalysis: Formation of C–O and C–N Bonds

The Sharpless Asymmetric Epoxidation (AE)

In the textbook we discussed Yamada's synthesis of kalihinene by the AE reaction. We said: 'reaction with the lithium derivative of the sulfone **52** gave **56** which could be converted into **49** by acetylation of the two OH groups, desilylation and Mitsunobu-style inversion of the alcohol with Ph_3P and CCl_4.'

Lithiation of the sulfone **52-OBn** gives **W1** that attacks the less hindered end of the epoxide with inversion at C-2 and retention (no change) at C-3. The product from this reaction is **W2**. Sodium amalgam reduces off the sulfone, making the stereochemistry at that centre irrelevant, to give **56** stereospecifically.[1]

The Mitsunobu-style reaction requires an unprotected OH at C-6 so the other free OHs must be protected by acetylation, giving **W3** and the silyl group is then removed with fluoride. The Mitsunobu-style reaction evidently occurs without allylic rearrangement (chapter 18) whereas direct reaction with HCl would certainly occur with rearrangement.

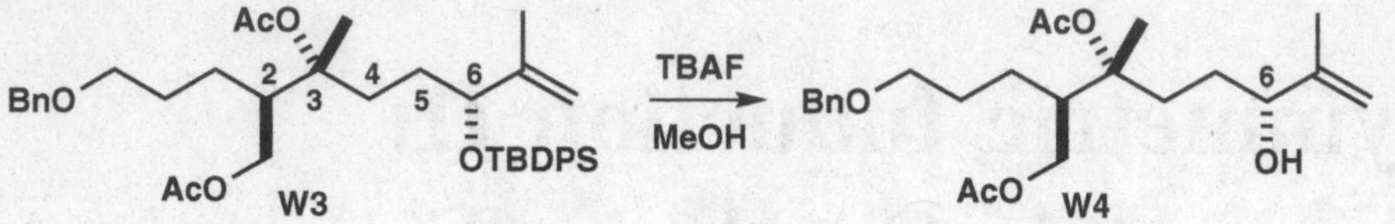

We say 'Mitsunobu-style' (as opposed to Mitsunobu itself) as the reaction does not include DEAD but there are very clear similarities – the alcohol is turned into a good leaving group (leaving as Ph_3PO) while the CCl_4 plays a similar role to the DEAD in that the $^-CCl_3$, anion, generated by reaction with Ph_3P, picks up a proton to form $CHCl_3$.

The Sharpless Asymmetric Epoxidation and the Synthesis of Diltiazem

The rest of the synthesis of diltiazem[2] appears here. Protection of the OH as an acetal **W5** is followed by reduction of the nitro group to the amine **W6**. Removal of the acetate with benzylamine and methylation by diazomethane gave the complete structure **W7** needing only cyclisation to form the seven-membered cyclic amide **58**.

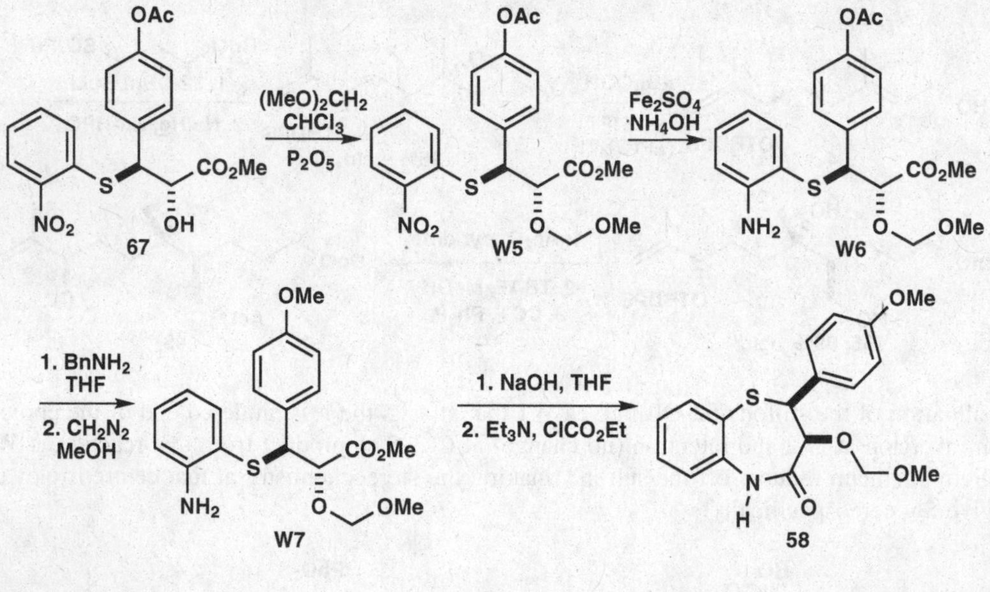

The Synthesis of Reticuline by Sharpless Asymmetric Dihydroxylation (AD)

We carried the synthesis of reticuline **117** up to intermediate **124**. Quite a lot remains to be done, the most serious reaction being the cyclisation by some sort of Friedel-Crafts reaction to complete the piperidine ring.

124; 65% yield from diol → 117; (*R*)-reticuline

First the free OH was acetylated and then the key step was to hydrolyse the acetal in strong acid so that the aldehyde released cyclised onto the benzene ring **W8**. All positions are activated but cyclisation occurs at the less hindered position. Cyclisation onto the other ring would give a seven-membered ring. All that remains is to acetylate the alcohol and hydrogenate all benzyl-O bonds.[3]

124; 65% yield from diol

1. Ac₂O NaOAc
2. HCl, H₂O acetone

W8; 87% yield

Ac₂O NaOAc

W9

H₂, Pd/C *i*-PrOH, H₂O

117; reticuline; 88% yield

Dihydroxylating Compounds With More Than One Double Bond
I – Regioselectivity

Problem 25.1: Suggest an enantioselective synthesis of the bicyclooctane **W10**. The aroma of beer is partly due to this volatile compound.

W10

Answer 25.1: The first thing to recognise is that this molecule is an acetal. The first disconnection should be fairly obvious (see The Disconnection Approach chapter 5 on 1,1-diX disconnections). This yields a molecule containing a diol and an aldehyde function **W11** and with one chiral centre only (notice, by the way, that the TM has *two* chiral centres. If you are in any doubt about how the second one is controlled in the acetal formation then try to make a model of the other possible diastereomer). Diol **W11** is drawn as closely as possible to the acetal (note the hydrogen at the back of the six-membered ring is going away from us so the carbon–carbon bond is thick). This is redrawn in the more extended way **W11a**. We recognise that the diol would have been made using a dihydroxylation reaction, but the aldehyde may be sensitive to the oxidative conditions and so it should be protected before the dihydroxylation. There are various ways that this may be done but one way is to protect it as a double bond. This may not sound sensible to the uninitiated, but a monosubstituted double bond is much less reactive than a trisubstituted double bond.

The first step of the synthesis is the regioselective mono dihydroxylation of a diene. The trisubstituted double bond is preferentially dihydroxylated (Guideline One). The carbonyl can then be revealed by ozonolysis and the acetal formed using catalytic tosic acid.[4]

Other bicyclic acetals can be made in a similar way.[5] These include (−)-frontalin[6], (+)-*exo*-brevicomin[7] and an apple aroma compound.[8]

The example explored above is fairly straightforward when it comes to predicting which double bond will react preferentially. In the textbook we said that predicting regioselectivity could be a bit of a minefield. In fact the four guidelines that are presented would allow you to predict the outcome on many compounds correctly but difficulty can arise if there is a conflict between them. Actually, the term regioselectivity is used very loosely in this context because there is only really an issue of *regioselectivity* when the double bonds are conjugated; choosing between one or other isolated double bond, as we did above, is a matter of *chemoselectivity*. **Problem 25.2**: Predict the 'regiochemical' outcome from a single dihydroxylation of these compounds.

Answer 25.2: The first two compounds are fairly simple. Guildeline One tells us that *trans* double bonds react faster than *cis* or terminal double bonds.

Hence **W18** is formed by the reaction of the only *trans* double bond in **W14**. The regio-selectivity with **W15** is not fantastic and **W19** is formed in a 2:1 ratio with its regioisomer. Things become a little more complicated with the next two examples as there is a conflict between Guidelines. With **W16** we have a *cis* double bond and a trisubstituted double bond. Guideline One would encourage us to react the trisubstituted double bond. However, we have here a conjugated system which Guildline Three says we should react to leave behind the most conjugated system. That would be achieved by reaction of the *cis* double bond and conjugation, it would seem, is more important. The same factors are in conflict in **W17** except we might suspect that the electron-withdrawing ester would deactivate the trisubstituted double bond.

Continuing with the theme of conjugated systems, an example is given where conjugated diene **141** reacts to leave the more conjugated system. The mnemonic device for the AD reaction features an 'attractive area' for aromatic groups. Increasing the size of the aromatic ring from phenyl in **141** to naphthyl in **W22** changes the selectivity. The attraction between the naphthyl ring and the 'attractive area' overcomes the selectivity for leaving behind the more conjugated system hence **W24** is the major product.

By contrast, as we see with **W25**, the binding of the aromatic ring can be discouraged though the use of *tert*-butyl groups to give more reaction at the remote double bond. **Problem 25.3**: Suggest a way to synthesise protected tetraol **W28** from **156**.

156   R=TBDMS   **W28**

Answer 25.3: The underlying tetraol **W29** is not a chiral molecule. There is a mirror plane running through the molecule. One of the 1,2-diols has an (R, R) configuration and the other an (S, S). Clearly we are going to have to use both enantiomers of AD reagent (one at a time!) to make this. Before we move on, let us note that although **W29** is not chiral, **W28** is. The chirality is precariously in place because of the different protecting groups used on each of the diols.

W29   achiral   **W30**   **W31**

On the face of it then the synthesis should be straightforward enough – react one alkene with one enantiomer of AD reagent, protect it and then react the other alkene with the other enantiomer of reagent. Easy. Well it is not quite so simple. As we saw in the textbook, the acetonide substrate **W30** is *matched* with the reagent that was used to generate it. This was all very well when we wanted compound **160** but it will not do here as it is mismatched with the pseudoenantiomeric reagents. Compound **W30** was made via reaction of **156** using (DHQD)$_2$PYR followed by the formation of the acetonide. Subsequent reaction of **W30** using (DHQ)$_2$PYR (the mismatched conditions) gave a poor diastereoselectivity although at least the compound we want **W33** is the major product.

W30   AD with (DHQ)$_2$PYR   **W32**   24 : 76   **W33**

The silyl version of protected diol **W30**, **W31**, is what we need for good selectivity (the matched case). And this works very well with selectivity at 98:2 except, of course with partially protected tetraol **W35** we are now heading towards the *wrong* enantiomer of protected tetraol **W28**.

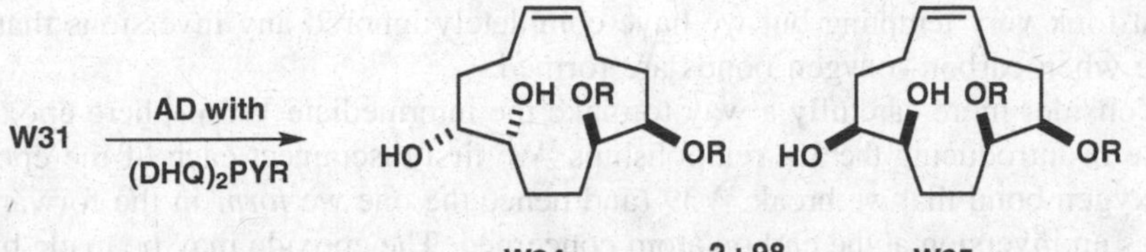

$$W31 \xrightarrow[\text{(DHQ)}_2\text{PYR}]{\text{AD with}} \quad W34 \quad + \quad W35 \qquad 2:98$$

At least that is all the tricky chemistry resolved. All we need to do to get it right (beginning with **156**) is to *start* with (DHQ)$_2$PYR, silylate, *then* apply (DHQD)$_2$PYR (which will be matched with the silylated substrate) and finally turn the free diol (the enantiomer of **W35**) into the acetonide.[9]

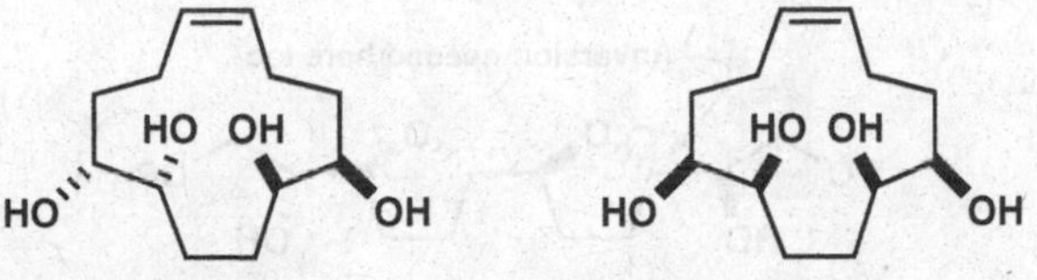

162   163

Employing both Asymmetric Epoxidation and Asymmetric Dihydroxylation

The diepoxide **W36** was needed for the synthesis of (+)-parviflorin **W37** – a natural product with cytotoxic activity.

W36

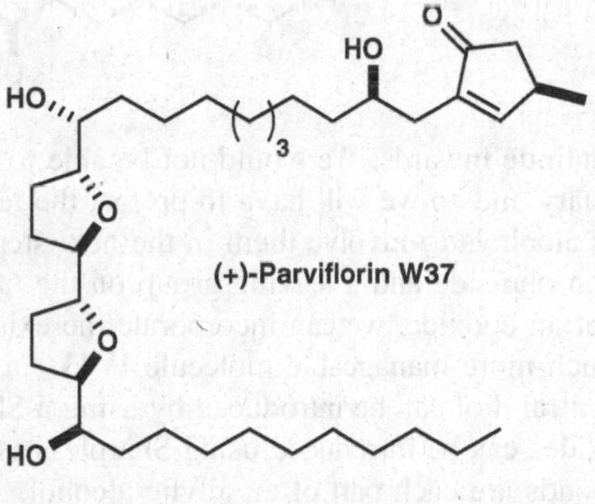

(+)-Parviflorin W37

Diepoxide **W36** has a C_2 axis of symmetry. We will keep this in mind with the rest of our analysis. The diepoxide has three sets of 1,2-related oxygen atoms that look as though they might have come from the corresponding diols.

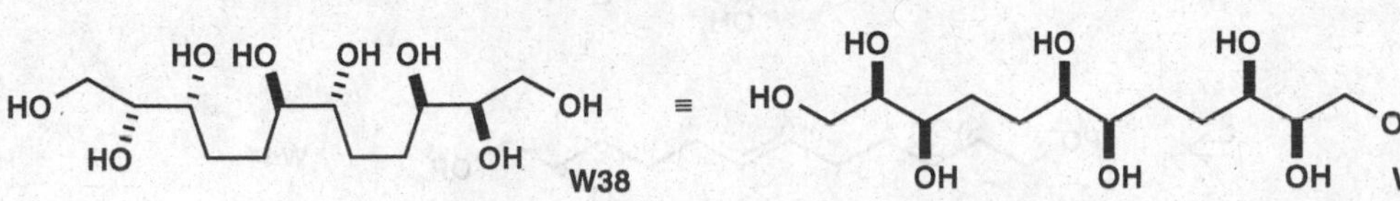

W38 ≡ **W38a**

If we were to view the stereochemistry of **W38** in an open chain we would see **W38a**. This octanol could conceivably come from the corresponding triene using three dihydroxylations.

This might look very tempting but we have completely ignored any inversions that may need to be done when carbon–oxygen bonds are formed.

Let us consider more carefully a way to make the intermediate **W36** – here epoxides play a crucial role in introducing the 1,2 relationships. We first disconnect each of the epoxides. The carbon–oxygen bond that we break **W39** (and hence the one we *form* in the forward reaction) necessitates an inversion at the carbon atom concerned. The epoxide may be made by attacking a tosylate with an alcohol.

We continue inwards. We would not be able to tosylate a secondary alcohol in the presence of a primary and so we will have to protect the terminal alcohols. The secondary alcohols we reveal as alcohols to involve them in the next step. The disconnection of the THF leads to an alcohol on one side and a leaving group on the other. And if we make that leaving group the oxygen of an epoxide, we can incorporate the existing functionality. We now have what looks like a much more manageable molecule **W43** – a bisepoxide with a diol in the middle.

The central diol can be introduced by using a Sharpless asymmetric dihydroxylation and the two epoxides can be introduced using Sharpless asymmetric epoxidation because the two outer double bonds are each part of an allylic alcohol.

The synthesis is that of Hoye and Ye.[10] and starts with triene **W45** which is converted to the diepoxide **W44** by Sharpless epoxidation. Notice that there are *two* enantioselective epoxidations in the same molecule at the same time. This will enhance the enantiomeric excess (see 'Some Principles' in chapter 20 and 'Double Methods' in chapter 28). The diepoxide **W44** was generated in about 97% ee and recrystallised to optical purity. The alcohols had then served their immediate function of directing the epoxidations and before any more alcohol functionality was introduced to the molecule they were protected – there can be no problem with regioselectivity if you protect your alcohols before others are even present. Asymmetric dihydroxylation was used to install the two chiral centres in the middle of the molecule **W43** which was then treated with trifluoroacetic acid to bring about the first two ring closures (of the five membered rings) and the first two ring openings (of the epoxides).

The two secondary alcohols were turned into tosyl groups and then the molecule desilylated with TBAF to liberate alkoxide anions which displace the tosylates and make the epoxides.

Problem 25.4: Now that we have seen how to make the target **W36** from triene **W45**, consider how you would modify the synthesis to make the same compound from isomeric triene **W46**.

Answer 25.4: The synthesis used by Hoye and Ye.[10] involved two inversions at each end of the molecule. The chiral centres of the original epoxide are both inverted (and at both ends of the molecule too). One is inverted in the formation of the five membered ring and the other in the epoxide formation. They need not be. A slight modification allows for the inversion of only *one* if we turn the reaction sequence into more of a 'cascade'.

As before we use epoxidation followed by dihydroxylation but this time we do not protect the primary alcohols as silyl ethers. Instead we tosylate them to turn them into leaving groups. Chemoselective tosylation of the primary alcohol groups should not pose a problem as the central alcohols are secondary and should not react nearly so quickly with tosyl chloride.

One of the alcohols attacks the near end of the epoxide. There is inversion at this centre. The alkoxide that results is a nucleophile (in contrast to the corresponding synthesis where this was turned into a tosylate) and displaces the tosylate at the end of the molecule. **Problem 25.5**: In the last couple of examples we have had more than one stereoselective step – usually an enantioselective step followed by a diastereoselective step. This problem (and problem 25.6b) test your maths for this sort of process. The terminal olefin **W48** is dihydroxylated, using (DHQD)$_2$PHAL as the ligand, in 84% ee. What would you expect the enantiomeric excess to be from the double dihydroxylation of **W49** under the same conditions?

Answer 25.5: The dihydroxylation of monosubstituted olefins is not as good as with many other substitution patterns and so **W48** is hydroxylated in only 84% ee. An 84% ee would correspond to a 92:8 mixture of enantiomers. We might expect the level of selectivity for each double bond to be similar in the case of the bis terminal olefin **W49**. Matters are complicated by there being two olefins in the same molecule but the maths behind this sort of thing is discussed in 'Some Principles' in chapter 20. The ratio of the two enantiomers will be $0.92^2 : 0.08^2$ which gives us an ee of 98.5%. There will be a proportion $(2 \times 0.92 \times 0.08)$ of the other diastereomer too. Exactly this material **W49** was used by Hoye[10] in his synthesis of parviflorin to overcome the problem of the low enantiomeric excesses that could be achieved by AD on terminal olefins. After a couple more manipulations the hydroquinone linker was disposed of. If you found this problem too easy have a look at Problem, 25.6b.

Problem 25.6a: How would you convert geraniol **W50** to the tetrahydrofuran **W51**?

Answer 25.6a: This conversion was done by Taber *et al.* in their synthesis of (+)–tuberine.[11] The product contains three chiral centres and the synthesis of the tetrahydrofuran from geraniol involves little more than the installation of those chiral centres. If you have just read chapter 25 in the textbook then you will probably feel that the allylic double bond is just *itching* to be subjected to the Sharpless epoxidation – and you would be right. Sharpless epoxidation yields the enantiomerically enriched epoxide **W52**. The alcohol is then activated with tosyl chloride. When the double bond in **W53** is dihydroxylated with AD mix β, the resulting diol **W54** reacts under the reaction conditions to form the target tetrahydrofuran **W51**. The epoxide is regioselectively opened to form the five-, rather than the six-, membered ring (or indeed the seven-membered ring by attack of the other hydroxyl group).

Problem 25.6b: There are two stereoselective steps in the above reaction (one enantioselective step and one diastereoselective step) but we have not mentioned *how* stereoselective they are. We know that they will both be very good. If the stereoselectivity of each reaction is the same, and for every 30 molecules of TM **W51** formed we get one molecule of the diastereomer **W52**, what can you deduce about the enantiomeric purity of the product? This is an interesting exercise as we are able to get an estimation of ee (and it is only as estimation as we have to make assumptions) without measuring it at all!

Answer 25.6b: First, we are going to assume that both stereoselective reactions are stereoselective to the same degree. In other words we assume that there is no chiral recognition of any kind. These are whopping great assumptions which is why the figures we will generate can only ever be estimations. In a reaction of a prochiral substrate, x is the mole fraction of one enantiomer formed and $(1 - x)$ is the mole fraction of the other enantiomer. If those two enantiomers are then reacted in a similar process, the product ratio of one enantiomer **W51**: both enantiomers of the diastereomer **W52** : the other enantiomer of **W51** will be x^2: $2x(1 - x)$: $(1 - x)^2$. In this case we know that both enantiomers of the product constitute 30 parts of the 31. In other words, $[x^2 + (1 - x)^2]$ is 30/31 [or, if we want to consider the diastereomer, that $2x(1 - x)$ is 1/31]. Now for the most demanding maths in the whole book, we have to solve a standard quadratic equation. Choose your method but we find $x = 0.9836$.

We can now do two things: work out the enantiomeric excess after the first reaction (the ratio of enantiomers being $x : 1 - x$) and work out the enantiomeric excess after the second reaction [the ratio of enantiomers being x^2: $(1 - x)^2$] The first is 96.7% ee and the second a staggering 99.94% ee.

In fact, if you do not want the bother of a quadratic equation, an approximation can be made which gets around this. Look at the diagram in chapter 20 where the Horeau Principle is explained to see proportions of different diastereomers represented with different boxes. With good selectivity (and 30:1 is easily good enough) the large square will account for almost all of the 30 parts of 31. The other square is so small it can be ignored so our equation simplifies to $x^2 = 30/31$ [i.e. we have deleted the $(1 - x)^2$ bit]. We find $x = 0.9837$. This gives us the same ees as we found above.

Aminohydroxylation

In the textbook we saw how the ligand could be used to control the regiochemistry of the aminohydroxylation. This is also seen in chapter 30 (in the synthesis of lactacystin) where the anthraquinone-based ligand controlled how the cinnamate ester reacted. Regiochemistry can be controlled in another way and that is to attach the nitrogen source to the molecule that is to be reacted – the tethered aminohydroxylation (TA). Thus **W53** is aminohydroxylated to give **W54**. The role of the chiral ligand here is merely to promote the reaction. It is not there for any chiral induction and the pseudo-enantiomer (or indeed i-Pr$_2$NEt) can be used instead.[12] There is good stereocontrol in this reaction too. A different result is seen with the corresponding eight-membered ring **W55** and the seven-membered ring gives a 3:1 mixture of the *syn*:*anti* products.

Problem 25.7: Suggest a synthesis for (−)-chloramphenicol that makes use of an aminohydroxylation.

Answer 25.7: There is more than one way that this might be approached. A FGI (reduction) and FGA (the AA reaction) disconnections reveal a cinnamate ester **W56**. This certainly looks tempting although we might worry that the nitro group would compete with the electron-withdrawing carbonyl and reduce regioselectivity (for good regioselection we would want to keep the ends of the double bond differentiated). A synthesis by Sudalai and colleagues.[13] adopted an alternative approach and used a tethered aminohydroxylation (TA) to ensure regioselection. Superficially it looks as though either hydroxyl in **W57** could have been used to control the regiochemistry we want but because the TA must also control the chirality of the new stereocentres, it has to be done from the *benzylic* hydroxyl group and not from the primary hydroxyl group. We thus need allylic alcohol **W58** and we need it in enantiomerically pure form.

The alcohol **W58** was obtained in 98% ee using a Sharpless asymmetric epoxidation – we are not interested in the epoxide here but rather the allylic alcohol left behind in this kinetic resolution. The alcohol is converted into the carbamate **W59** which is the nitrogen source for the aminohydroxylation to cyclic carbamate **W60**. Hydrolysis and reaction of the nitrogen with methyl dichloroacetate gives the target compound.

Desymmetrisation

Problem 25.8: Suggest how a desymmetrisation might be employed to make a single enantiomer of amino acid **W61**. This compound is an anti-fungal agent and has been prepared on a pilot plant scale by such a route.[14]

Answer 25.8: We saw acid anhydrides being used in the textbook and so the carboxylic acid in **W61** is a big clue to substrate **W62**. One of the carboxylic acids needs to be completely removed however and replaced with nitrogen (and this needs to be done with retention of configuration). The acid anhydride **W62** is made in six steps from commercially available butanetetracarboxylic acid. We shall not go into details but although a mixture of the *cis* and *trans* diacid is made on the way, *both* isomers are converted to *meso* anhydride **W62**. Stoichiometric amounts of quinine and – 15 °C were needed for good enantiomeric excess and the cinnamyl alcohol was the best nucleophilic alcohol the workers found for yield, enantiomeric excess and crystallization. Following the reaction, crystalline racemic ester (±)-**W63** could be formed and removed to leave enantiomerically pure ester **W63** in the filtrate in >97% ee. Curtius rearrangement introduced the nitrogen atom with retention and gave the isocyanate which then reacted to form the carbamate **W64**. Finally, removal of ester and carbamate protecting groups from **W64** using $Pd(OAc)_2$, Ph_3P and morpholine gave amino acid **W61** in 85% yield and >99.5% ee.

References

1. H. Miyaoka, H. Shida, N. Yamada, H. Mitome and Y. Yamada, *Tetrahedron Lett.*, 2002, **43**, 2227.
2. J. Saunders, *Top Drugs: Top Synthetic Routes*, Oxford University Press, Oxford, 2000.
3. R. Hirsenkorn, *Tetrahedron Lett.*, 1990, **31**, 7591.
4. G. A. Crispino and K. B. Sharpless, *Synlett*, 1993, 47.
5. H. C. Kolb, M. S. VanNieuwenhze and K. B. Sharpless, *Chem. Rev.*, 1994, **94**, 2483.
6. B. Santiago, J. A. Soderquist, *J. Org. Chem.*, 1992, **57**, 5844; J. A. Turpin and L. O. Weigel, *Tetrahedron Lett.*, 1992, **33**, 6563.
7. J. A. Soderquist and A. M. Rane, *Tetrahedron Lett.*, 1993, **34**, 5031; D. N. Kumar and B. V. Rao, *Tetrahedron Lett.*, 2004, **45**, 2227.
8. M. A. Brimble, D. D. Rowan and J. A. Spicer, *Synthesis*, 1995, 1263.
9. U. Hugger and K. B. Sharpless, *Tetrahedron Lett.*, 1995, **36**, 6603.
10. T. R. Hoye and Z. Ye, *J. Am. Chem. Soc.*, 1996, **118**, 1801.
11. D. F. Taber, R. S. Bhamidipati and M. L. Thomas, *J. Org. Chem.*, 1994, **59**, 3442.
12. T. J. Donohoe, P. D. Johnson, A. Cowley and M. Keenan, *J. Am. Chem. Soc.*, 2002, **124**, 12934.
13. S. George, S. V. Narina and A. Sudalai, *Tetrahedron*, 2006, **62**, 10202.
14. J. Mittendorf, J. Benet-Buchholz, P. Fey and K.-H. Mohrs, *Synthesis* 2003, 136.

26

Asymmetric Induction III: Asymmetric Catalysis: Formation of C–H and C–C Bonds

Privileged Structures in Asymmetric Synthesis

In the textbook we explained that 'The first place in catalytic hydrogenation nowadays is taken by Rh or Ru complexes of BINAP.' BINAP **W1** and other binaphthols belong to the small group of 'privileged structures' that should be our first resort when trying to put asymmetry into some chemistry. We have met others: the *cinchona* alkaloids such as quinine **W2**, much simpler compounds that may be C_2 symmetric like tartaric acid **W3** (or the TADDOLS derived from it) or the diamine **W4**, proline **W5** and even the very simple amine **W6** and the salens derived from it (chapter 25). Quinine, tartaric acid and proline are natural products but the others are resolved, making both enantiomers available.

Preparation of BINOL and BINAP by Resolution

BINAP **W1** has axial chirality as the naphthalene rings cannot rotate past each other. The minimum blockage needed to stop rotation is about the size of an OH group so that enantiomers of BINOL are stable. BINOL is very easily made by $FeCl_3$-catalysed dimerisation of α-naphthol and it can be converted into BINAP so resolution of BINOL is attractive. The *cinchona* alkaloid salt N-benzylcinchonidinium chloride **W10** forms a crystalline complex with (R)-BINOL **W8**: refluxing in methanol liberates[1] 85–88% yield of **W8** having 99.8% ee. The enantiomer **W9**

Workbook for Organic Synthesis: Strategy and Control Paul Wyatt and Stuart Warren
© 2008 John Wiley & Sons, Ltd

can be recovered from the solution with 99% ee. Though there are many published resolutions of BINOL, this is the one that works best from our experience.

W7; (±)–BINOL   **W8; (R)-(+)–BINOL**   **W9; (S)-(–)–BINOL**   **W10**

BINAP can be made from either enantiomer of BINOL by triflation and a Ni-catalysed coupling with Ph_2PH without loss of enantiomeric excess.[2] If you prefer, racemic BINOL can be converted into the *bis*-phosphine oxide BINAPO **W12** by a Grignard sequence.[3]

The phosphine oxide **W12** forms a crystalline complex with dibenzoyl tartaric acid **W13**. One diastereoisomer crystallises from solution. Treatment with base liberates the phosphine oxide **W14** that can be reduced to BINAP with trichlorosilane.

The next section refers directly to chapter 26 in the textbook and it may help you to have that open while you look at it.

Asymmetric Reduction of 1,3-Dicarbonyl Compounds with Ru-BINAP

We gave examples of Ru-BINAP catalysed reductions of alkenes and a few of reductions of carbonyls such as **30** or even **43** with kinetic resolution.[4] Catalytic reduction of carbonyls is difficult as the C=O bond is much stronger than the C=C bond. **Problem 26.1**: Since all the good examples involve 1,3-dicarbonyl compounds, can you suggest an alternative explanation that does not involve carbonyl reduction?

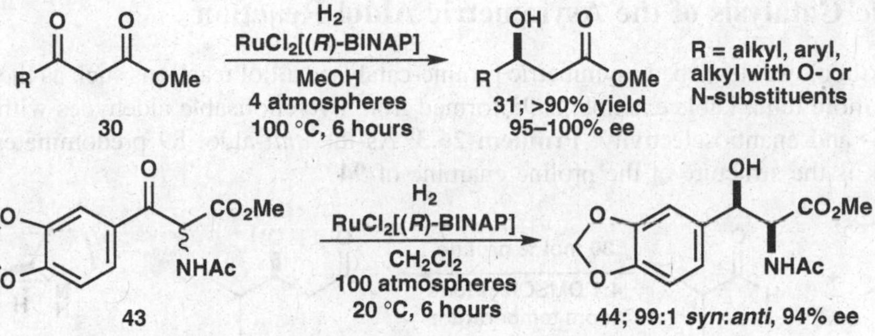

$R = $ alkyl, aryl, alkyl with O- or N-substituents

30

31; >90% yield 95–100% ee

H₂
RuCl₂[(R)-BINAP]
MeOH
4 atmospheres
100 °C, 6 hours

43

H₂
RuCl₂[(R)-BINAP]
CH₂Cl₂
100 atmospheres
20 °C, 6 hours

44; 99:1 *syn:anti*, 94% ee

Answer 26.1: As 1,3-dicarbonyl compounds easily from enols, is it possible that the C=C bond of the enol **W16** or **W17** is reduced rather than the carbonyl itself? There are problems with this interpretation: in the 'kinetic resolution' of **43** the enol would have to exist in the geometry shown **W17** but this is not unlikely. What do you think?

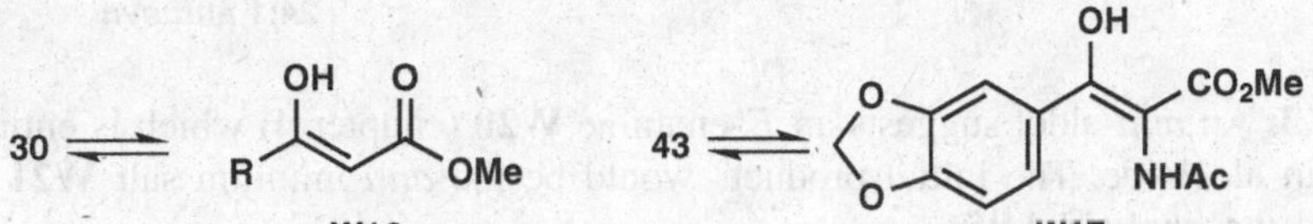

$30 \rightleftharpoons$ **W16**

$43 \rightleftharpoons$ **W17**

Noyori's Asymmetric Synthesis of Menthol

A key step in the asymmetric synthesis of menthol described in the textbook is the conversion of myrcine **46** into the allylic amine **47**. **Problem 26.2**: Propose a mechanism for this step.

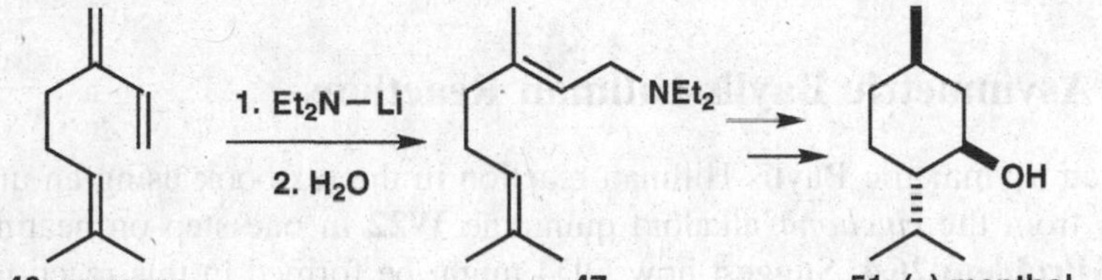

46; myrcene

1. Et₂N—Li
2. H₂O

47

51; (–)-menthol

Answer 26.2: Addition to the less substituted end of the diene **W18** gives a lithium allyl species **W19** that collects a proton at the remote centre to give the more stable trisubstituted alkene.[5]

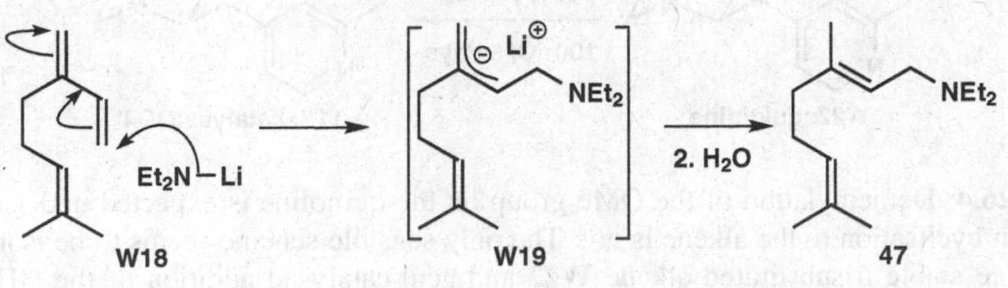

Et₂N—Li

W18

W19

2. H₂O

47

Organic Catalysis of the Asymmetric Aldol Reaction

In the textbook we described asymmetric proline-catalysed aldol reactions such as the formation of **90**. A more remarkable example is **95** formed from two enolisable aldehydes with very high diastereo- and enantioselectivity. **Problem 26.3**: As the *anti*-aldol **89** predominates, what do you think is the structure of the proline enamine of **94**?

Answer 26.3

Answer 26.3: An *anti*-aldol suggests an *E*-enamine **W20** (chapter 4) which is entirely reasonable from an aldehyde. The initial product[6] would be the *anti*-iminium salt **W21** that would hydrolyse to the final aldol **95**.

Catalysed Asymmetric Baylis-Hillman Reactions

We reported an asymmetric Baylis-Hillman reaction in the textbook using an unusual catalyst QD4 derived from the *cinchona* alkaloid quinidine **W22** in one step on heating in KBr and 85% H_3PO_4. **Problem 26.4**: Suggest how QD4 might be formed in this reaction.

Answer 26.4: Demethylation of the OMe group on the quinoline is expected under these conditions but cyclisation to the alkene is not. The only sensible scheme seems to be isomerisation to the more stable trisubstituted alkene **W23** and acid-catalysed addition of the OH group to the tertiary cation. The yield (60%) is reasonable under these harsh conditions.[7]

The catalyst **113** acts in the usual way for a Baylis-Hillman reaction, addition in a conjugative fashion, aldol reaction, and elimination. **Problem 26.5**: What is the mechanism of the elimination?

Answer 26.5: Like most reversible eliminations in basic solution, this must be an E1cB mechanism **W25** via the enolate.

Asymmetric Cyclopropanation Using the Box Ligands

We reported in the textbook that the antidepressant tranylcypromine **145** was made using the Evans box ligand **133**. A bulky ester gave a good enantiomeric excess but only moderate diastereoselectivity.

The diastereoisomers could be separated since the *trans* ester hydrolyses faster than the *cis* so the free acid was essentially pure *trans*. Further reactions including a Curtius rearrangement with retention gave tranylcypromine **145**. **Problem 26.6**: Why does the *trans* ester hydrolyse faster? Why are such vigorous conditions needed for the hydrolysis? Why does no epimerisation occur during the hydrolysis? Suggest reagents for the Curtius rearrangement and comment on the stereochemistry of this reaction.

149; *trans:cis* 88:12
trans: 97% ee

150; 79% yield
trans:cis 130:1

151
tranylcypromine

Answer 26.6: The rate-determining step in ester hydrolysis is the addition of hydroxide ion to the carbonyl group. With a bulky esterifying group (CHR$_2$) this step will be slow if there is a *cis*-phenyl group next door on the three-membered ring. The bulky group also explains the need for vigorous conditions.[8] Epimerisation does not occur because enolates *exo* to a three-membered ring are strained as the trigonal carbon wants an angle of 120° and gets 60°. A good reagent for the Curtius rearrangement is diphenylphosphoryl azide **W26** that gives an acid azide **W27** which loses nitrogen to give an unstable nitrene. Migration of the cyclopropane from C to N gives the isocyanate that is hydrolysed to **151**. Such rearrangement occurs with retention of configuration as the C–C bonding orbital (σ not σ*) is used.

150   W26   W27   W28   W29   151

Asymmetric Synthesis of Chiron's MIV-150

Chiron's anti-HIV drug MIV-150 **W30** contains a disubstituted cyclopropyl amine and, in view of the box-catalysed cyclopropanations we have just discussed, might be made[9] from the carboxylic acid **W31** and the styrene **W32**. The cyclopropanation would have to be *cis* selective unlike the route to **149**.

W30; Chiron's MIV-150   W31   W32

In the lab route, the cyclopropanation of protected **W32** used ligand **133** and CuOTf and achieved >98% ee but only in about 50% yield with a *cis:trans* ratio of 3:1. The yield of *cis*-**W31** in this step was never more than 45%.

W33   1. CuOTf / N$_2$CHCO$_2$Et / 2. HCl, H$_2$O / 3. LiOH, H$_2$O   W34

However, addition of the aminopyridine **W36** to the rest of the molecule was made easier as the isocyanate **W35**, formed when **W34** was treated with diphenylphosphoryl azide **W26**, could be coupled directly to the amine **W36** without hydrolysis and decarboxylation to give the urea in **W37**.

So there were good and bad things in this synthesis. New and old chemistry were combined in the development synthesis. The racemic *cis*-iodo amide **W41** could be made[10] from the amide **W39** by a deprotonation to give **W39** and then equilibration to the more stable *cis* **W40** (more stable because of coordination with Mg).

The free acid **W42** can be resolved with the amine **W43** (compound **6** in chapter 22) and esterified without loss of enantiomeric excess ready for coupling to the left-hand aromatic ring.

A Negishi coupling between the zinc derivative **W46** prepared by *ortho*-lithiation between the MeO and F directors (chapter 5) of **W45** and (+)-**W44** gave exclusively *cis*-**W47** in good yield and the Curtius rearrangement and coupling to the aminopyridine **W36** followed.

Ketene Cycloadditions

We discussed the formation and hydrolysis of β-lactones like **164** in the textbook. There is some confusion in the literature about the stereochemistry and mechanism of the hydrolysis of the lactone **164** to give malic acid **165**. At first the absolute stereochemistry of **164** was not known and did not seem to matter very much as it gave one enantiomer of malic acid.

It was later found that the overall process of NaOH hydrolysis of **164** to give malic acid went with inversion and that the inversion occurred during the hydrolysis of the CCl_3 group and not the lactone opening as the free acid **W48** could be made and hydrolysed with the same results. In addition, hydrolysis of a CCl_3 group to CO_2H is usually slow but here it is fast because participation **W49** drives out the first chloride, cyclisation **W50** to the new β-lactone **W51** explains the inversion and the second chloride is lost by elimination rather than substitution **W51**. The two reactions occurring in **W51** are independent.[11]

Asymmetric Synthesis of Enalapril by Ketene Cycloadditions

In the textbook we said that: 'the hydroxy-ester **180** is prepared for coupling to an enantiomer-ically pure amine by conversion to the triflate **181**. The coupling goes with inversion to give enalapril **182**.' **Problem 26.7**: Which amine is required for the synthesis of enalapril from **181**, and how would it be made?

Answer 26.7: The primary amine **W52** is needed and is prepared from proline by coupling with alanine (chiral pool strategy). Suitable protection of the amino group of alanine and the carboxyl group of proline and a suitable coupling agent will of course be required.[12]

Asymmetric Addition of Allyl Metals to Carbonyl Compounds

In this section of the textbook we referred to: 'Recent improvements include the replacement of toxic allyl tin with the allyl silane **203**. Addition to the alkoxyaldehyde **202** gives **204** that can be transformed into the half-protected dialdehyde **205**.' **Problem 26.8**: Suggest a mechanism for the formation of **204** and how the transformation into **205** might be achieved. R is an acyl group.

204; 97% yield, 91% ee
(97% after recrystallisation)

Answer 26.8: This is an ene reaction in which a silyl group is transferred rather than a proton **W53**. Conversion into **204** requires oxidative cleavage of the alkene and protection at C-5, protection at C-3, and deprotection and oxidation at C-1. The trick is to choose the order of events so that the two ends of the molecule are never the same.[13]

Carreira's solution[13] was to ozonise the alkene and protect the new aldehyde as an acetal **W55**. Then he deprotected the primary alcohol and oxidised with the stable radical TEMPO using NaOCl (bleach) as the stoichiometric oxidant. TEMPO is selective for primary alcohols.

Asymmetric Allylation of Enolates

In the textbook we briefly described Helmchen's use of the manganese-containing phosphine **244** for alkylation of malonates with racemic allylic chloride **246** to give the adduct **247** that is used in the synthesis of the lactone **248**.

Now we can continue the synthesis[14] to get **248**. Hydrolysis of all the esters and decarboxylation of the malonic acid gave approximately a 1:1 ratio of epimers of **W56**. The key step was iodolactonisation (chapter 17) giving mostly (ratio about 3:1) the epimer **W57**. Investigation revealed that the unwanted epimer cyclises more slowly. Elimination gave a mixture of unsaturated lactones **248** that could be separated by chromatography. It is clear that **W57** is the kinetic product as the OH group is on the inside of the folded molecule.

Organic Catalysis: Robinson Annelation

The starting material for Corey's synthesis[15] of dysidiolide was a single enantiomer **W60** of the product of an asymmetric Robinson annelation, but not a catalytic one. The trione **W58** was cyclised with an equivalent of phenylalanine **W59** and half an equivalent of camphor sulfonic **W61** acid to give the bicyclic ene-dione in 79% yield and 89% ee, raised to 100% ee by recrystallisation. Since the reagents are cheap, the fact that they are used in stoichiometric amounts is no problem though it is interesting to note that catalytic proline does better (100% yield, 98% ee) on a slightly different compound (see compound **78** in the chapter). This intermediate was also used in a synthesis[16] of the steroid biosynthesis inhibitor **W62**.

Organic Catalysis of Conjugate Addition

In the textbook we reported that asymmetric conjugate addition occur with *N*-methyl pyrrole **104** and a simple enal **W63** under the influence of catalyst **102** to give the monoadduct **105** in good yield and enantiomeric excess.

With crotonaldehyde double addition[17] occurs with the same pyrrole **104** to give the 2,4-disubstituted pyrrole **W64**. Each C–C bond forming event is independent and catalyst controlled as the C_2 symmetric compound (93% ee) predominates over the much less interesting *meso* compound. The two new centres have a remote 1,5-relationship. There are related examples in chapter 25. Presumably crotonaldehyde is more reactive both electronically and sterically than **W63**.

W64; 83% yield; 93% ee, 9:1 C_2:*meso*

Asymmetric Diels-Alder Reactions

Other good catalysts for the Diels-Alder reaction include box ligands **W66** based on the *cis*-amino-indanol[18] **W65** described in chapter 25.

W65: *cis*-amino-indanol   W66   W67

Simple Diels-Alder reactions such as cyclopentadiene and the acyl oxazolidinone using the box ligand **W66** and copper triflate gave adduct **W69** in reasonable yield and enantiomeric excess. The heterocyclic amide is necessary to hold the copper complex **W70** in a fixed conformation and hence distinguish the lone pairs on the acyl oxygen atom. What was particularly interesting in this work was a study of the effect of the 'bite angle' of the catalyst (θ in **W67**). This angle was varied by changing the angle ϕ in the ligand. Ligand **W66**, with $\phi = 104.7$, gave 82.5% ee but ligand **W67**, with $\phi = 110.6$, gave 96.3% ee. This work from Merck[18] uses calculations to back experimental results in the best modern manner.

W68   W69   W70

Alkynyl Zinc Reactions

We described Carreira's catalysed addition of alkynyl-Zn species to aldehydes in the textbook. The catalyst is *N*-methylephedrine **212** acting as a ligand for zinc.

210   211; 88% yield, 99% ee   212; *N*-methylephedrine

An interesting extension to Carreira's work is to follow the addition of propargyl acetate to an aldehyde by protection and then Pd-catalysed acetate transposition to give first an acetal **W74** and then an *E*-enal **W75** with the newly introduced chiral centre intact.[19] This in effect adds an unsaturated d^3 reagent asymmetrically to the aldehyde.

W71 → W72; 95% yield, 96% ee → W73; 90% yield → W74; 85% yield → W75; 94% yield

Palladium Allyl Cation Complexes

In the textbook chapter we mentioned asymmetric reactions of allylic acetates catalysed by palladium with the C_2-symmetric ligand **236**. The question of monoalkylation with bis allylic acetates did not then arise but we now want to show that it is indeed possible.

ligand 236

It is demonstrated by the remarkable reactions of the six-membered ring compound **W76**. Reaction with malonate gives the monosubstituted product **W77** in good yield and enantiomeric excess and conversion to a cyclopropane **W78** follows. Meldrum's acid, however, gives a lactone **W80**. **Problem 26.9**: Suggest mechanisms for the formation of the cyclopropane and the lactone.

W76 → W77; 81% yield, 99% ee → W78; 90% yield, 94% ee

Meldrum's acid → W79; 60% yield, 98% ee → W80; 75% yield, 98% ee

Answer 26.9: The desymmetrisation step to give **W77** or **W79** is a straightforward choice made by the ligand for displacement of one or other of the enantiotopic benzoates by palladium as described in the textbook. The mechanism of the cyclisation of **W77** to the cyclopropane **W78** looks straightforward too as only one Pd-allyl cation **W81** can form and the cyclisation **W82** must give the observed geometry as the tether delivers the enolate from the bottom face and the Pd requires the same approach.[20]

W77 → W81 → W82 → W78; 90% yield, 94% ee

The complication is clearer when the second reaction is studied. Why switch from enantiomerically pure ligand to racemic ligand for the second step? The reason is simple. The stereochemistry required for the cyclisation **W83** is a mismatch with the stereochemistry of the catalyst. As **W79** is already enantiomerically pure (98% ee) racemic catalyst is fine and inevitably provides both enantiomers. The first cyclisation product **W84** can give the lactone **W80** by various combinations of hydrolysis and decarboxylation.

W79 → W83 → W84 → W80; 75% yield, 98% ee

Asymmetric Phase-Transfer Catalysis

Another remarkable discovery was made during the synthesis of the uricosuric indacrinone **W88** at Merck.[21] The synthesis is simple, and deals with achiral intermediates until almost the last stage when a phase-transfer methylation of **W85**, catalysed by trialkylbenzylammonium salts, completes the carbon skeleton. This base-catalysed reaction goes through an achiral enolate so **W86** is the first chiral compound having an inviolable chiral centre. The base catalyst is a phase-transfer catalyst, a tetra-alkyl ammonium salt and the alkylation is carried out in a mixture of water and toluene. The base (NaOH) dissolves only in the water and the compound **W85** only in the toluene. The catalyst sits on the solvent boundary and carries small amounts of reagents into the other layer.

W85; chiral → (MeCl, toluene; 50% NaOH/H_2O; Phase Transfer Catalyst e.g. $PhCH_2NEt_3^{\oplus}$ $Cl^{\ominus}$) → W86; chiral

1. $AlCl_3$ → (±)-W87 → (2. NaI, K_2CO_3; EtO_2C–Cl; 3. NaOH) → (±)-W88; 63% overall yield

Initial work with *N*-alkyl cinchoninium cations, prepared from cinchonine **W89** by simple alkylation, showed that the *N*-benzyl compounds **W91** were the best. The unsubstituted *N*-benzyl compound **W91** (R = H) gave only 20–30% ee but improvements by careful choice of conditions eventually gave 96% ee. A Hammett plot showed that electron-withdrawing substituents such as X = F gave better results and that X = CF$_3$ **W91** was the best.

W89; cinchonine   **W90**   **W91**

These refinements give an salutary example of how to 'tune' a reaction towards excellence and so we give them in some detail. The variables were:

- solvent;
- concentration of organic reagents, PT catalyst and NaOH;
- physical conditions like speed of stirring;
- temperature;
- leaving group, i.e. MeCl, MeBr, MeI.

These variables are not of course independent and improvement in enantiomeric excess by changing one may then require fine tuning in the adjustment of another. In the event the course of development went something like this: Nonpolar solvents like benzene and toluene gave higher ees than more polar ones such as CH$_2$Cl$_2$ or *t*-BuOMe. High dilution helped. Higher concentrations of NaOH improved the ee and a maximum was reached at 50% NaOH concentration. Excellent agitation was critical. The reaction mixture is in two phases and agitation mixes these better ensuring that the catalyst is present in the alkylation step. Increasing the catalyst concentration accelerated the reaction but *did not affect the ee*. The poorer leaving groups were better, i.e. MeCl>MeBr>MeI.

The final adjustments[21] were made with MeCl under the best conditions as defined above. At a constant MeCl concentration of 14 mol per mol of **W85**, lower temperatures helped (Table 26.1).

Table 26.1   Variation of exantiomeric excess with temperature in the PTC methylation of **W85**

Temperature	25 °C	15 °C	5 °C
Enantiomeric excess	78%	84%	90%
Rate	fast	reasonable	too slow

A compromise at 15 °C seemed best and reducing the MeCl concentration to 7 mol per mol of **W85** raised the reasonable 84% ee to an excellent 96% ee at 40% conversion. These conditions were modified to give the best enantiomeric excess and most convenient operations at 100% conversion (Table 26.2).

Table 26.2 Optimum conditions for asymmetric PTC methylation

Amounts of organic reagents	0.61 g indanone (**56**), 0.7 g MeCl
Solvent	Toluene, 25 ml
Base	50% aqueous NaOH, 5 ml
Catalyst	(**58**; X = CF₃), 0.11 g
Temperature	20 °C
Time	18 hours
Yield of **57**	95%
Enantiomeric excess of **57**	92%

The proposed transition state **W93** binds the four aromatic rings in appropriate pairs and uses Coulombic attraction between N^+ and O^- to hold the enolate rigid while MeCl adds to the free front face.

W85 → (+)-**W88**; 95% yield ~ 96% ee

MeCl, toluene
50% NaOH/H₂O
~10% CHIRAL PTC W89
20 °C, 18 hours

W92; asymmetric phase-transfer catalyst made by alkylation of cinchonine

W93

Catalytic Asymmetric Induction on a Large Scale

The vitronectin receptor antagonist SB-273005 is being developed by GSK as a promising drug for a range of conditions.[22] The reaction chosen for large scale asymmetric induction was catalytic hydrogenation over a rhodium catalyst and unsaturated acid **W96** looked ideal for this purpose.

W94; GSK's SB-273005 **W95**

In the event, 50 kg batches could be hydrogenated in methanol in the presence of Cy₂NH (dicyclohexylamine) to give the acetal **W97** in 84% yield as the dicyclohexylammonium salt. The substrate for the hydrogenation is probably the dicyclohexylammonium salt of the acetal of **W96**. Conversion to the azepinone **W98** by reductive amination and cyclisation was straightforward.

References

1. D. Cai, D. L. Hughes, T. R. Verhoeven and P. J. Reider, *Org. Synth.*, 1999, **76**, 1.

2. D. Cai, J. F. Payack, D. R. Bender, D. L. Hughes, T. R. Verhoeven and P. J. Reider, *Org. Synth.*, 1999, **76**, 6.

3. H. Takaya, S. Akutagawa and R. Noyori, *Org. Synth.*, 1988, **67**, 20.

4. M. Kitamura, M. Tokunaga, T. Ohkuma and R. Noyori, *Org. Synth.*, 1993, **71**, 1.

5. S. Akutagawa and K. Tani, in I. Ojima, *Catalytic Asymmetric Synthesis*, chapter 3.

6. A. B. Northrup and D. W. C. MacMillan, *J. Am. Chem. Soc.*, 2002, **124**, 6798.

7. Y. Iwabuchi, M. Nakatani, N. Yokoyama and S. Hatakeyama, *J. Am. Chem. Soc.*, 1999, **121**, 10219.

8. F.-C. Shu and Q. Zhou, *Synth. Commun.*, 1999, **29**, 567.

9. S. Cai, M. Dimitroff, T. McKennon, M. Reider, L. Robarge, D. Ryckman, X. Shang and J. Therrien, *Org. Process Res. Dev.*, 2004, **8**, 353.

10. M.-X. Zhang and P. E. Eaton, *Angew. Chem., Int. Ed.*, 2002, **41** 2169.

11. H. Wynberg and E. G. J. Staring, *J. Chem. Soc., Chem. Commun.*, 1984, 1181; P. E. F. Ketelaar, E. G. J. Staring and H. Wynberg, *Tetrahedron Lett.*, 1985, **26**, 4665; C. E. Song, J. K. Lee, S. H. Lee and S. Lee, *Tetrahedron: Asymmetry*, 1995, **6**, 1063.

12. H. Yanagisawa, S. Ishihara, A. Ando, T. Kanazaki, S. Miyamoto, H. Koike, Y. Iijima, K. Oizumi, Y. Matsushita and T. Hata, *J. Med. Chem.*, 1987, **30**, 1984.

13. J. W. Bode, D. R. Gauthier and E. M. Carreira, *Chem. Commun.*, 2001, 2560.

14. S. Kudis and G. Helmchen, *Tetrahedron*, 1998, **54**, 10449.

15. E. J. Corey and B. E. Roberts, *J. Am. Chem. Soc.*, 1997, **119**, 12425.

16. H. Hagiwara and H. Uda, *J. Org. Chem.*, 1988, **53**, 2308.

17. D. W. C. MacMillan, *J. Am. Chem. Soc.*, 2000, **122**, 4243; N. A. Paras and D. W. C. MacMillan, *J. Am. Chem. Soc.*, 2001, **123**, 4370.

18. I. W. Davies, L. Gerena, L. Castonguay, C. H. Senanayake, R. D. Larsen, T. R. Verhoeven and P. J. Reider, *Chem. Commun.*, 1996, 1753; I. W. Davies, C. H. Senanayake, R. D. Larsen, T. R. Verhoeven and P. J. Reider *Tetrahedron Lett.*, 1996, **37**, 1725.

19. E. El-Sayed, N. K. Anand and E. M. Carreira, *Org. Lett.*, 2001, **3**, 3017.

20. B. M. Trost, S. Tanimori and P. T. Dunn, *J. Am. Chem. Soc.*, 1997, **119**, 2735.

21. U. H. Dolling, P. Davis, and E. J. J. Grabowski, *J. Am. Chem. Soc.*, 1984, **106**, 446.

22. M. D. Wallace, M. A. McGuire, M. S. Lu, L. Goldfinger, L. Liu, W. Dai and S. Shilcrat, *Org. Process Res. Dev.*, 2004, **8**, 738.

27

Asymmetric Induction IV: Substrate-Based Strategy

Asymmetric Synthesis of a β-Lactam

An Evans-style chiral auxiliary can be used to make β-lactams **W3** by the route outlined below.[1] The product is formed 'completely regio- and enantio-selectively'. **Problem 27.1**: Suggest a mechanism and explain the sense of the induction.

Answer 27.1: To get a four-membered ring we need a $2 + 2$ cycloaddition and that needs a ketene from the acid chloride. Such reactions are well known:[2] elimination of chloride from the enolate **W4** and cycloaddition **W5** gives the β-lactam **W3**.

The regioselectivity is evident from the structure of the product and the relative and absolute stereochemistry with the chiral auxiliary were discovered by Evans and Sjogren.[3] They do not suggest a structure for the transition state but some arrangement such as **W6** gives the right answer **W7** (note the marked Hs).

Asymmetric Synthesis of Amino Acids

We gave a synthesis of the 'fat' amino acid **16** in the textbook. The amino-nitrile **13** was rather unstable but **14**. HCl and **15** were crystallised. In neither case could the diastereomeric excess or the enantiomeric excess be determined in those early days. However, **16** was crystallised and then recrystallised from aqueous MeOH to constant rotation, salts with quinine and ephedrine made and recrystallised and taken back to **16** without loss of rotation.[4] Hence they estimated the enantiomeric excess of **16** as 100%. **Problem 27.2**: How would we proceed nowadays?

Answer 27.2: We should use NMR to determine the diastereomeric excess of **13** or **14** before the chiral auxiliary was removed and we should use a method such as chiral HPLC to determine the enantiomeric excess of **15** and **16** and not rely on rotations.

Chiral Enolates from Amino Acids

Schöllkopf's Bislactim Ethers

In the textbook we revealed that the lithium enolate **48** of the lactim **47** reacts with electrophiles on the face opposite the branched isopropyl group. Hydrolysis of **49** requires only dilute aqueous acid as it is an easily protonated imine. **Problem 27.3**: Given that the product of hydrolysis is the methyl ester **W8** suggest a mechanism for the hydrolysis. Why are amides unlikely intermediates?

Answer 27.3: Protonation at nitrogen gives **W9** attacked by water to give the tetrahedral intermediate **W10**. Proton transfer to nitrogen allows the two oxygen atoms to push the protonated nitrogen out **W11** and reveal the methyl ester of half of the product **W12**. The same steps on the other imine give **W8** and the methyl ester of valine.[5] An amide could be formed by the loss of methanol from **W10** but is an unlikely intermediate as amides are very difficult to hydrolyse and the product would have free acids rather than methyl esters.

We also described a synthesis of the unnatural amino acid **63** by conjugate addition of the interesting pyridine derivative **(E)-51**. **Problem 27.4**: Suggest a synthesis of **(E)-51**.

Answer 27.4: No synthesis is given in the paper[6] but it should be easily made from the available pyridine-4-aldehyde by a Wittig or aldol style of reaction. There are many possibilities–perhaps a HWE olefination (chapter 15) is most appealing.

Williams's Asymmetric Glycine Equivalent

We also described the synthesis of the unusual amino acid **55** by alkylation with the silylated bromide **61**. You might notice the radical bromination step giving only moderate yields. **Problem 27.5**: Suggest a mechanism for the conversion of **60** to **61**.

Answer 27.5: NBS generates a low concentration of bromine and the reaction is initiated by hydrogen abstraction with bromine radicals **W13** and capture of bromine by the resulting conjugated radical completing the radical chain.[7] Where does the bromine come from? The HBr released in the first step reacts by an ionic mechanism **W15** with NBS **W14** to give bromine and succinimide. By this means a low concentration of bromine is maintained.

Removing the chiral auxiliary has to be done in two stages: the first cleaves both benzylic C–O bonds and releases **65** and the second cleaves the C–N bond. **Problem 27.6**: What happens to the NCO$_2$Bn group?

Answer 27.6: The O–Bn group is hydrogenated to release the carbamic acid **W16** that decarboxylates rapidly **W17** in its zwitterionic form.

Seebach's Relay Chiral Centre in Synthesis with Hydroxy Acids

We described the synthesis of the cyclic compound **79** from mandelic acid **78** and the alkylation of the enolate **80**. **Problem 27.7**: Explain why the cyclisation gives *cis*-**79** but the alkylation gives *trans*-**81**. You may find conformational drawings helpful.

Answer 27.7: The cyclisation is acetal formation and it is under thermodynamic control. Most five-membered rings have a flexible 'envelope' conformation, such as **W18** in which any atom may be the one on the 'flap' of the envelope, as in **W19**. In **79** the ester group is planar and only the other oxygen can be the 'flap' **79A**. Hence the *cis* compound is more stable as *t*-Bu and Ph are both equatorial in **79a**. The enolate **80** is simply flat and alkylation is under kinetic control so the reagent (PrI) prefers to attack on the diastereotopic face not occupied by the *t*-butyl group.

W18	W19	79a
envelope conformation	alternative envelope	

The Evans Chiral Auxiliary Approach to Asymmetric Alkylation of Enolates

We showed in the textbook that adding the alkylating agents in one order (R^1 first and R^2 second) gave **96** and adding them in the reverse order (R^2 first and R^1 second) gave *ent*-**96**. Full flexibility demands that both R^1 Br and R^2 Br are able to do S_N2 reactions with lithium enolates.

An alternative strategy for *ent*-**96** was to construct the oxazolidine **93c** from *nor*-ephedrine from the chiral pool (chapter 25) and alkylate in the first order. This route too can be used to make either enantiomer.

Strangely sodium enolates made with $(Me_3Si)_2NNa$ give better selectivity. Thus ethylation of **W20** gives a 96:4 ratio of **W21** and **W23** while methylation of **W22** gives a 93:7 ratio of **W23** to **W21**. Because the chiral auxiliary is still attached, separation of the diasteroisomers allows high ees after the auxiliary is removed.[8]

Adding Chiral Centres to an Evans Alkylation Product

We described in the textbook how **101** can be made in high yield and excellent enantiomeric excess by an Evans alkylation. **Problem 27.8**: Suggest how **101** might be developed into both diastereoisomers **W24** and **W25** of a methylated product.

Answer 27.8: Two molecules of LDA make the lithium salt of the carboxylic acid and the enolate of the *t*-Bu ester **W26**. Methylation gives selectively **W25** in a good ratio. If the enolate is regenerated and reacted with a proton, the proton shows the same diastereoselectivity as the methyl group did (i.e. reaction on the bottom face as drawn) to give **W24**. As expected, the proton of MeOH is less selective than the methyl group of MeI, being smaller, but again we are dealing with diastereoisomers and they can be separated.[9]

Asymmetric Synthesis of Pumiliotoxin via the Diels-Alder Reaction

We described a synthesis of pumiliotoxin **143** in the textbook from **138**, the product of an asymmetric Diels-Alder reaction. **Problem 27.9**: Explain what is going on in the remaining steps.

Answer 27.9: The Horner-Wadsworth-Emmons reaction with a stabilised anion is expected to give the *E*-enone **141** though this is unimportant in view of what happens next. Catalytic reduction of **141** first reduces the two alkenes and removes the Cbz group from nitrogen (Problem 27.6). The free amine can now form a cyclic imine **W27** with the ketone and that is hydrogenated too.[10] The imine is like a *cis*-decalin **W27a** having *exo* and *endo* faces and the *exo* face is the only one that can lie flat on the catalyst.

Asymmetric Synthesis of Shikimic Acid via the Diels-Alder Reaction

Problem 27.10: Explain the stereoselectivity in the next three steps to give **149** and **150**. Only relative stereochemistry needs to be controlled as the starting material **146** is already a single enantiomer.

Answer 27.10: Iodolactonisation must give a 1,3-diaxial bridge as the opening of the iodonium ion **W28** must give the *trans* diaxial product. Of the two β-Hs only the axial one has the right *anti-peri*-planar geometry for an E2 reaction **W29a**. Now the diaxial bridge blocks the top face of the alkene in **149a** and epoxidation must occur from the bottom face as drawn.[11] Both iodination and epoxidation are of course stereospecifically *cis*.

The next few steps are more straightforward. The elimination to give **151** can again occur only on that side as an 'up' proton is needed. Epoxidation occurs again on the bottom face: not only because the diaxial bridge is still there but because allylic alcohols generally deliver peracids intramolecularly (chapter 21). The last step is described in the textbook.

A Chiral Pool Synthesis of Shikimic Acid (Revision of Chapter 23)

It is appropriate to outline a chiral pool synthesis of shikimic acid from a sugar by Fleet[12] to remind you that there is no single 'right' answer to a synthesis problem. A few simple disconnections (mainly two aldols) took Fleet back to a very sugar-like molecule **W32** with oxygen atoms on each of five carbons.

148; shikimic acid   W30   W31   W32

However, the two terminal aldehydes need to be distinguished. Fortunately, the sugar literature revealed a derivative of mannose **W33** that could easily be converted into **W35** whose similarity to **W32** is emphasised by numbered atoms. Two of the OH groups are protected as an acetal, one OH and one CHO are tied up as a cyclic hemiacetal and the other aldehyde is a free primary alcohol.[13]

W33   W34   W35

In fact Fleet decided to use an HWE reaction rather than an aldol for the main C–C bond-forming step so a phosphonate **W36** was made from **W35** and converted in three steps without isolation of intermediates to methyl shikimate **W37**. **Problem 27.11**: What are the intermediates and what is happening?

W36; 74% yield, 1:1

W37
methyl shikimate
62% yield from W36

Answer 27.11: Hydrogenation removes the benzyl group revealing a hemiacetal **W38** that is in equilibrium with the aldehyde. Treatment with methoxide forms the phosphonate anion **W39** and catalyses the opening of the hemiacetal so that the HWE reaction can occur. The undefined chiral centre in **W36** disappears in **W39**. As the HWE is intramolecular there is no debate over the geometry of the new alkene. The dotted arrow shows the atoms that are linked. Acid removes the acetal.

If you compare the strategies of these two syntheses of shikimic acid you will see that they are dominated by the choice of asymmetric method: asymmetric Diels-Alder reaction in one case and chiral pool in the other. The reactions used in the synthesis are subordinate to the strategy.

Asymmetric Conjugate Addition (Michael Reactions)

The examples[14] we gave of conjugate addition to **176**, acids **180** with an aryl and **181** with an alkyl nucleophile in >99% ee in both cases may have given you a doubt: why is **180** (*S*) and **181** (*R*) when any sensible person would say that they have the same stereochemistry? The answer is trivial: they do have the same stereochemistry, it is just that the rules give priority to Ph over CH_2CO_2H in **180** but priority to CH_2CO_2H over Bu in **181**. There is no fundamental significance in *R* and *S*.

We used a π-stacking description of the induction in these and other chiral auxiliaries in the textbook and you may like to read a good discussion of π-stacking effects in asymmetric synthesis by Jones and Chapman.[15]

We referred briefly in the textbook to the same auxiliary giving high enantioselectivity in Diels-Alder reactions with cyclopentadiene. This very early work deserves more comment.[16] It was the first report of a chiral auxiliary and the language Corey uses betrays this; 'an improved method for the preparation of a key prostaglandin intermediate in optically pure form *without resolution* by a process which utilizes a new, readily accessible, recyclable, and efficient chiral controlling group.' The best definition of a chiral auxiliary we have read anywhere.

Sulfoxides as Chiral Auxiliaries

The sulfoxide **206**, used by Posner[17] in his asymmetric synthesis of podorhizone, is made from propargyl alcohol in three simple steps. **Problem 27.12**: Give mechanisms for the reactions and explain the stereochemical control.

Answer 27.12: The first few steps are straightforward: hydrostannylation of alkynes is described in chapter 16: the initially formed *Z*-**W44** equilibrates to *E*-**W44** in a radical process. exchange of tin for iodine and iodine for lithium go with retention as expected for electrophilic substitution on a trigonal centre. The first interesting step is the reaction of *E*-**W45** with **W42** and we need to explore the background to this.

Menthyl toluene-*p*-sulfinate is made from menthol and racemic toluene-*p*-sulfinyl chloride in pyridine.[18] After recrystallisation about 32% (of 50%) of crystalline **W42** can be isolated. Up to this point the menthol has been a covalently bound resolving agent. There is no need to remove it in a separate step as it is established[19] that reaction of **W42** with organolithium compounds goes with inversion at sulfur, either by an S_N2 reaction (at sulfur) or by an addition-elimination *via* four-valent sulfur. This is how (−)-**206** is formed. Both **W42** and **W43** are now commercially available.

Iodolactone Reactions

In the textbook we revealed that the iodolactone **223**, made by asymmetric Birch reduction, can be fragmented in two different ways by different reagents to give the lactones **224** or **225** that bear little resemblance to the original benzene ring. **Problem 27.13**: Suggest mechanisms for the two fragmentations. (*Hint*: In one case the nucleophile attacks the lactone carbonyl and in the other the ketone.)

224   **223**   **225**

Answer 27.13: Attack at the more electrophilic ketone by hydroxide fragments the molecule **W47** by a reverse Claisen ester condensation to give **W48** that loses HI in an E1cB reaction. The stereochemistry at R is lost in the fragmentation but this does not matter as it is destroyed by the elimination. The drawing of **225** below is the same as above but inverted horizontally and vertically. An alternative is fragmentation in one step **W49**.

The other product is derived from transesterification of the lactone **W50** followed by fragmentation by attack of the released oxyanion on the ketone **W51**. If this reaction happened with hydroxide, the oxyanion would be quenched by the new CO_2H group. Schultz[20] gives no mechanistic details but says 'it is noteworthy that [all products] are formed without racemisation.'

W50   **W51**   **224**

Schöllkopf's Bislactim Ethers Revisited

More light was shed on the conjugate addition of Schöllkopf's bislactim ethers during the BAYER synthesis of the hepatoprotective agent clausenamide.[21] Planning to make this from a bislactim ether, they needed to disconnect back to a glycine anion equivalent. So removal of the OH group with oxidation of an enolate in mind (chapter 33) and removal of the bonds in **W53** gave a simple amino acid that could be made from **47** by conjugate addition.

W52; (+)-clausenamide   W53   W54

The lithium derivative of the bislactim ether **W55**, the enantiomer of **47**, gave good stereo-selectivity in the expected sense on the ring (new group added *anti* to the i-Pr group) and, if **Z-56** is use, the right stereochemistry in the side chain.

W55   Z-W56   W57; 73% yield, 100% ee   W58

They had previously tried *E*-**56** but this gave the wrong stereochemistry in the side chain. So there was an unexpected dependence on alkene geometry in the Michael acceptor which they explain by involving both oxygens of the ester with the lithium atom on nitrogen. The reagents have to approach in parallel planes **W60** or, in three dimensions, something like **W61**.

W59   W60   W61

The future development of new chiral auxiliaries will depend on such careful analysis of interactions between them and the compounds from which new chiral centres are to be created.

References

1. T. W. Hudyma and R. A. Partyka, US Patent 5,106,842 1992 to Bristol-Meyers Squibb, *Chem. Abstr.*, 1992, **117**, 48215.
2. F. H. van der Steen and G. van Koten, *Tetrahedron*, 1991, **47**, 7503.
3. D. A. Evans and E. B. Sjogren, *Tetrahedron Lett.*, 1985, **26**, 3783, 3787.
4. K. Q. Do, P. Thanei, M. Caviezel and R. Schwyzer, *Helv. Chim. Acta*, 1979, **62**, 956.
5. U. Groth and U. Schöllkopf, *Synthesis*, 1983, 37; D. Pettig and U. Schöllkopf, *Synthesis*, 1988, 173.
6. U. Schöllkopf, D. Pettig, U. Busse, E. Egert and M. Dyrbusch, *Synthesis*, 1986, 737.
7. S. R. Schow, S. Q. DeJoy, M. M. Wick and S. S. Kerwar, *J. Org. Chem.*, 1994, **59**, 6850.
8. D. A. Evans, *Morrison*, **3**, p. 1.
9. R. P. Beckett, M. J. Crimmin, M. H. Davis and Z. Spavold, *Synlett*, 1993, 137.
10. L. A. Overman, G. F. Taylor, C. B. Petty and P. J. Jessup, *J. Org. Chem.*, 1978, **43**, 2164.
11. P. A. Bartlett and L. A. McQuaid, *J. Am. Chem. Soc.*, 1984, **106**, 7854.
12. G. W. J. Fleet and T. K. M. Shing, *J. Chem. Soc., Chem. Commun.*, 1983, 849.

13. J. S. Brimacombe, F. Hunedy and L. C. N. Tucker, *J. Chem. Soc.*, 1968, 1381.

14. W. Oppolzer and H. J. Löher, *Helv. Chim. Acta*, 1981, **64**, 2808.

15. G. B. Jones and B. J. Chapman, *Synthesis*, 1995, 475.

16. E. J. Corey and H. E. Ensley, *J. Am. Chem. Soc.*, 1975, **97**, 6908.

17. G. H. Posner, T. P. Kogan, S. R. Haines and L. L. Frye, *Tetrahedron Lett.*, 1984, **25**, 2627.

18. M. Hulce, J. P. Mallamo, L. H. Frye, T. P. Kogan and G. H. Posner, *Org. Synth.*, 1986, **64**, 198.

19. G. Solladié, *Synthesis*, 1981, 185.

20. A. G. Schultz, *Chem. Commun.*, 1999, 1263.

21. W. Hartwig and L. Born, *J. Org. Chem.*, 1987, **52**, 4352.

28

Kinetic Resolution

Standard Kinetic Resolution Reactions

In chapter 16 we saw how kinetic resolution using the Sharpless asymmetric epoxidation (see chapter 25) with L-(+)-di-isopropyl tartrate removed the unwanted enantiomer as the epoxide **59** and left the required enantiomer (−)-**58** for transformation into (+)-grandisol **60**. The enantiomeric excess of **58** depends upon the level of conversion. The 94% ee was obtained at 59% completion although >98% ee could be obtained at 80% completion.[1]

(±)-cis-58 → **59** + **(−)-cis-58; 94% ee 31% yield** → **60; (+)-grandisol**

Problem 28.1: Epoxide **W2** had been prepared by Mori in 87% ee with a Sharpless dihydroxylation (which made **W1**) to introduce the chirality.[2] The epoxide was used to synthesise the sex pheromone **W3** of *Janus integer* but 87% ee was simply not good enough. Starting with **W2** of 87% enantiomeric excess, how might the ee of **W2** be improved by kinetic resolution?

Answer 28.1: You might think that with 87% enantiomeric excess an easy way to improve enantiomeric excess would be to recrystallise. This *was* possible with diol **W1** but in such poor yield that an additional asymmetric method was applied. In the case of grandisol **60**, epoxidation left behind the compound we wanted. Here we already have epoxide and we want

Workbook for Organic Synthesis: Strategy and Control   Paul Wyatt and Stuart Warren
© 2008 John Wiley & Sons, Ltd

to remove the epoxide we do not want. The answer is hydrolytic kinetic resolution (HKR). The catalyst is very similar to the chromium salen complex we saw in textbook chapter 28 (which was used for a dynamic kinetic resolution). Here the Jacobsen *cobalt* salen[3,4] **W4** catalyses an ordinary kinetic resolution and hydrolyses the wrong enantiomer. Epoxide **W2** is thus produced in 96% ee and 72% yield (and diol **W1** is produced by the hydrolysis in 20% yield and 49% ee).

W2, 72%, 96% ee

Other kinetic resolutions had been tried. For example diol **W1** was treated with a lipase and vinyl acetate to try to make the monoacetate but this was not successful in improving the enantiomeric excess.

Cobalt salen complexes are used in more traditional kinetic resolutions. The protected catechol **W5** reacts with an excess of racemic epichlorohydrin which is kinetically resolved in the presence of a cobalt salen. A special oligomeric form of the salen (represented diagrammatically by **W7**) is tremendously efficacious (compared with monomeric forms) and gives the product in 99% yield and 98% ee.[5]

Problem 28.2: Suggest how ester **W8** might be prepared by kinetic resolution of (±)-**W8**.

Answer 28.2: The ester can be prepared by removing the enantiomer we do not want from the racemic mixture. This can be done by conjugate addition of an optically pure lithium amide to racemic **W8**.[6] The advantage of generating the optically pure compound by taking what is left behind (rather than what is generated) is two-fold here. First, as explored in the textbook, the enantiomeric excess can be as high as you like by running the reaction further. Secondly, we do not have to worry about things like the extra chiral centres that are generated when **W10** is formed. There are several diastereomers of **W10** possible but they matter not one jot. We shall see more reactions of this sort of substrate in the section Parallel Kinetic Resolutions and there, since both enantiomers react, we *will* be concerned with diastereoselectivity as well as enantioselectivity.

(±)-W8 → W10 + W8

Problem 28.3: Acid **W13** is needed in the synthesis of roxifiban which is an antagonist of a platelet glycoprotein receptor.[7] Suggest (i) a synthesis of acid (±)-**W13** from nitrile oxide **W12**, (ii) a way to kinetically resolve acid **W13** and (iii) a way to recycle the wrong enantiomer.

W11, roxifiban

W12

W13

Answer 28.3: Part (i) is easy, although we will be making the *ester* of **W13**. Cycloaddition with *iso*-butyl ester **W14** gives the isoxazóle (±)-**W15** in 90–95% yield. For part (ii) kinetic resolution is conducted using an enzyme in phosphate buffer to give, after recrystallisation, the acid **W13** in >99% ee and 38% yield. The enzyme hydrolyses one of the esters to give the acid and leaves the other ester alone. These reactions were performed on pilot plant 40 kg scale. Enantiomerically pure **W13** had been previously prepared by preparative chiral HPLC and by fractional crystallization of its cinchonidine salts.[8]

W12 + **W14** → (±)-**W15**

1. phosphate buffer / Amano PS30 lipase
2. EtOH recrystallisation

W13, 38% yield, >99% ee

(S)-W14

The final thing to do, part (iii), is recycle (*S*)-**W14** by turning it into (±)-**W14**. This is done in base (KO*t*-Bu) though it is not immediately obvious why base works because the chrial centre is not acidic. However, once **W14** is deprotonated in base it can undergo a retro-Michael addition reaction which destroys the chiral centre leading to a racemic mixture of **W14**. If it were possible to find a way to racemise **W14** *while* the kinetic resolution were taking place then we would have a dynamic kinetic resolution. We shall come onto this.

Dynamic Kinetic Resolutions

The complication with dynamic kinetic resolutions, over ordinary kinetic resolutions, is getting the two enantiomers of starting material to interconvert and thus produce a *dynamic* version and the potential 100% yield. **Problem 28.4**: Alcohol **W17** is a required intermediate in the synthesis of taranabant, an inverse agonist on a cannabinoid receptor with potential in the treatment of obesity.[9] Racemic alcohol (±)-**W17** can be made diastereoselectively from racemic ketone **W15** by stereoselective reduction with L-selectride.

Starting from acid **W16**, suggest a stereoselective synthesis of **W17** which incorporates a dynamic kinetic resolution.

Answer 28.4: The key intermediate is ketone **W15**. The dianion of acid **W16** can be reacted with the benzyl chloride **W18** to give acid **W19**. Formation of the Weinreb amide and reaction with methyl Grignard gave ketone **W15**.

It is with ketone **W15** that the 'dynamic' part of the reaction happens and this is made relatively easy by the fact that the chiral centre is both benzylic and at the α-position making it nice and acidic. Racemisation via deprotonation is thus achieved by adding base. It is important that the rate of racemisation is fast relative to the rates of reaction so that there is no build up of one enantiomer of starting material. A reagent that reacts selectively (ie $k_1 \gg k_2$) with one enantiomer of the ketone is needed. Of course, in order to produce the diastereomer we want, the reducing agent must also distinguish the diastereotopic faces of the ketone.

The reducing agent is hydrogen using Noyori's catalyst with xyl-BINAP ligand and the DIAPEN additive (textbook chapter 26).

This strategy (a Noyori catalyst and base) for DKR has been applied to other substrates which include, on the pilot scale, the production [from (±)-**W20**] of alcohol **W21** which is an intermediate in the synthesis of drug target **W22**.[10]

(±)-W20 → W21, 97% ee ; W22

Problem 28.5: Suggest how acetate **W24** might be made by dynamic kinetic resolution of alcohol **W23**.

W23 → W24

Answer 28.5: Contrast this Problem with the previous one. Racemisation of the ketone **W15** was fairly easy because the proton at the stereogenic centre is so acidic. This is certainly not the case here and racemisation of alcohol **W23** is not merely achieved by adding the right base. The racemisation is achieved by adding a ruthenium catalyst **W25**.[11] Studies into the mechanism of racemisation suggest the active species is a ruthenium hydride and alcohols are racemised via their corresponding ketones.[12] Catalyst **W25** has an advantage over some others because it does not need to be used in anaerobic conditions. The enantioselection is done with an enzyme and, as we are accustomed to seeing, a vinyl acetate (textbook chapter 29).

W25 ; W23 → W24, 98% >99% ee ; 4 mol% W25, K₃PO₄, Novozym 435

Problem 28.6: We have previously encountered acid **W13** in a kinetic resolution above (Problem 28.3). Suggest a way that acid **W13** may be made in enantiomerically pure form using a dynamic kinetic resolution.

(*R*)-W13

Answer 28.6: The clue to this was in the last part of Problem 28.3–the racemisation of **W14**. So, add base perhaps? The trouble with this is the lipase and its tolerance (or lack of) to high pHs. Although the workers on this problem tried putting electron-withdrawing groups on the substrate this did not improve matters. The answer was to move to a thiolester, **W26**. The pK_a of the α-proton on a thiolester is lower than with ordinary esters. A host of reaction conditions was tried but ultimately the enzyme used was the same one as that in the ordinary kinetic resolution and, when the reaction was performed at pH 9.25 with trimethylamine, a DKR ensued. Conversion was >99% and the product was formed in 97.6% ee. The process was then run successfully on multi-kilogram scale.

W26 → phosphate buffer / Amano PS30 lipase, 2 eq Me₃N → W13, 99% conversion, 97.6% ee

Problem 28.7: The synthesis below shows a synthesis of racemic aminoquinuclidine **W32**. This compound was needed as a precursor to a substance P antagonist. Suggest a way that the synthesis might be modified to include a dynamic kinetic resolution and thus generate enantiomerically pure **W32** without wasting 50% of any intermediate.

Answer 28.7: Several approaches have been made to this target and these have been summarized.[13] The routes have included an ordinary resolution of **W32** using tartaric acid. The enantiomeric excess is good but the yield, naturally, is less than 50% and the resolution happens on the final product which is the worst place to resolve! The best place to resolve is with the first compound that is chiral–ketone **W29**. There is a potential problem though, and that is the chiral centre may be quite unstable during the formation of imine **W30**. It is this instability that also gives us the opportunity to introduce a *dynamic* element into the resolution. L-Tartaric acid itself is quite good at selectively crystallising with the (*S*)-enantiomer of **W29** and it was found that the ketone remaining in solution was racemic which meant that it must be racemising. The addition of acetic acid is to speed up the racemisation. The free enantiomerically enriched ketone can be liberated by the careful addition of base to the tartrate salt.

(±)-W29 → 1. L-tartaric acid / AcOH 2. Toluene / NaOH (aq) → (*S*)-W29 90% yield 96% ee

Once ketone (*S*)-**W29** has been formed as a single enantiomer we need to keep the stereochemistry intact. The workers wanted to speed things up by not isolating the imine but doing a reductive amination. Many methods for the formation of amine **W31** by reductive amination of **W29** using various borohydrides led to either racemisation of **W29** or **W30** or poor *cis/trans* ratios of product **W31**. Finally the use of hydrogen in combination with Ti(O*i*-Pr)₄ was able to provide good *cis/trans* selectivity (9:1) and not erode the enantiomeric excess (≥92% for *cis* diastereomer). Recrystallisation took the enantiomeric excess to 99% and debenzylation gave the target (*S*,*S*)-**W32**.

(*S*,*S*)-W31
61% yield
99% ee, 99% de

(*S*,*S*)-W32
95% yield
99% ee

Parallel Kinetic Resolutions

With an ordinary kinetic resolution in its simple form one enantiomer reacts and the other one does not. This can be difficult enough to achieve but the extraordinary thing about parallel kinetic resolutions is that the other enantiomer, rather than not reacting or reacting slowly, does a different reaction! In a synthesis of annonaceous acetogenins, phosphonates **W34** and **W36** could be used in both a desymmetrisation reaction (of **W33**) and a parallel kinetic resolution. We explore the parallel kinetic resolution below.[14]

Problem 28.8: Consider what reaction might occur between racemic aldehyde **W35** and phosphonate **W36** under basic conditions.

Answer 28.8: Were this an ordinary kinetic resolution we might expect a result with one enantiomer reacting and the other not. In this case each enantiomer reacts to give different products—one gives the *cis* **W37** and the other the *trans* olefin **W38** in a 40:60 ratio. Diastereoselectivity is excellent in both cases. Remember that, since the esters are the 8-phenylmenthyl esters that the enantiomeric excess of every product will be as good as the original enantiomeric excess of the 8-phenylmenthol. This parallel kinetic resolution is unusual in that there are *not* two pseudo-enantiomers of reagent. Remarkably there is just (one enantiomer of) one reagent and it reacts differently with each enantiomer of substrate.

W37; 98:2 dr

W38; 96:4 dr

The parallel kinetic resolution over, a cunning part of the synthesis followed. Compounds **W37** and **W38** were not separated but carried through the synthesis together to give phosphinates **W39** and **W40**. In a lovely example of stereoconvergence, these two diastereomers react differently (one with retention and the other with inversion) in a palladium-catalysed process to give the same diastereomer (and same enantiomer) of compound **W41**.[14]

It is extraordinary to reflect that both enantiomers of a substrate **W35** were transformed into the same enantiomer of product **W41** through the use of just one enantiomer of reagent. **Problem 28.9**: In Problem 28.2 we saw ester **W8** resolved by ordinary kinetic resolution. How might the similar reaction of ester **W42** be transformed into a parallel kinetic resolution? Consider what conditions must be fulfilled for the parallel kinetic resolution to work.

Answer 28.9: The obvious answer is to find another similar optically pure lithium amide that can react with the other enantiomer of **W42**. This might be lithium amide **W44** or **W45**. We would hope then to achieve a reaction of the kind below.

This does indeed work and the authors explain how they carefully arrived at this reaction.[6] The conditions needed are (i) that the two reactions do not interfere with one another, (ii) that they proceed at approximately the same rate and (iii) that we also get good diastereoselectivity. If the reactions proceed at different rates there will be a build up of one of the enantiomers. Experiments using racemic amines gave no indication that the reactions interfered with one another (see the Eames example using racemic oxazolidinone in textbook chapter 28) but lithium amide **W45** reacts at about twice the rate of **W9**. Although this can be compensated for by adding twice as much of the slower-reacting nucleophile, matters are simpler using **W44** and **W9** which react at about the same rate. Here is the overall reaction.

The hindered phenol is important to get good diastereoselection in the protonation of the product from the conjugate addition (something we were not concerned about in Problem 28.2). For example, diastereomers **W47** and **W48** both result from the same enantiomer of ester reacting with **W44** (of which there is only the one enantiomer anyway) but differ because protonation has occurred on opposite sides of the enolate. The fraction of **W48** is greatly reduced when the hindered phenol is used for protonation of the enolate. Finally, pseudo-enantiomers **W46** and **W47** are easily separated and removing the benzyl groups (from the nitrogen atom that is) gives the *genuine* enantiomers of **W49**.

References

1. D. P. G. Hamon and K. L. Tuck, *J. Org. Chem.*, 2000, **65**, 7839.
2. K. Mori, *Eur. J. Org. Chem.*, 2005, 2040.
3. M. Tokunaga, J. F. Larrow, F. Kakiuchi and E. N. Jacobsen, *Science*, 1997, **277**, 936.
4. E. N. Jacobsen, *Acc. Chem. Res.*, 2000, **33**, 421.
5. J. M. Ready and E. N. Jacobsen, *J. Am. Chem. Soc.*, 2001, **123**, 2687.
6. S. G. Davies, A. C. Garner, M. J. C. Long, A. D. Smith, M. J. Sweet and J. M. Withey, *Org. Biomol. Chem.*, 2004, **2**, 3355.
7. J. A. Pesti, J. Yin. L. Zhang, L. Anzalone, R. E. Waltermire, P. Ma, E. Gorko, P. N. Confalone, J. Fortunak, C. Silman, J. Blackwell, J. C. Chung, M. D. Hrytsak, M. Cooke, L. Powell and C. Ray, *Org. Process Res. Dev.*, 2004, **8**, 22.
8. J. Wityak, T. M. Sielecki, D. J. Pinto, G. Emmett, J. Y. Sze, J. Liu, E. Tobin, S. Wang, B. Jiang, P. Ma, S. A. Mousa, R. A. Wexler and R. E. Olson, *J. Med. Chem.*, 1997, **40**, 50.

9. C. Chen, L. F. Frey, S. Shultz, D. J. Wallace, K. Marcantonio, J. F. Payack, E. Vazquez, S. A. Springfield, G. Zhou, P. Liu, G. R. Kieczykowski, A. M. Chen, B. D. Phenix, U. Singh, J. Strine, B. Izzo and S. W. Krska, *Org. Process Res. Dev.*, 2007, **11**, 616.
10. M. Scalone and P. Waldmeier, *Org. Process Res. Dev.*, 2003, **7**, 418.
11. N. Kim, S. B. Ko, M. S. Kwon, M. J. Kim and J. Park, *Org. Lett.*, 2005, **7**, 4523.
12. J. H. Choi, Y. K. Choi, Y. H. Kim, E. S. Park, E. J. Kim, M. J. Kim and J. Park, *J. Org. Chem.*, 2004, **69**, 1972.
13. T. C. Nugent and R. Seemayer, *Org. Process Res. Dev.*, 2006, **10**, 142.
14. D. Strand, P. O. Norrby and T. Rein, *J. Org. Chem.*, 2006, **71**, 1879.

29

Enzymes: Biological Methods in Asymmetric Synthesis

Epothilone Synthesis

In the textbook we said that the lithium enolate of the lactone **176** was allylated with the expected stereoselectivity to give **177** and then, by a series of reactions described in the workbook, the aldehyde **178**.

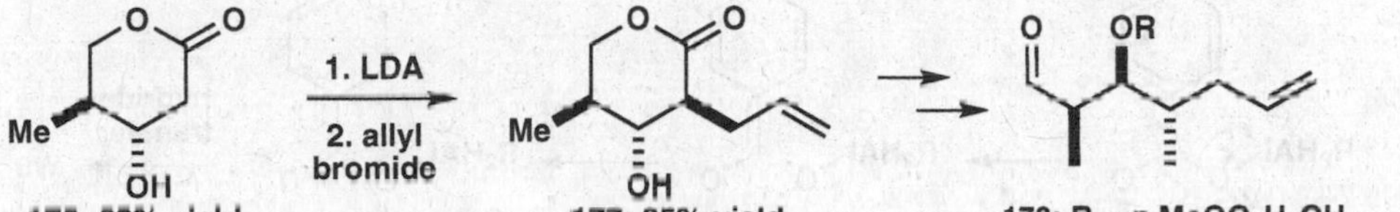

176; 62% yield   177; 85% yield   178; R = p-MeOC$_6$H$_4$CH$_2$

The lactone **177** is first opened by methoxide ion to give **W1** and hence the diol ester **178** without change in stereochemistry – **W1** simply needs to be rotated and redrawn.

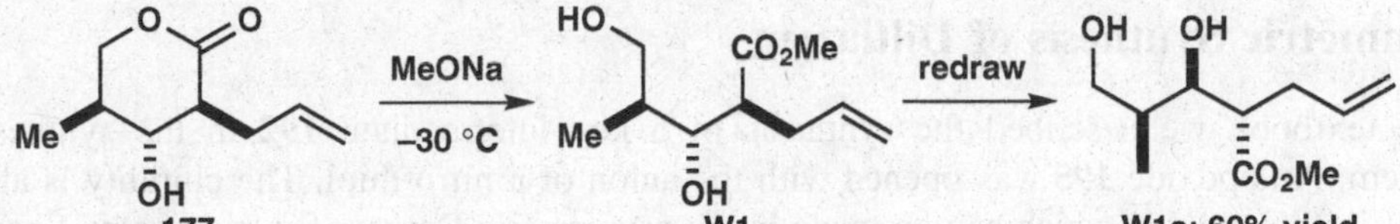

177   W1   W1a; 60% yield

Now come the more interesting reactions. Acetal exchange with the dimethylacetal of anisaldehyde **W2** gives the more stable cyclic acetal **W3**. The ester group is reduced, mesylated and reduced again to give the methyl group in **W5** without loss of stereochemistry. Now we come to the point of the PMP protection. Because of the electron-donating methoxy group, cleavage by the Lewis acidic reducing agent DIBAL gives the mono-protected diol **W6** chemo-selectively. Dess-Martin oxidation provides the aldehyde **178** needed for the next step.[1]
Problem 29.1: Draw a detailed mechanism for the DIBAL reaction.

W2 → W1 (camphor sulfonic acid) → W3; 95% yield → LiAlH₄ (0 °C to room temperature) → W4; 90% yield → 1. MsCl, Et₃N 94% yield / 2. LiAlH₄ 88% yield → W5 → DIBAL toluene → W6; 93% yield → Dess-Martin → 178 96% yield

Answer 29.1: The DIBAL presumably adds to the less hindered oxygen atom to give **W7** that cleaves with help from both the other oxygen atoms to give a cation **W8** that can also be drawn as **W8a**. Presumably the oxygen atom from the secondary alcohol is better able to stabilise the positive charge than the other.

Asymmetric Synthesis of Diltiazem

In the textbook we described the synthesis of a key intermediate **192** in the synthesis of diltiazem. An epoxide **195** was opened with the anion of a nitro thiol. The chirality is already present in the epoxide which can be made by an asymmetric Darzens reaction or by Sharpless epoxidation. You may have noticed that the opening of the epoxide is, strangely, a *cis* opening.[2]

192; Ar = *p*-MeO-phenyl   193   194   195

The answer is that two inversions are required. The epoxide **W9** was first opened with chloride ion from pyridinium chloride to give **W10** – no doubt the workers were pleased to see that the S_N2 reaction happened next to the aromatic ring and not the ester group (both are activating). The second S_N2 reaction, also at the benzylic position, was with the anion of *o*-nitrobenzene thiol. It should be easy to see how **193** can be made from **W11**.

In fact there is more to it than this. Detailed mechanistic work[3] showed that the *anion* of *o*-nitrobenzene thiol reacts with the *p*-methoxy analogue **W12** with clean inversion, giving *anti*-**W13** in 80% yield while reaction with the thiol itself and a Lewis acid catalysts gives mostly *syn*-**W13** (4.2:1 *syn:anti*) presumably via an S_N1 mechanism.

Cypermethrin Synthesis

In the textbook we reported the coupling of enantiomerically pure acid chloride **W14** with enantiomerically pure cyanohydrin prepared by kinetic resolution with an enzyme to give the right enantiomer of the right diastereoisomer **200** of cypermethrin.[4]

The preparation of the acid **W14** follows. Reduction of ketone **W15** with the ephedrine complex of $LiAlH_4$ gave the enantiomerically enriched alcohol **W16** (72% ee). Acylation with ketene dimer gave the acetoacetate **W17**. Reaction with the sulfonyl azide **W18** gave the diazoester and treatment with aqueous NaOH removed the now redundant acetyl group to give **W19**. Carbene formation with Cu(II) catalysis led to stereospecific insertion into the alkene from the face opposite the CCl_3 group **W20** and elimination of Cl and ester created the dichlorovinyl group in **W14**. Coupling with enantiomerically pure cyanohydrin and purification removes the minor isomer.[5]

An Asymmetric Synthesis of Salbutamol (GSK's Ventolin)

In chapter 7 of this workbook we described a synthesis of salbutamol **W22** in which the last two steps were reductions. **Problem 29.2**: How might one make this synthesis asymmetric?

Structures: **W21** (left) and **W22; salbutamol (GSK's Ventolin)** (right). Reagents: 1. LiAlH₄ 2. H₂, Pd/C

Answer 29.2: Since **W25** is achiral, the obvious way is to use an asymmetric reducing agent from chapter 24 or 26 or an enzyme to carry out a kinetic resolution on the alcohol before deprotection or even of **W26** itself.

Problem 29.3: Now suggest an alternative synthetic strategy using an oxynitrilase or some related enzyme.

Answer 29.3: The active enantiomer happens to be (R)-**W22** that could clearly be derived from the cyanohydrin **W23** by reduction to a primary amine and alkylation.

Structures: (R)-**W22** (left) and **W23** (right). Arrow labelled reduction / alkylation.

There are obviously many ways to do this and we shall describe just one published method.[6] It was necessary to protect the two alcohols and the cyclic acetal **W24** was ideal. In this synthesis, the enzyme (R)-oxynitrilase was used to make the cyanohydrin (R)-**W25** rather than in a kinetic resolution. Thus a 93% conversion was used to give 97% ee.

Structures: **W24** (left) and (R)-**W25** (right). Reagents: (R)-oxynitrilase, HCN, i-Pr₂O. 97% ee at 93% conversion.

Now how can one alkylate at nitrogen with a tertiary alkyl group? The Ritter reaction is the answer and a nitrile is perfect for that. So acylation followed by reaction with t-BuOH in strong acid gives the amide **W27** without loss of enantiomeric excess and reduction with LiAlH₄ reduces the amide to the amine and removes the acetate. Sadly they were now unable to remove the acetal protection. So if your method is different, it may lack this sting in the tail.

Structures: (R)-**W25** → (R)-**W26; 92% yield** → (R)-**W27; 62% yield**. Reagents: Ac₂O / pyridine; then t-BuOH, H₂SO₄, HOAc.

Lotrafiban

In the textbook chapter we described a synthesis of lotrafiban by the chiral pool strategy. The fragment in the frame in **W29** was made from aspartic acid. This synthesis was abandoned in favour of a kinetic resolution by an enzyme because a loss of stereochemical integrity occurred during the closure of the seven-membered ring **W29** from precursor **W28**. **Problem 29.4**: Suggest a mechanism for this reaction and an explanation for the partial racemisation.

Answer 29.4: The mechanism is aromatic nucleophilic substitution by the normal addition-elimination mechanism. The amine can add to the ring **W30** because of the ester carbonyl and the fluorine atom and the loss of fluoride completes the process.[7] Racemisation is presumably because of enolate formation in **W28** or **W29** catalysed either by the oxyanion in **W31** or by fluoride – a relatively strong base as it is not solvated by the polar aprotic solvent DMSO.

Asymmetric Synthesis with Baker's Yeast

At the start of the textbook chapter we revealed that baker's yeast reduction of hydroxyacetone **7** gives the useful diol **8**. In more detail, it has been used in an asymmetric synthesis of ofloxacin **W32**. We described a chiral pool synthesis[8] of this antibiotic in chapter 23.

Since we know that ofloxacin can be made (see the C–N and C–O disconnections on **W32a**) from **W33–35** by a series of nucleophilic substitutions, we simply have to convert enantiomerically pure **8** into **W35** – the only chiral starting material. When the enantioselectivity of the drug was being studied[9] both enantiomers of **W39** were needed. **Problem 29.5**: Which enantiomer of **W35** is easier to make from **8**? Suggest a synthesis of that enantiomer from **8**.

Answer 29.5: A little manipulation reveals that (*S*)-**W35** unfortunately has the same absolute chirality **W37** as **8** so a simple nucleophilic substitution will not give (*S*)-**W35**: we would need nucleophilic substitution with *retention*. That enantiomer was the one made from alanine (it is *S*-alaninol) in chapter 23 and turns out to be the active one.

However, the other enantiomer (*R*)-**W35** can easily be made from **8**. You might have considered protecting the primary alcohol with, say, a bulky silyl group, converting the secondary alcohol into a leaving group, say a sulfonate ester, displacing with a suitable nitrogen nucleophile, say azide, with inversion, and converting into the required amino group (by reduction if you used azide). In fact, both enantiomers of **W35** are available commercially.

Palladium-Allyl Cation Chemistry

In the textbook we described a desymmetrisation of the bis-acetate **37** made by the Pd-catalysed attack of AcOH on the monoepoxide (±)-**35** to give the racemic monoester **36**. **Problem 29.6**: Explain the stereo- and regioselectivity of the formation of **36**.

Answer 29.6: Palladium forms a π-complex with the alkene that reacts by expulsion of the epoxide **W38** to give a palladium-allyl cation complex **W39**. The palladium adds from the bottom face to displace the epoxide with inversion.[10] The anion formed is basic enough to deprotonate acetic acid and acetate then adds to the palladium-allyl cation again with inversion **W40** to give the π-complex of the product **W41**. Attack at the other end of the allyl cation would have to be uncomfortably close to the OH group.

Polymer-Supported Enzymes

We reported in the textbook how racemic products from polymer supported reagents **43** could be used in the synthesis of a key intermediate (+)-**46** for the bryostatins after kinetic resolution with an enzyme. **Problem 29.7**: Suggest a mechanism for the formation of **44**.

Answer 29.7: Aliphatic nitro-compounds are dehydrated by PhNCO with the formation of nitrile oxides that do 1,3-dipolar cycloadditions **W42** on alkenes. The regioselectivity may look strange as we should expect the oxyanion end of the nitrile oxide to have the largest HOMO coefficient and to attack the other end of the enone. However, nitrile oxides are 1,3-dipoles and it is the HOMO of the enone that attacks the LUMO of the nitrile oxide.

The dehydration step produces CO_2 and $PhNH_2$ as by-products. We must assume that PhNCO attacks one of the oxygen atoms of the nitro group **W43** and that two proton transfers end in elimination **W45**.

The second chiral centre is introduced by L-selectride reduction of the racemic ketone **44**. The product was a racemic mixture of exclusively *syn*-diols **45**. **Problem 29.8**: Suggest an explanation for the high diastereoselectivity.

Answer 29.8: Probably the best explanation is a Felkin style conformation **44a** with the bulky reducing agent approaching alongside H and opposite the electronegative oxygen atom (chapter 21). The dotted line shows that O and CH_2 are actually joined in a ring and the dotted arrow shows the line of approach of the hydride donor.[11]

The Stereochemistry of Inositol Derivatives

We described in the textbook how the acetonide **70** has been used[12] to make both enantiomers of pinitol **73**. You will have seen that pinitol is a derivative (methyl ether) of one stereoisomer of inositol **W52**. There are obviously many stereoisomers of inositol. **Problem 29.9**: If the methyl group were removed from pinitol, would that isomer still be chiral?

Answer 29.9: Isomer **W47** is what we get and it has no plane or centre of symmetry so is still chiral as would be any of its monomethyl ethers. Isomers **W48** and **W49** also have three OH groups up and three down. Isomer **W50** has four up and two down.

Problem 29.10: Is any of these chiral?

Answer 29.10: No. Each has a plane of symmetry. Amazingly, **W47** is the only inositol isomer that is chiral. Is Nature just showing off making pinitol?

Enzymatic Baeyer-Villiger Rearrangements

In the textbook we gave the almost incredible example of the Baeyer-Villiger rearrangement of racemic **111** where one enantiomer gives **110** and the other **112**. The latest news on this reaction is that recombinant whole cells of *Escherichia coli* expressing cyclohexanone mono-oxygenase are even more efficient.[13]

With careful engineering the recombinant whole cells can be used[14] on a 210 g scale in a flow reactor industrial process giving 174 g of mixed lactones. A portion (24 g) of the two products were separated by chromatography and crystallisation to give 5.3 g of **110** and 8.3 g of **112**.

Asymmetric Synthesis of Pyrrolidines by Aldolases

We described an aldolase controlled synthesis of the azide **146** in the textbook and said that reduction of the azide gave the trihydroxy pyrrolidine **147** with excellent control over a third chiral centre, less than 10% of the epimer being formed. **Problem 29.11**: Explain the formation of **147**.

Answer 29.11: Reduction of the azide gives the amine **W51** that cyclises to the imine **W52**. Hydrogenation occurs from the top face (as drawn) because this gives the most stable all *trans* (and all pseudo-equatorial) product.[15]

Asymmetric Synthesis of Epothilone by Aldolases

Both halves of epothilone were made in the textbook from aldolase-derived compounds. One half was made from **172** by formation of the protected keto-aldehyde **183** followed by successive stereoselective Wittig-Horner and Wittig reactions to put in the two alkenes with control over stereochemistry. The Wittig-Horner is under thermodynamic control giving the trisubstituted *E*-alkene **185** while the Wittig is under kinetic control giving the *Z*-iodoalkene **175**; R = Ac (chapter 15). **Problem 29.12**: Draw the structures of the intermediate from **172** to **175** explaining the roles of the various reagents. You do not need to draw a mechanism for the Swern oxidation.

Answer 29.12: Acetylation of both OHs and deacetylation of the anomeric ester gave selectively protected hemiacetal **W54** and hence the dithian **182** by reaction of the open chain aldehyde with $HS(CH_2)_3SH$. Swern oxidation selectively gives the aldehyde without affecting the sulfur atoms. The lithium derivative **W55** of the phosphine oxide **184** gives the *E* alkene **185** selectively via the reversible formation of a mixture of diastereoisomers of adduct **W56**. This is the thermodynamically controlled version of the kinetically controlled Horner-Wittig reaction we met in chapter 15. The dithian is removed with Hg(II) catalysis and the slightly unusual Wittig process completes the synthesis[16] of **175**.

The final lactonisation uses the famous Yamaguchi process.[17] The hydroxy acid, represented by **W58**, is converted to a mixed anhydride **W60** with the acid chloride **W59**. Normal concentrations can be used for this stage: 1 mmol in 10 ml THF. The formation of a macrolactone demands dilution so 500 ml toluene are added and **W60** is added to a refluxing solution of DMAP in 100 ml toluene. Yields are usually good (85% in this case) but are in any case better than other methods.

Asymmetric Synthesis of the Schering-Plough Anti-Fungal Agent SCH51048

In the textbook we reported an enzyme catalysed acylation of the prochiral diol **213** was cleanly monoacetylated by Novozyme (Novo Nordisk) SP 435 (98% yield, 98% ee) and mentioned that iodo-etherification gave the required diastereoselectivity **214** of a key intermediate.[18] **Problem 29.13**: Suggest a mechanism for the iodoetherification and explain the stereochemistry.

Answer 29.13: In the same way as an iodolactonisation, iodine may attack either face of the alkene but the cyclisation will predominate **W61** that gives the most stable product. The iodolactone **215** either has the largest group (Ar) pseudo-equatorial **215a** or else two groups pseudoequatorial **215b**. These are just conformers of the same diastereoisomer.

Problem 29.14: An earlier attempt was the base-catalysed cyclisation of the bis-tosylate **208**. With a variety of bases, the best result was 60:40 in favour of the unwanted diastereoisomer *anti*-**209**. Why do you suppose that the selectivity (84:16 the right way, in favour of **215**) of the cyclisation of **214** is so much better than that of **208**?

Answer 29.14: The reactions are completely different. The cyclisation of **208** is an S_N2 reaction under kinetic control. The basic mechanism **W62** conceals the fact that the two CH_2OTs groups are diastereotopic: cyclisation to one **W62a** gives *syn*-**209** and to the other **W62b** gives *anti*-**209**. There is very little difference between these two; the chiral centre already present is too far away to have much influence.[19]

W62   W62a gives *syn*-209   W62b gives *anti*-209

A Dramatic Enzymatic Desymmetrisation

The amlodipine type of dihydropyridine drugs are chiral but only just so. In the textbook we pointed out that the symmetrical compound **234** can be easily made by the Hantzsch pyridine synthesis and described the remarkable desymmetrisation using the Amano company's own lipase. **Problem 29.15**: How can the symmetrical compound be made?

HO$_2$C ... CO$_2$Pr-*i*   **Amano lipase AH**   *i*-PrO$_2$C ... CO$_2$Pr-*i*   **Amano lipase AH**   *i*-PrO$_2$C ... CO$_2$H

i-Pr$_2$O saturated with water   cyclohexane saturated with water

Me   Me   Me   Me   Me   Me
(*S*)-233   (±)-234   (*R*)-233

Answer 29.15: The Hantzsch pyridine synthesis allows the one pot assembly of symmetrical compounds like **234** from two molecules of an acetoacetate ester **W64**, an aromatic aldehyde **W66** and ammonia. This tandem process (chapter 36) is very efficient.[20]

i-PrO$_2$C ... CO$_2$*i*-Pr   *i*-PrO$_2$C ... CO$_2$*i*-Pr

2x C–N
enamines

Me   N   Me   Me   O O   Me
H
(±)-234   W63   1,5-diCO

i-PrO$_2$C ... CO$_2$*i*-Pr   aldol   *i*-PrO$_2$C   CHO   CO$_2$*i*-Pr
W66

Me O   O Me   Me O   NH$_3$ O Me
W64   W65   W64   W64

References

1. J. Liu and C.-H. Wong, *Angew. Chem., Int. Ed.*, 2002, **41**, 1404.

2. J. Saunders, *Top Drugs,* page 23.

3. T. Hashiyama, H. Inoue, M. Konda and M. Takeda, *J. Chem. Soc., Perkin Trans. 1,* 1984, 1725; T. Hashiyama, H. Inoue, M. Takeda, K. Aoe and K. Kotera, *J. Chem. Soc., Perkin Trans. 1,* 1985, 421.

4. J. Roos, U. Stelzer and F. Effenberger, *Tetrahedron: Asymmetry,* 1998, **9**, 1043.

5. C. E. Hatch, J. S. Baum, T. Takashima and K. Kondo, *J. Org. Chem.,* 1980, **45**, 3281.

6. F. Effenberger and J. Jäger, *J. Org. Chem.,* 1997, **62**, 3867.

7. T. C. Walsgrove, L. Powell and A. Wells, *Org. Process Res. Dev.,* 2002, **6**, 488.

8. H. Egawa, T. Miyamoto and J. Matsumoto, *Chem. Pharm. Bull.,* 1986, **34**, 4098.

9. L. A. Mitscher, P. N. Sharma, D. T. W. Chu, L. L. Shen and A. G. Pernet, *J. Med. Chem.,* 1987, **30**, 2283.

10. D. R. Deardorff and D. C. Myles, *Org. Synth.,* 1989, **67**, 114.

11. J. A. López-Pelegrín, P. Wentworth, F. Sieber, W. A. Metz and K. M. Janda, *J. Org. Chem.,* 2000, **65**, 8527.

12. T. Hudlicky, J. D. Price, F. Rulin and T. Tsunoda, *J. Am. Chem. Soc.,* 1990, **112**, 9440.

13. P. A. Bird, C. A. Sharp and J. M. Woodley, *Org. Process. Res. Dev.,* 2002, **6**, 569.

14. S. D. Doig, P. J. Avenell, P. A. Bird, P. Gallati, K. S. Lander, G. J. Lye, R. Wohlgemuth and J. M. Woodley, *Biotechnol. Prog.,* 2002, **18**, 1039.

15. R. R. Hung, J. A. Straub and G. M. Whitesides, *J. Org. Chem.,* 1991, **56**, 3849; K. K.-C. Liu, T. Kajimoto, L. Chen, Z. Zhong, Y. Ichikawa and C.-H. Wong, *J. Org. Chem.,* 1991, **56**, 6280; *J. Am. Chem. Soc.,* 1991, **113**, 6187.

16. J. Liu and C.-H. Wong, *Angew. Chem., Int. Ed.,* 2002, **41**, 1404.

17. J. Inanaga, K. Hirata, H. Saeki, T. Katsuki and M. Yamaguchi, *Bull. Chem. Soc. Jpn,* 1979, **52**, 1989.

18. B. Morgan, D. R. Dodds, A. Zaks, D. R. Andrews and R. Klesse, *J. Org. Chem.,* 1997, **62**, 7736.

19. A. K. Saksena, V. M. Girijavallabhan, R. G. Lovey, R. E. Pike, H. Wang, A. K. Ganguly, B. Morgan, A. Zaks and M. S. Puar *Tetrahedron Lett.,* 1995, **36**, 1787.

20. Y. Hirose, K. Kariya, I. Sasaki, Y. Kurono, H. Ebiike and K. Achiwa, *Tetrahedron Lett.,* 1992, **33**, 7157.

30

New Chiral Centres from Old: Enantiomerically Pure Compounds and Sophisticated Syntheses

Both in chapters 30 and 27 in the textbook we described a clever way to remove the chiral auxiliary after asymmetric Birch reduction. Reduction of the azide **6** to the amine with Ph₃P leads to the imine **7** by spontaneous ring closure. **Problem 30.1**: Draw mechanisms for **5** going to **6** going to **7**, explaining the stereochemistry.

Answer 30.1: Reversible formation of the iodonium ion allows cyclisation only if the iodine is on the lower face **W1** so that *trans*-diaxial opening **W1** is possible. The resulting iminium ion **W2** is easily hydrolysed by water. Ph₃P reduces the azide to a primary amine that cyclises onto the ketone to give **7**.

Acylation with **8** moves the imine to an enamine **9** and opening the lactone with the lithium alkoxide of benzyl alcohol gives the epoxide **10** ready for the key cyclisation. **Problem 30.2**: Show how the lactone **9** gives the epoxide **10** again explaining the stereochemistry.

Answer 30.2: The oxyanion opens the lactone **W3** to give the anion of the *trans*-diaxial iodo-hydrin that closes **W4** to the epoxide[1] **10**.

Induction in the Synthesis of 8-Phenylmenthol

The only new chiral centre introduced in the asymmetric synthesis of the chiral auxiliary 8-phenylmenthol **21** occurs during a thermodynamically controlled reduction with *i*-PrONa in *i*-PrOH. This reduction is totally stereospecific giving the equatorial alcohol in both cases. An 87:13 mixture of *trans:cis* **24** gives an 87:13 mixture of **21** and **25**. The two alcohols are easily separated as their chloroacetyl esters. **Problem 30.3**: Suggest a mechanism for the reduction that explains why it is thermodynamically controlled.

Answer 30.3: Thermodynamic control suggests a reversible reduction of the ketone, quite unlike, say, $LiAlH_4$. The *i*-PrO anion can donate a hydride in a similar way, however, **W5** (using cyclohexanone in place of **24**) forming acetone and the new alkoxide that can reverse the reaction. Of the compounds present, acetone is the most volatile (boiling point of acetone 56 °C, boiling point of *i*-PrOH 82 °C) and so it distils out driving the reaction across.[2]

Diastereoselective Alkylation of 2-Hydroxy Enolates

We described the diastereoselective alkylation of enolate **66** of the hydroxy-lactone **65** in the textbook. **Problem 30.4**: Suggest how **65** could be made from **64**.

Answer 30.4: The problem is to differentiate the two acids. One simple way is to reduce both acids to the diol **W7**, form the acetonide **W8** as acetone greatly prefers five-membered cyclic acetals, and oxidise the remaining alcohol to the acid **W9** with Jones reagent. Releasing the diol leads to spontaneous cyclisation.[3]

An Unusual Brominating Agent

In the textbook we described a cyclisation using the 'convenient' brominating agent **74** to create three new chiral centres in one operation **W75**.

This brominating agent is made from phenol by exhaustive bromination. **Problem 30.5**: Draw the mechanism for the introduction of the fourth bromine atom. How does **74** act as a Br$^+$ donor?

Answer 30.5: The fourth bromination occurs not at a *meta* position but for a second time at the *para* position **W12** even though this means disrupting the aromaticity. In reaction with a nucleophile it is one of the *para* bromine atoms that is removed to give the anion **W15** of **W11** and this can be recycled to **74** if required.

Folded Molecules in Diastereoselective Reactions

In the textbook we described a bicyclic system **82** used in the synthesis of astericanolide. It starts with a vinyl-lithium reagent derived from **79** (chapter 16): reaction with a single enantiomer of menthyl *p*-tolylsulfinate gave the sulfoxide **80**. Michael addition of the propargyl alcohol **81** was not very efficient but could be persuaded to give one diastereoisomer of the adduct **82** in 38% yield. Michael addition of the alcohol has been followed by Michael addition of the resulting anion to the acetylenic ester. **Problem 30.6**: Draw out the mechanism in detail.

Answer 30.6: Conjugate addition of the alcohol to the bottom face of **80** to keep out of the way of the tolyl group on sulfur **W16** puts in one new centre.[4] The intermediate anion has no stereochemistry but adds in a second conjugate addition **W17** to the acetylenic ester also from the bottom face because the electrophile is tethered or, if you prefer, a *cis*-fused 5/5 ring system is much more stable. Note that the p-orbitals of the enolate in **W18** are orthogonal to the alkene and in the plane of the ring – hence the high *E*:*Z* selectivity.

Stereochemical Transmission by Cyclic Transition States: Sigmatropic Rearrangements

The sequence used in the textbook to make **98** from proline looks involved but in fact only three intermediates are isolated: **92**, **96**, and one between **97** and **98**. **Problem 30.7**: Suggest how proline can be converted into **91** and how **97** can be converted into **98**.

Answer 30.7: One suitable way starts with the ethyl ester of Boc protected proline **W19**, reduction to the aldehyde **W20** and, without isolation, immediate conversion by a HWE reaction (chapter 15) into the unsaturated ester **W21**. It proved difficult to reduce the ester to the alcohol **91** without conjugate reduction but a combination of DIBAL and the Lewis acid BF_3 worked very well.[5]

In converting **97** to **98** you must cleave the alkene, reduce the amide and remove the benzyl group. Mulzer did this by ozonolysis, reduction to the alcohol **W22**, reduction of the amide to the amine (which was not purified as it existed as the borane complex **W23**) and hydrogenation.[6] Other reasonable methods are obviously acceptable providing they do not put either chiral centre at risk.

Between these two stages an Ireland-Claisen rearrangement on the *E*-enolate of **92** creates the new chiral centres in **95**. The mechanism is a standard [3,3]-sigmatropic rearrangement **94** and occurs suprafacially across the backbone of the molecule through a chair-like transition state. We hope you agree with the relative stereochemistry of the new centres. **Problem 30.8**: Explain the stereochemical outcome in detail.

Answer 30.8: Mulzer makes a list[6] of important factors in this reaction:

(a) The deprotonation of **92** leads to a chelated enolate **93**.

(b) A chair-like transition state is adopted, as normal for acyclic substrates.

(c) With respect to the diastereofacial induction, *Houk's model* is applicable, resulting in a minimum of inside crowding.

(d) The *N*-Boc group exerts a strong *anti*-peri-planar effect.

He gives two diagrams for the rearrangement **W24** and **W25** which we (roughly) reproduce and hope your attempts resembled these. Note that **W25** shows what he means by point (d).

An Asymmetric Synthesis of Hastanecine

The synthesis of hastanecine we featured in the textbook began with a reaction between the anhydride **100** and the amine **101**. **Problem 30.9**: Suggest a synthesis for the amine **101**.

Answer 30.9: This is not an asymmetric synthesis as the amine **101** is achiral. There are obviously many solutions using nitriles or amides but Hart chose[7] to use Claisen and Curtius rearrangements.

Problem 30.10: What is **W30** and what reactions are going on between **W29** and **101**?

Answer 30.10: Azide ion reacts with **W29** to give the acid azide **W31** that does the Curtius rearrangement to give the isocyanate **W30**. Capture of the isocyanate with *t*-BuOH gives the carbamate **W32** which is hydrolysed and decarboxylated to give **101**. We have discussed these last two steps before.

Felkin-Anh Control

We start with a simple but important example. The HIV protease inhibitor viracept **W33** disconnects simply into three pieces. The central open chain portion **W35** has two adjacent chiral centres neither of which will be disturbed during amide coupling with **W34** or S_N2 reaction with **W35**.

In the original synthesis of **W33** by Aguron Pharmaceuticals[8] this key intermediate **133** was made from serine **131** *via* the simple ketone **132**. Reduction of **132** with $NaBH_4$ gave mostly the stereochemistry required **133**. The Felkin conformation is **134** with the large (and electronegative) NHCbz group orthogonal to the carbonyl group and nucleophilic hydride approaching at the Bürgi-Dunitz angle alongside the small H atom. In the long run, this synthesis is not the best, but it makes the point.[9]

Asymmetric Synthesis of an Anti-Cancer Drug

The search for a manufacturing method for Janssen's anti-cancer drug **W37** as one enantiomer of one diastereoisomer in high yield led to the racemic amino-ketone **W40** easily made from **W38** via a Friedel-Crafts acylation and a substitution with dimethyl amine.[10]

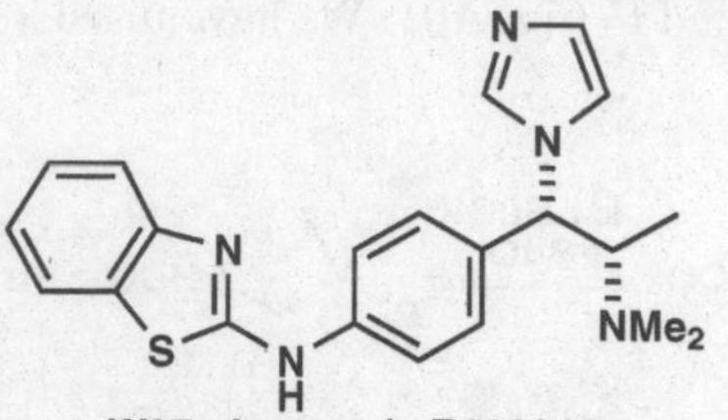

W37; Janssen's R116010

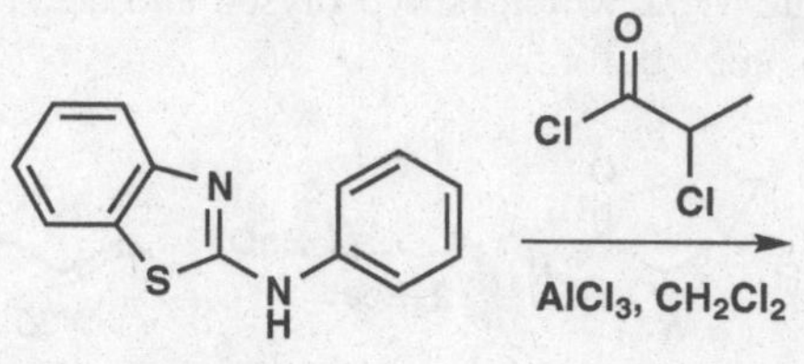

W38

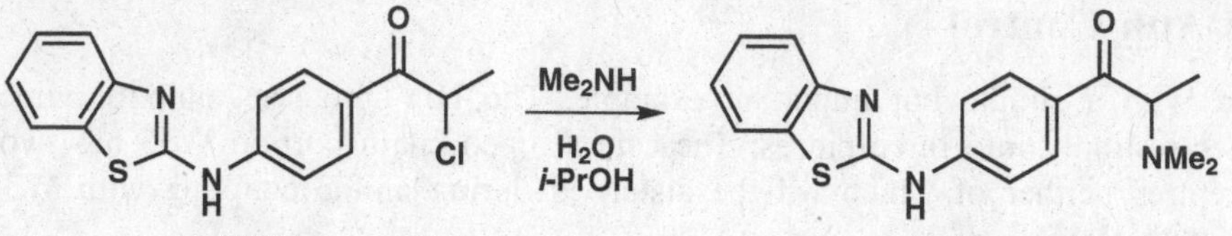

(±)-W39

(±)-W40

Dynamic kinetic resolution (chapter 28) of **W40** with (+)-di-*p*-toluoyltartaric acid equilibrated the easily enolisable chiral centre and precipitated a 90% yield of the salt of the (*S*)-amine having 80% ee. Reduction with $NaBH_4$ in *i*-PrOH was very diastereoselective giving 99:1 *syn:anti*-**124**.

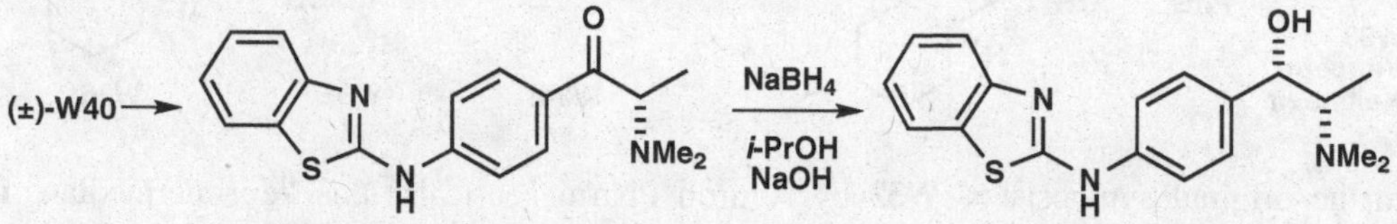

(*S*)-W40 as di-*p*-toluoyl-tartrate

(*S,S*)-W41; 80% yield, 80% ee

This is Felkin-Anh control: the NMe_2 group sits at right angles to the carbonyl group and nucleophilic hydride approaches at the Bürgi-Dunitz angle alongside the H atom **W42**. The synthesis is completed by displacement of OH with an excess of imidazole using CDI (carbonyl-di-imidazole) **W43** to activate the alcohol. Notice that this displacement happens with retention of configuration. **Problem 30.11**: Explain this stereochemical outcome.

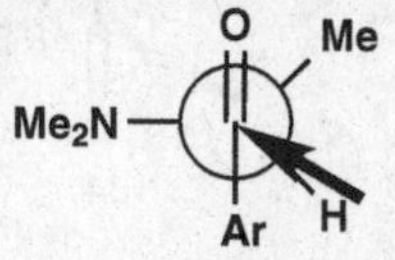

W42

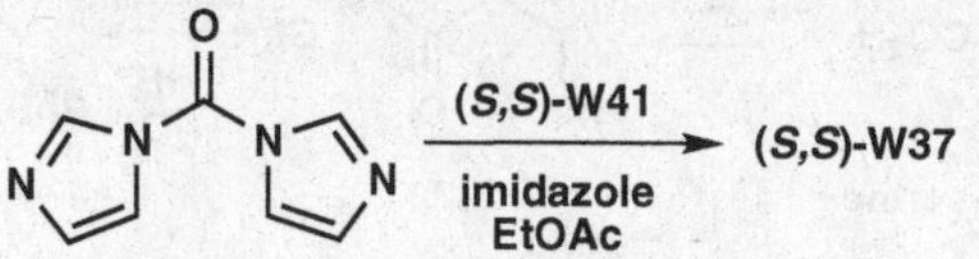

W43

Answer 30.11: The imidazole anion is a good leaving group as it is aromatic and stabilised by two nitrogen atoms. Acylation of **W41** gives an intermediate that reacts by N participation **W44** to give CO_2 and an aziridinium ion that is attacked by the imidazole anion **W45** to give **W37** with retention as a result of two inversions.

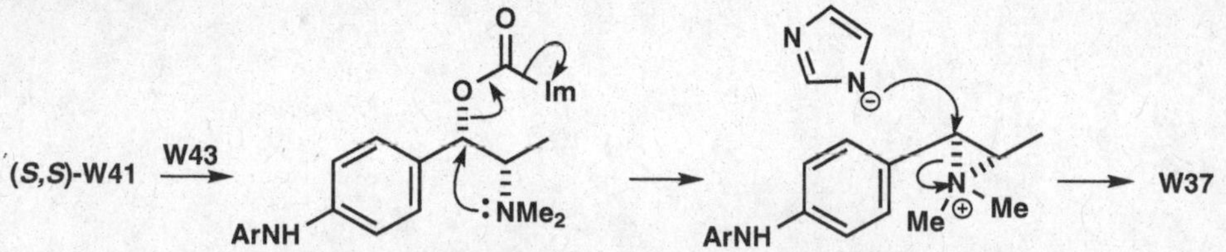

(*S,S*)-W41

W44; Im = imidazole

W45

W37

Houk Control

In the textbook we described the alkylation of an aspartic acid derivative to give either pyrrolidine dicarboxylic acid **141** or **142** depending on whether Houk or chelation control was used.

Now we shall describe some related chemistry that can be used to make either *syn*- or *anti*-piperidine dicarboxylic acid **W46**. Aspartic acid is easily protected as **W47** and the enolate made, at the carbon not having the amino group, with metal hexamethyldisilazides [$MN(SiMe_3)_2$ where M is Li or K]. Reaction with allyl halides gives a mixture of *anti*-**W48** and *syn*-**W48**. The lithium enolate gives a 3:2 ratio of *anti*-**W48**:*syn*-**W48** but the potassium enolate gives a 6:1 ratio.

The explanation is that the (E)-lithium enolate **W48** can adopt a Houk conformation **W48a** with little difference between top and bottom faces of the enolate but the (Z) potassium enolate is chelated **W49**. **Problem 30.12**: How might *anti*-**W48** be converted into *anti*-**W46**?

Answer 30.12: The published route[11] uses hydroboration and oxidation to make the aldehyde **W51**. Hydrogenation removes the benzyl groups and completes reductive amination to give **W52** easily hydrolysed to *anti*-**W46**.

Asymmetric Synthesis of Piperidine 3,4-Dicarboxylic Acids

A related reductive amination strategy was used to make piperidine 3,4-dicarboxylic acid **W55**. This time an allyl group was oxidatively cleaved to give the aldehyde **W54** for reductive amination. The protecting groups are different but all except the *t*-butyl ester are again removed by catalytic reduction. How do we make **W53**?

With no suitable natural product from the chiral pool, the substrate-based strategy (chapter 27) of alkylation of an Evans' enolate was used to make the first chiral centre but the second was inserted by open-chain enolate control after the chiral auxiliary had been removed. Alkylation of the lithium enolate of the protected acyl oxazolidinone **W56** with a bromoacetate and removal of the auxiliary gave the acid **W58** as a single enantiomer.[12]

Enolate formation (LDA) from **W59** and alkylation with allyl iodide gave a single diastereoisomer of **W53**. The scheme shows that the alkylation has occurred in a *syn* fashion and suggests that the lithium enolate is chelated this time. Earlier work on such alkylations shows that a chelated lithium enolate gives the best explanation.[13]

Reactions of Allyl Silanes

The typical reaction of an allyl silane with an electrophile (textbook chapters 12 and 30) is at the remote atom of the alkene with loss of the silyl group. This transfers, but does not create, chirality. The chirality is transferred during the reaction of the allyl silane to the other end of the allylic system. We can now explain how chirality is created either by asymmetric Pd-catalysed reactions or by enzymes. The Pd-catalysed reactions use the asymmetric addition of a racemic Grignard reagent to vinyl bromides catalysed by a Pd(II) complex **W60** held together by a ferrocene.[14] Racemisation of the Grignard is faster than the coupling so this is a dynamic catalytic kinetic resolution. There is no loss of stereochemical (*E* or *Z*) integrity of the vinyl group but *E*-alkenes give much higher enantiomeric excess values than *Z*-alkenes.

The enzymatic method goes like this.[15] Platinum-catalysed hydrosilylation of the alkene **W63** gives only *E*-**W64** as the H and $SiMe_2Ph$ groups are added *cis* (chapter 16). Note that the catalyst **W65** is achiral and the product *E*-**W64** is racemic.

The allylic alcohol **W64** is first resolved by the lipase Amano AK using vinyl acetate as the esterifying agent. You will recall from chapter 29 that this drives the equilibrium forward as the released vinyl alcohol immediately becomes acetaldehyde and is no longer a substrate for the enzyme. This kinetic resolution is very efficient: nearly 50% of each compound can be isolated by chromatography and the ester **W66** can easily be converted without loss of enantiomeric excess to (*R*)-**W64** by reduction with $LiAlH_4$.

Now each enantiomer of the vinyl silane **W64** must be converted into an allyl silane. Treatment of either enantiomer with methyl *ortho*-acetate and catalytic propionic acid does the job giving (*S*)- or (*R*)-**W67** in good yield. Both **W64** and **W66** are formed in >95% ee by these processes. **Problem 30.13**: Explain the formation of **W67** and its stereochemistry.

Answer 30.13: This is an aliphatic Claisen rearrangement (chapter 19). Acid-catalysed acetal exchange gives a new *ortho*-ester that undergoes a suprafacial (i.e. the migrating group remains on the same surface of the allylic part of the molecule) [3,3]-sigmatropic rearrangement **W69** giving **W67** drawn in a more helpful way. You are to be commended if you also drew a chair transition state for this process.

More Remote Asymmetric Induction

In the textbook we described two consecutive hydroxyl-directed epoxidations and THF ring closures, the first being the formation of **231**. We said: 'the hydroxyl group now directs the epoxidation of the nearer alkene using a vanadium directed delivery of *t*-BuOOH.' **Problem 30.14**: Draw the structure of the epoxide and show how it is opened by the OH group to give **231**.

228   **231; 75% yield, 8:1 diasts**

Answer 30.14: The drawings here come from Kishi's explanation of the stereochemistry. In order for the hydroxyl group to direct the epoxidation, the alkene must be close to the OH group in a conformation like **228a** and the epoxide is **W70**. This must be rotated about bond 'a' to give a conformation that can cyclise **W71** to give a redrawn **231a**. The second epoxidation and cyclisation go the same way.[16]

228a   **W70; 8:1**   **W71**   **231a**

A [2,3]-Sigmatropic Rearrangement

We described [2,3]-rearrangements in the textbook. Such a reaction was used in Kallmerten's synthesis of rapamycin. The starting material **W72** was prepared from D-glucose (chiral pool strategy) and rearranged to **W73** with BuLi. **Problem 30.15**: Describe this in detail.

W72   **W73; 3:1 *E:Z***

Answer 30.15: BuLi removes the proton from next to the oxazoline and a [2,3]-sigmatropic (Wittig) rearrangement **W74** on the anion gives the more stable oxyanion **W75**. The oxazoline unit is transferred suprafacially across the top face of the allylic system but the new centre next to the oxazoline is less obvious. Avoiding the temptation to draw a thick bond between the two new centres, it is better to mark the hydrogens (chapter 20). Diagram **W74a** should make all clear.[17]

W74   **W75**   **W74a**

References

1. A. G. Schultz, M. A. Holoboski and M. S. Smyth, *J. Am. Chem. Soc.*, 1996, **118**, 6210.
2. O. Ort, *Org. Synth.*, 1987, **65**, 203.
3. H.-M. Shieh and G. D. Prestwich, *J. Org. Chem.*, 1981, **46**, 4319; A. R. Chamberlin and M. Dezube, *Tetrahedron Lett.*, 1982, **23**, 3055; H.-M. Shieh and G. D. Prestwich, *Tetrahedron Lett.*, 1982, **23**, 4643.
4. L. A. Paquette, J. Tae, M. P. Arrington and A. H. Sadoun, *J. Am. Chem. Soc.*, 2000, **122**, 2742.
5. T. Moriwake, S. Hamano, D. Miki, S. Saito and S. Torii, *Chem. Lett.*, 1986, 815.
6. J. Mulzer and M. Shanyoor, *Tetrahedron Lett.*, 1993, **34**, 6545.
7. D. J. Hart and T.-K. Yang, *J. Chem. Soc., Chem. Commun.*, 1983, 135; *Tetrahedron Lett.*, 1982, 2761.
8. S. W. Kaldor, V. J. Kalish, J. F. Davies, B. V. Shetty, J. E. Fritz, K. Appelt, J. A. Burgess, K. M. Campanale, N. Y. Chirgadze, D. K. Clawson, B. A. Dressman, S. D. Hatch, D. A. Khalil, M. B. Kosa, P. P. Lubbehusen, M. A. Muesing, A. K. Patick, S. H. Reich, K. S. Su and J. H. Tatlock, *J. Med. Chem.*, 1997, **40**, 3979.
9. M. Ikunaka, *Chem. Eur. J.*, 2003, **9**, 379.
10. W. Aelterman, Y. Lang, B. Willemsens, I. Vervest, S. Leurs and F. De Knaep, *Org. Proc. Res. Dev.*, 2001, **5**, 467.
11. C.-B. Xue, X. He, J. Roderick, R. L. Corbett and C. P. Decicco, *J. Org. Chem.*, 2002, **67**, 865.
12. C.-B. Xue, M. E. Voss, D. J. Nelson, J. J.-W. Duan, R. J. Cherney, I. C. Jacobson, X. He, J. Roderick, L. Chen, R. L. Corbett, L. Wang, D. T. Meyer, K. Kennedy, W. F. DeGrado, K. D. Hardman, C. A. Teleha, B. D. Jaffee, R.-Q. Liu, R. A. Copeland, M. B. Covington, D. D. Christ, J. M. Trzaskos, R. C. Newton, R. L. Magolda, R. R. Wexler and C. P. Decicco, *J. Med. Chem.*, 2001, **44**, 2636.
13. R. P. Beckett, M. J. Crimmin, M. H. Davis and Z. Spavold, *Synlett*, 1993, 137.
14. T. Hayashi, M. Konishi, H. Ito and M. Kumada, *J. Am. Chem. Soc.*, 1982, **104**, 4962.
15. R. T. Beresis, J. S. Solomon, M. G. Yang, N. F. Jain and J. S. Panek, *Org. Synth.*, 1998, **75**, 78.
16. T. Fukuyama, B. Vranesic, D. P. Negri and Y. Kishi, *Tetrahedron Lett.*, 1978, 2741.
17. N. Sin and J. Kallmerten, *Tetrahedron Lett.*, 1993, **34**, 753.

31

Strategy of Asymmetric Synthesis

We have tried in this chapter to set a few problems that require decisions based on strategy as well as giving further examples and explanations of material from the textbook. It is not a simple matter to devise genuine problems on strategy that can be solved reasonably easily.

Problem 31.1: Suggest asymmetric syntheses of lineatin **35**, filifolone **36**, or raikovenal **37** using methods from part III of the chapter if you wish.

35; (+)-lineatin   36; (−)-filifolone   37; raikovenal

Answer 31.1: Lineatin: The main thing to notice about lineatin **35a** is that it is an acetal and therefore we should start with a 1,1-diX disconnection revealing the hydroxyaldehyde **W1**. To get back to an intermediate from the grandisol synthesis we need an oxidation change to the lactone **W2** and imagine making that by a Baeyer-Villiger rearrangement on the cyclopentenone **W3**.

35a $\xrightarrow[\text{acetal}]{\text{1,1-diX}}$ W1 $\xrightarrow[\text{reduction}]{\text{FGI}}$ W2 $\xrightarrow{\text{B–V}}$ W3

There are a multiplicity of possible ways to do all this but we shall give the synthesis from the review by chemists from Bologna.[1] The chemistry used in the chapter for grandisol can be adapted to make the key intermediate for lineatin and filifolone. This synthesis uses the same intermediate **W5** as used in the textbook for grandisol. Racemic **W5** can be reduced to **W6** and then resolved with camphanic acid chloride **66**.

Workbook for Organic Synthesis: Strategy and Control   Paul Wyatt and Stuart Warren
© 2008 John Wiley & Sons, Ltd

W4 → (±)-W5; 82% yield → (R)-W6 | 66; (1S)-(−)-camphanic acid chloride

Protection with t-BuMe$_2$SiCl, hydroboration from the *exo*-face, oxidation and methylation give the bicyclic ketone **W9**.

(R)-W6 → W7; 70% yield → W8; 78% yield → W9; 82% yield

Desilylation and a Baeyer-Villiger rearrangement with the more substituted centre migrating gives lactone **W2**. Reduction with DIBAL gives the masked aldehyde **W10** that rearranges in acid into the acetal (+)-lineatin **35** with loss of water.[2] The stereochemistry of the acetal centre is fixed by the '*endo*' nature of the OH group that can reach only the underside of the molecule as drawn.

W3; 83% yield → W2; 73% yield → W10

Filifolone: A small change in the ketene-based synthesis of **W5** gives filifolone directly. Compound **W11** lacks one methyl group compared with **W5** and $2 + 2$ cycloaddition to the ketene derived from it gives the ketone **W12**. Double methylation gives racemic filifolone.[3] If a single enantiomer is required, resolution of the corresponding alcohol **W13** with **66** would answer.

W11 → W12; 82% → (±)-36; 72% yield → W13

Raikovenal: The key intermediate, ketone **W14**, can be made in good yield by two routes from **W12** differing only in the order of events.[4]

W14; 75% yield ← 36 72% yield ← W12 → W14; 95% yield → W14 75% yield

The side chain was extended by a Peterson reaction (chapter 15) using the lithium derivative of **W15**. The mixture of enol ethers **W16** was hydrolysed to a mixture of aldehydes (*exo:endo* 4:1) **W17**.

A Horner-Wadsworth-Emmons olefination (chapter 15) added the remaining carbon atoms **W19** but the *E:Z* ratio was an unexpectedly disappointing 2:3. Reduction to the diol **W20** at last allowed separation of the pure *E-exo*-diol that could be converted into *E*-raikovenal **37** by oxidation of the allylic alcohol with MnO_2. This route is not as bad as it sounds as the other isomers of **W20** can be recycled to **W17** by oxidative cleavage of the alkene.

Problem 31.2: Discussing the synthesis[5] of the collagenase inhibitor Trocade in the textbook we said: 'this method depends on the availability of optically pure hydroxyacids.' Suggest how enantiomerically pure hydroxy acid **82** might be made.

82; 2-(*R*)-hydroxyacid

Answer 31.2: This problem was to give you a chance to assess the merits of the various strategies we have unfolded in the last section of the textbook. There are many answers, such as:

(1) *Resolution* (chapter 22): This compound is ideal for resolution as it has a carboxylic acid and some salts with enantiomerically pure amines will be separable by crystallisation.

(2) *Asymmetric reagents* (chapter 24): Asymmetric reduction of the ketoacid **W21** with one of the reagents such as BINAL-H or Ipc_2BCl looks promising as does the asymmetric oxidation of the enolate of **W22** by one of Davis's oxaziranes described in chapter 33.

(3) *Asymmetric catalysis* (chapter 26): Catalytic asymmetric hydrogenation of **21** with catalysts such as BINAP-Rh or Ru complexes looks a good bet.

(4) *Substrate strategy* (chapter 27): Alkylation or oxidation of an enolate of a derivative of an Evans's auxiliary such as **W23** or **W24** offers another alternative.

W21   W22   W23   W24

(5) *Kinetic resolution and enzymes* (chapters 28 and 29): Enzymatic acylation of **82** or hydrolysis of one of its esters look pretty secure. The wrong enantiomer can be recycled (both here and after resolution) by oxidation to **W21** and reduction with $NaBH_4$.

Problem 31.3: What types of selectivity are demonstrated in the epoxidation of **94**?

t-BuOOH, Ti(O*i*-Pr)$_4$

1.5 equiv
(+)-di-isopropyl
tartrate, −20 °C

94

(*R*)-(+)-94
23–38% yield 72% ee

(−)-95
27–33% yield >95% ee

Answer 31.3: The epoxidation is regioselective (only the allylic alcohol reacts), stereospecific (the *anti* epoxide is formed) and enantioselective (only one enantiomer is formed).[6]

Problem 31.4: How might the Grignard reagent **104**, or the equivalent lithium derivative, used in the synthesis of ifetroban, be prepared as a single enantiomer?

104   W25

Answer 31.4: You might be attracted to *ortho*-lithiation of **W26** using the heterocycle as the directing group but this has a reasonably acidic hydrogen atom at the benzylic position that could be removed instead. They actually preferred[7] to react 2-bromo-benzaldehyde with ephedrine **W27** and then treat with BuLi (halogen-lithium exchange) and exchange with $MgBr_2$.

W26   BuLi? → W25   2-bromo-benzaldehyde   W27; (1*S*, 2*R*)-(+)-ephedrine   W28

Problem 31.5: Suggest reagents for converting **166** into **167**, a 'key late intermediate' in the synthesis of lactacystin. It should be obvious to you that direct protection of **166** with anything does not occur at the right place.

166 **167**

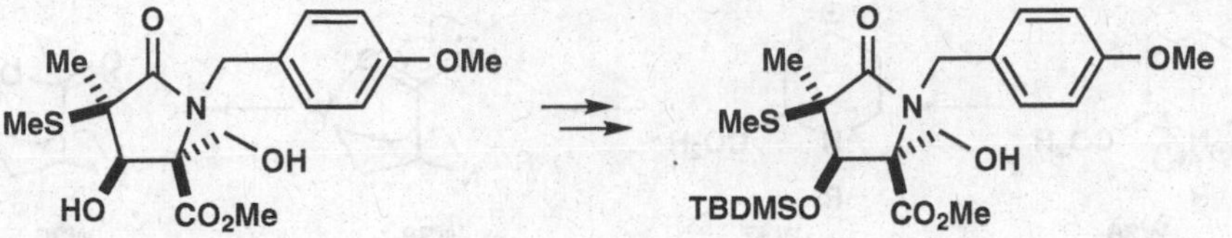

Answer 31.5: Any OH protection goes on the primary alcohol first so some preliminary protection with a group 'orthogonal' to TBDMS is needed. You may have made various suggestions but one principle should be that it should be a large group to minimise double protection. Corey chose pivaloyl and the sequence becomes esterification, protection of the secondary alcohol and doubly protected **W30** can be transesterified to **167** with methoxide in methanol in 92% yield.[8]

W29; 85% yield **W30; 98% yield**

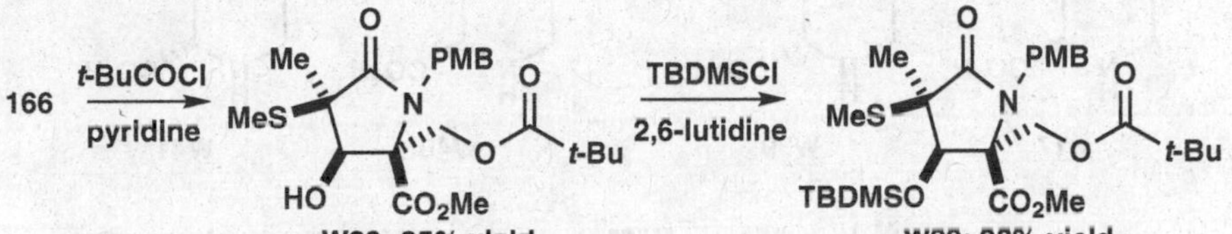

The Mannich/Prins Reaction

In the textbook we described the synthesis[9] of **26** from **24** and glyoxylic acid **25** as 'a cross between a Mannich and a Prins reaction' and we should new explain what we mean.

24 **26**

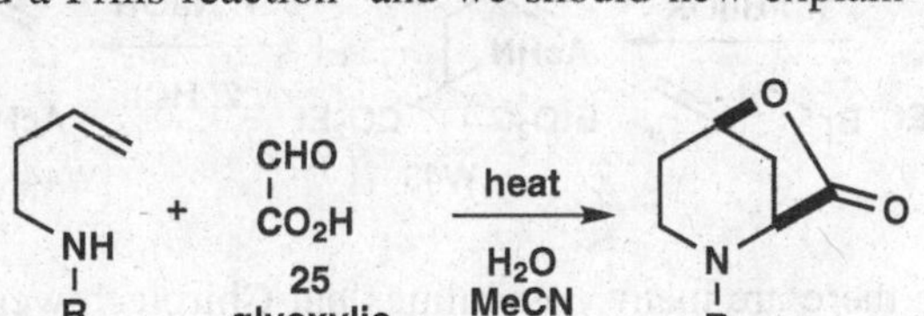

The Mannich reaction occurs when an enol attacks an imine (usually an iminium salt) to form a new C–C bond as in **W31** to **W32**. A Prins reaction occurs when an alkene attacks a carbonyl group and some nucleophile collects the cation **W34** so formed. Both occur in acidic solution.

W31 **W32** **W33** **W34** **W35**

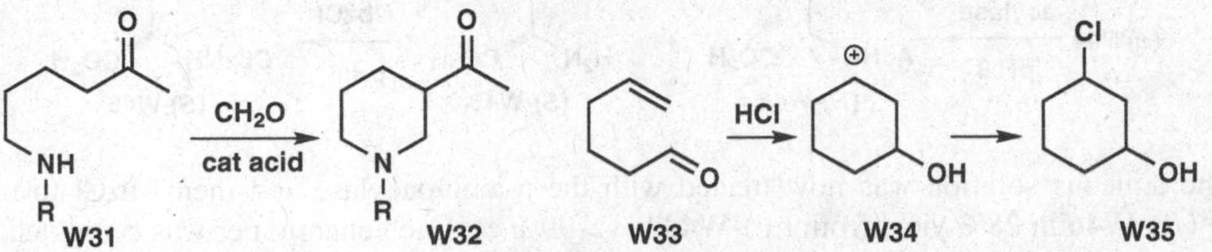

Here the alkene attacks the iminium salt **W36** formed from **24** and glyoxylic acid. The resulting cation is captured intramolecularly **W38** by the carboxylic acid to give **26**. You should see aspects of the Mannich and the Prins in this sequence.

This enantiomer of the lactone **26** was needed to make one enantiomer of 4-hydroxypipecolic acid **17**. This is one enantiomer of the *syn*-diastereoisomer: the other is **W39**. The other (*anti*-) diastereoisomer also has two enantiomers: **W40** and **W41**. More recently workers at Chirotech[10] set out to make all four enantiomers as 'chiral scaffolds for drug discovery.'

They chose to start with *N*-allyl glycine **W44** easily made (in racemic form of course) from the amino-malonate **W42**. They plan to make the ring by inserting formaldehyde between the alkene and the nitrogen atom in a similar Mannich/Prins style to the reactions we have just seen. This sequence will not disturb the amino acid chiral centre. **Problem 31.6**: Suggest a suitable strategy for moving forward.

Answer 31.6: Of course there are many possibilities but Chirotech were keen to promote two of their own enzymes. These are a recombinant *Thermococcus litoralis* L-aminoacylase and *Alcaligenes* sp. D-aminoacylase. Using the L-aminoacylase first the (*S*)-enantiomer of **W44** was hydrolysed to the amino acid (*S*)-**W45** and protected as the Cbz derivative (*S*)-**W46** that could be extracted into toluene and isolated in 36% yield and >99% ee leaving unreacted (*R*)-**W44** in the aqueous solution in 49% yield but only 80% ee.

The aqueous solution was now treated with the D-aminoacylase and then CbzCl to make pure (*R*)-**W46** in 28% yield from (±)-**W44** but >99% ee. Each enantiomer was converted into its methyl ester, e.g. **W46**. The two enantiomeric series have been separated. Each must now be converted into two diastereoisomers.

The Mannich/Prins cyclisation is not stereoselective in the way of **W38** but that is all right as they wanted both diastereoisomers **W48** and **W49**. A third enzyme, the commercial lipase AY30 hydrolysed the formate ester of *anti*-**W48** leaving *syn*-**W49** untouched.

Separation was achieved by reaction of *anti*-**W50** with phthalic anhydride and separation of the two compounds by solvent partition between saturated aqueous ammonium carbonate, that dissolved the half acid (*S*)-*anti*-**W52**, and toluene, that dissolved the diester (*S*)-*syn*-**W48**. The same process on the other enantiomeric series led to all four isomers.

The Synthesis of Pfizer's Sampatrilat – a Metalloproteinase Inhibitor

Pfizer's sampatrilat **W53** has valuable activity against various enzymes: it is an ACE inhibitor as well as a zinc metalloproteinase inhibitor. **Problem 31.7**: Comment in general terms about the stereochemical difficulties in designing a synthesis for sampatrilat and suggest by what strategies these might be overcome. You may like to suggest some preliminary disconnections.

Answer 31.7: The compound has three chiral centres but each is well separated from any other so controlling the relative stereochemistry appears impossibly difficult. The only practical strategy is to disconnect the molecule into three pieces each containing one chiral centre and make each of these pieces separately in enantiomerically pure form.[11] The obvious disconnections are of the amide bonds. Two pieces come directly from the chiral pool (chapter 23): **W56** is tyrosine and **W54** is a sulfonamide of lysine. The remaining piece **W55** will have to be made.

The laboratory synthesis started with straightforward chemistry to build the unsaturated ester **W60** from diethyl malonate. Reaction with the C_2 symmetric amine **W61** gave an intermediate **W62** ready for the formation of the vital chiral centre. This is the substrate-based strategy (chapter 27) with the amine **W61** acting as the chiral auxiliary.

Conjugate addition of the dilithium enolate of **W63** to **W62** gave **W64** with good diastereoselectivity (though the reaction was unreliable) and the key intermediate **W65** could be made from it. This was a key intermediate as it was 'extremely crystalline' and made purification simple. However most intermediates were not crystalline and chromatography was repeatedly used. Even worse the 15 steps required led to only 2% overall yield. And in addition the allylic bromide **W60** was a dangerous lachrymator. In spite of all this it was used to make 500 g of sampatrilat.

The First Development Route

The substrate-based strategy was abandoned in favour of what appears at first sight to be a simple resolution of a carboxylic acid. This was attractive as a resolution of **W66** was already being used to make another drug and the unwanted enantiomer from that project could be used for sampatrilat. However, the first step was an elimination to give achiral **W67**. Asymmetry was induced by conjugate addition of the simple amine **W68** (reagent-based strategy, chapter 24). The reaction worked but the asymmetric induction was erratic and fell as the reaction proceeded.

The Second Development Route

The difficulties with the successful laboratory route were overcome by using the allylic chloride **W71** made from *t*-butyl acrylate by the Baylis-Hillman reaction. The catalyst was quinuclidinol giving 64% yield on a 40 kg scale.

The rest of the synthesis was essentially the same as the laboratory route but the typical −78 °C of a laboratory LDA reaction could be replaced by −10 to 0 °C giving 88% yield of **W64**. There are other aspects of carrying out well known reactions on a large scale that are dealt with in detail in this revealing paper.

A Final Example of a Choice of Strategy

Reboxetine is a selective drug for neuropathic pain marketed by Pfizer in racemic form.[12] However the (*S*,*S*)-enantiomer is much more active than the racemate. **Problem 31.8**: Suggest how you might achieve an asymmetric synthesis.

W72; (*S*,*S*)-reboxetine

Answer 31.8: It is not obvious where to start disconnecting but the core of the molecule is a three carbon chain **W72a** bearing heteroatoms at each carbon and both chiral centres. This might be derived from epichlorohydrin **W73**, a member of the 'new chiral pool' (chapter 23) but would require asymmetric induction at C-1 when the phenyl group is introduced. However, Sharpless AE (chapter 25) is ideal as both chiral centres are present in **W74**, the epoxide of readily available cinnamyl alcohol **W75**. This is the strategy adopted by Pfizer.

The first stages were straightforward. The AE reaction gave 92% ee on a large scale upgraded to >98% by crystallisation. Epoxide opening with the phenol **W76** occurred at the benzylic position giving **W77** essentially as one enantiomer of one diastereoisomer.

The next essential selectivity is attack by a nitrogen nucleophile at the primary alcohol. This they did by making the epoxide **W78**. Though a sequence of protection, activation, deprotection and cyclisation was needed, this was evidently efficient as the amino alcohol **W79** was formed in 60% yield over the five steps.

Two carbon atoms now need to be inserted to make the ring and it was convenient to acylate at nitrogen first and do the more difficult *O*-alkylation as an intramolecular reaction. Reduction of the amide by Vitride, better known as REDAL NaAlH$_2$(OCH$_2$CH$_2$OMe)$_2$ gave reboxetine as the (*S,S*)-enantiomer.

References

1. E. Marotta, P. Righi and G. Rosini, *Org. Process Res. Dev.*, 1999, **3**, 206.
2. K. Mori, T. Uematsu, M. Minobe and K. Yanagi, *Tetrahedron*, 1983, **39**, 1735.
3. G. Confalonieri, E. Marotta, F. Rama, P. Righi, G. Rosini, R. Serra and F. Venturelli, *Tetrahedron*, 1994, **50**, 3235.
4. G. Guella, F. Dini, F. Erra and F. Pietra, *J. Chem. Soc., Chem. Commun.*, 1994, 2585.
5. M. J. Broadhurst, P. A. Brown, G. Lawton, N. Ballantyne, N. Borkakoti, K. M. K. Bottomley, M. I. Cooper, A. J. Eatherton, I. R. Kilford, P. J. Malsher, J. S. Nixon, E. J. Lewis, B. M. Sutton and W. H. Johnson, *Bioorg. Med. Chem. Lett.*, 1997, **7**, 2299.
6. G. C. Crawley, M. T. Briggs, R. I. Dowell, P. N. Edwards, P. M. Hamilton, J. F. Kingston, K. Oldham, D. Waterson and D. P. Whalley, *J. Med. Chem.*, 1993, **36**, 295.
7. S. D. Real, D. R. Kronenthal and H. Y. Wu, *Tetrahedron Lett.*, 1993, **34**, 8063.
8. E. J. Corey, W. Li and T. Nagamitsu, *Angew. Chem., Int. Ed.*, 1998, **37**, 1676.
9. S. J. Hays, T. C. Malone and G. Johnson, *J. Org. Chem.*, 1991, **56**, 4084.
10. R. C. Lloyd, M. F. B. Smith, D. Brick, S. J. C. Taylor, D. A. Chaplin and R. McCague, *Org. Process Res. Dev.*, 2002, **6**, 762.
11. P. J. Dunn, M. L. Hughes, P. M. Searle and A. S. Wood, *Org. Process Res. Dev.*, 2003, **7**, 244.
12. K. E. Henegar and M. Cebula, *Org. Process Res. Dev.*, 2007, **11**, 354.

E
Functional Group Strategy

32

Functionalisation of Pyridine

This first section deals mainly with further exploration of chemistry from the textbook chapter. It will help if you have that chapter open while you deal with the problems. **Problem 32.1**: Comment of the choice of the order of events in the Flupirtine synthesis.

Answer 32.1: The nitration is carried out first because two electron-donating groups are needed for the reaction to work. Then the nucleophilic substitution must be done next while the nitro group is still there to help. Reduction of nitro to amine must be done before acylation as it is that amino group which is to be acylated.

Problem 32.2: In the final step of the Flupirtene synthesis, compound **W1** is acylated to give a single amide (carbamate) by acylation at one of four amines in **W1**. Why is only that one amino group acylated? Comment on the detailed mechanism of the reaction.

Answer 32.2: Only that amine is acylated because it is the most nucleophilic. The pyridine nitrogen and the other two amines form a single conjugated system **W2a** and **W3a** which would acylate at the pyridine nitrogen atom if at all. The amine which is acylated is conjugated with the pyridine ring but not with the pyridine nitrogen atom.

Workbook for Organic Synthesis: Strategy and Control   Paul Wyatt and Stuart Warren
© 2008 John Wiley & Sons, Ltd

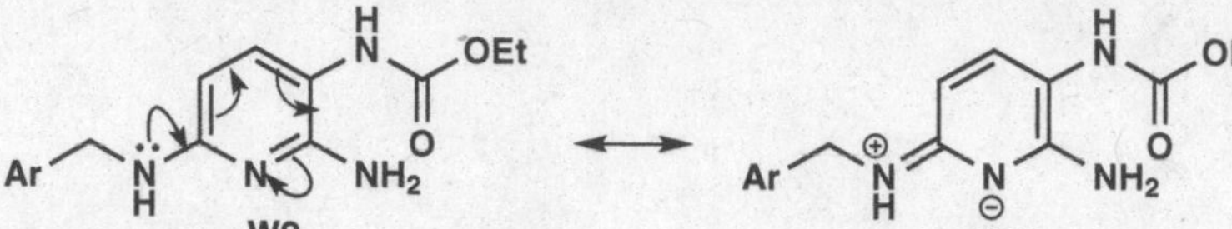

W2 ⟷ W2a

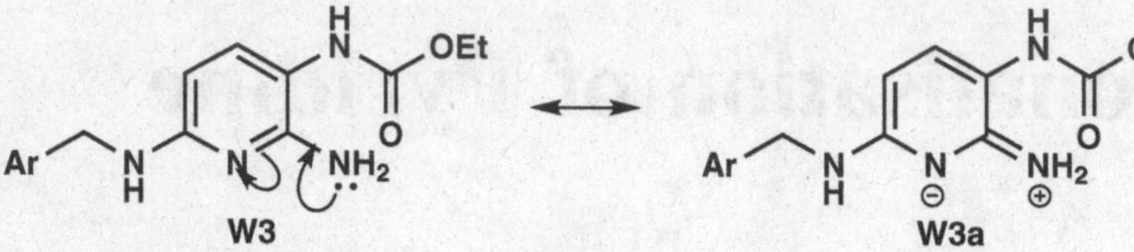

W3 ⟷ W3a

The pyridine nitrogen atom might well get involved in the reaction in a DMAP style of nucleophilic catalysis. Acylation at the pyridine nitrogen atom of one molecule **W4** would create an acylation agent which could acylate the 3-NH$_2$ group of another molecule **W5**.

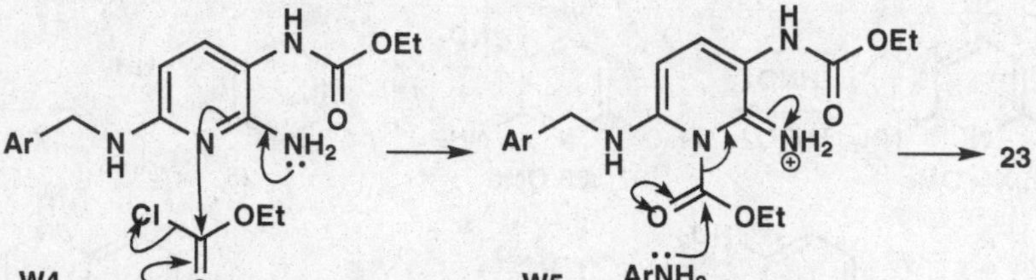

W4   W5   ArNH$_2$ 23

Traditional Electrophilic Nitration of Acridine

References for electrophilic substitution on simple heterocycles are found in Joule and Mills. Nitration of acridine **W6** gives a mixture of at least four nitro derivatives containing the 2-nitro compound as the major product. **Problem 32.3**: Explain this result and suggest how you might prepare pure 9-nitro-acridine **W7**.

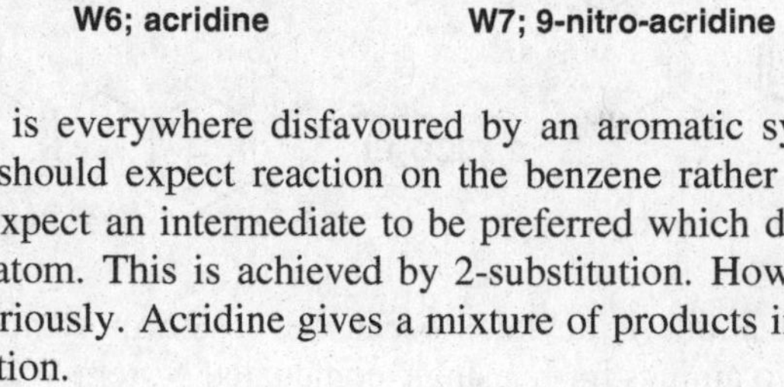

W6; acridine   W7; 9-nitro-acridine

Answer 32.3: Nitration is everywhere disfavoured by an aromatic system which includes a pyridine ring. First we should expect reaction on the benzene rather than the pyridine rings and second we should expect an intermediate to be preferred which does not put the positive charge on the nitrogen atom. This is achieved by 2-substitution. However, we must not take this last argument too seriously. Acridine gives a mixture of products in poor yield. This is not a practically useful reaction.

The synthesis of 9-nitro-acridine **W7** is most obviously tackled *via* the *N*-oxide. Pyridine-*N*-oxides tend to react in the '*para*' position and quinolines should do so even more **W10** as the intermediate **W11** has two 'real' benzene rings. A phosphine removes the oxygen atom to give **W7**. Anthracene is nitrated in the same position for the same reason.

Substitution on Quinolines

Problem 32.4: Give mechanisms for and comment on the selectivity in these substitutions on quinolines.

Answer 32.4: Notice the contrast between electrophilic and nucleophilic substitution: electrophiles attack the benzene ring but nucleophiles attack the pyridine ring. That was painless but exact explanations of the ratios are more tricky. In the nitration, we might have expected product **W14** to be favoured as the charge in the intermediate is not delocalised onto nitrogen whereas in intermediate **W18** it is. Another argument might be that the quinoline nitrogen is protonated which deactivates both rings but the pyridine more so that the pyridine acts as an electron-withdrawing and therefore *meta*-directing substituent on the benzene ring. But this does not work either: **W13** is nitrated *meta* to N but **W14** is *para* to N.

W17   W18   W19 → W13

However, these arguments clearly do not hold water as the two products are formed in equal amounts. We do much better if we ignore the pyridine ring **W20** with its lower energy HOMO and say that what is left of the benzene ring behaves as a butadiene structure having the largest coefficients in the HOMO at the ends **W21**. Reaction occurs at both ends of this system. Maybe the main message is: Beware! Any electrophilic substitution on any aromatic system including a pyridine ring must be referred to the literature. Our simplistic mechanistic arguments are not reliable in these situations.

W20   W21

It is much easier to explain the nucleophilic substitution – the molecule is naturally electrophilic and its LUMO is more easily understood than the HOMO of quinoline. Tosylation occurs at oxygen and then the aniline adds **W22** or **W24** to the two electrophilic atoms in the pyridine ring. The 2-position **W22** is more electrophilic (larger coefficient in the LUMO: nearer to N^+) and more reaction occurs here. Finally a proton is lost **W23** from the 2- (or 4-) position to regenerate the aromatic ring.

W22   W23   W24

Reactions of Activated Pyridines

The major product in the nitration of **W25** is used in a synthesis of porphobilinogen. **Problem 32.5**: Why does the nitration go in such high yield? Suggest how the yield of the major isomer **W26** might be improved.

W25   W26; 70% yield   W27; 25% yield

Answer 32.5: The reaction goes well because of the activating amino group, and to a lesser extent, methyl group. The 5-nitro compound **W26** is what we expect as the reaction occurs *ortho* and/or *para* to the amino group and in spite of the pyridine nitrogen. We expect and get more **W26** than **W27** because of steric hindrance so perhaps the best way to improve the yield would be to bulk up the amino group with, say, a benzyl group or even two.

The nitration product **W29** is used in a synthesis of pyridoxine **W30**. **Problem 32.6**: What is the role of the acetic anhydride in the nitration step? Why was this particular starting material **W28** chosen for the synthesis in spite of the poor yield in the nitration step? How might the yield of the nitration product be improved? How would you finish the synthesis or, if you prefer, suggest an alternative pathway from the same starting material?

Answer 32.6: Acetic anhydride reacts with nitric acid to give NO_2^+ AcO^- as a solution in acetic acid, thereby avoiding strongly acidic conditions.

The starting material **W28** was chosen because it is so easy to make in large quantities by a typical pyridone **W33** condensation reaction:

The electron-withdrawing cyanide and carbonyl groups in the ring do not help the nitration and one suggestion might be to convert the pyridone temporarily to a 2-OMe or 2-Cl derivative such as **W34** as these can be easily hydrolysed back to the pyridone afterwards. As pyridoxine **W30** has no carbonyl group (it is a pyridine not a pyridone) the 2-Cl derivative is particularly appealing because it can be reduced catalytically to 2-H, though the 2-OMe gives more activation for the nitration step. Alternatively the cyanide could be reduced before nitration.

This synthesis is the original and ancient one by Folkers[1] so we ought to be able to do a bit better today! It went on like this:

W28 →PCl₅/POCl₃→ W36; 40% yield →H₂/Pd, HOAc→ W37; 40% yield →1. NaNO₂, HCl / 2. 48% HBr / 3. AgCl, H₂O→ W30

The nitro group is introduced with the idea of transforming it later into an OH group by diazotisation. No step in the synthesis goes in good yield, though it was a major achievement for 1939, and we might want to rethink the whole thing. Suggestions are made in the chapter for the direct introduction of oxygen rather than nitrogen into an aromatic ring.

Working Towards Bakke Nitration

Problem 32.7: Suggest mechanisms for the successful sulfonation[2] **162** and bromination[3] **163** of pyridine in solutions containing SO_3, using the explanation for successful nitration if you wish.

152; 70% yield ←very concentrated H₂SO₄ / catalytic HgSO₄ 220 °C, 24 hours← →Br₂, 130 °C / very concentrated H₂SO₄→ 153; 86% yield

Answer 32.7: The hint was well meant. Though this explanation has not, as far as we are aware, been published, it seems sensible to attack at nitrogen with the electrophile **W38**, use SO_3 or something else as a nucleophile, and then a [1,5]-sigmatropic Br or SO_2OH shift like this **W39**.

→Br₂→ W38 →SO₃→ W39 →[1,5]Br→ W40 → 86

The Nitration of Benzene with *N*-Nitro Heterocycles

Problem 32.8: Draw mechanisms for the formation of **14** and for the nitration of benzene with **15**.

13 →HNO₃, Ac₂O / HOAc→ 14 →BF₃→ [15] →benzene→ 89% yield

Answer 32.8: As we explained above, acetic anhydride reacts with nitric acid to give NO_2^+ AcO^- as a solution in acetic acid. The nitrating agent is NO_2^+ in a solution that does not protonate the pyrazole.[4] Either nitrogen atom could react, e.g. **W41**.

Iodination of 3-Hydroxypyridines

In the textbook we discussed the halogenation of **33**. Though it had been suggested that **32** was the product, in fact **34** was formed, and the same isomer **35** was the result of chlorination.

Iodination of **35** went *para* to the OH group **W44** as did a useful reaction with aqueous formaldehyde **W45**. Only with the 2- and 6-positions blocked did reaction finally occur[5] at the 4-position **W46**. It appears that reaction next to nitrogen is indeed preferred.

Electrophilic Substitution on Pyridine *N*-Oxides

We discussed reactions of **53** with electrophiles to give **54** in the textbook and mentioned that the oxygen may be removed with trivalent phosphorus compounds. **Problem 32.9**: Suggest a mechanism for this step, bearing in mind that PCl_3 reacts more rapidly than PPh_3. The other product is $POCl_3$.

Answer 32.9: You could start by nucleophilic attack of the phosphine on the oxyanion, but that does not look right. It seems better to use the oxyanion to attack the phosphorus atom **W47**. Then electrons can flow the other way **W48** to break the weak N–O bond and make the strong P=O bond. The best mechanism is probably to combine the two **W49** with a partial negative charge on the phosphorus atom in the transition state.

W47   W48   55 + POCl$_3$ W49

Lithiation and the Halogen Dance

In the textbook chapter we concentrated on controlling lithium/halogen exchange by controlling the site of halogenation. However, there is increasing evidence that selective lithiation of, for example, dibromopyridines is possible. It was already known that available 2,5-dibromopyridine **W50** gave the 5-Li derivative **W51** with BuLi in ether and reaction with electrophiles gave **W52** in good yield.[6] Aldehydes **W53**; R = H (64% from DMF) and ketones **W53**; R = t-Bu (71% from t-BuCN) and R = Me (82% from MeCONMe$_2$) can be made using the methods of chapter 8.

W50   BuLi Et$_2$O   W51   E$^\oplus$ W52   W53

New work from Merck[7] revealed that a simple change of solvent to toluene caused a reversal of selectivity: the 2-Li derivative **W54** and hence compounds such as the alcohol **W56** were formed in good yield. The aldehyde **W55** was unstable and was not isolated. Other electrophiles include aldehydes, ketones, disulfides and silyl halides.

W50   BuLi PhMe   W54   Me$_2$NCHO   W55   NaBH$_4$ W56; 78% yield

The Synthesis of Mappicine

In the synthesis of mappicine in the textbook, each component **125** and **126** is made by lithiation of a heterocycle. The quinoline component **125** was made from **127** by a simple sequence of reactions. **Problem 32.10**: Give full mechanisms for all these reactions explaining any selectivity.

127   1. LDA 2. CH$_2$O   128   PBr$_3$ 125

Answer 32.10: Lithiation with LDA normally occurs by *ortho*-lithiation rather than halogen/lithium exchange.[8] The only directing group in **127** is the chlorine atom so the electrophile is introduced *ortho* to it. Reaction with PBr$_3$ replaces the benzylic OH with Br in an S$_N$2 reaction and bromide replaces chloride by the standard mechanism of S$_N$Ar on pyridines. The other component **126** is made from **122**.

Problem 32.11: Give full mechanisms for all these reactions explaining any selectivity.

122   **129; 84% yield**   **130; 84% yield**   **126; 92% yield**

Answer 32.11: This time BuLi is used and halogen/lithium exchange with the iodide is preferred to *ortho*-lithiation. The rest is straightforward but we might note that the last step is a hydrolysis of the 2-fluoro-pyridine **129** rather than substitution by Cl. Addition of the weak nucleophile water **W57** is specific acid catalysed and assisted by the electronegative fluorine atom but *not* by the carbonyl group. The loss of the bad leaving group fluoride is fast anyway as it restores aromaticity.

W57   **W58**   **W59**

In the synthesis of camptothecin[9] given in the textbook, an unexplained step was the conversion of intermediate **144** on acidic work-up into **135**. **Problem 32.12**: Explain what is happening.

143   **144**   **135**   **145 TCC alcohol**

Answer 32.12: Acid protonates the lithium alkoxide and hydrolyses the acetal giving the triol **W60**. Only one OH group is well placed to cyclise onto the ester to give **135**.

144   **W60**   **135**

The Baeyer Anti-Cancer Drug BAY 43-9006

In the synthesis of the Baeyer anti-cancer drug BAY 43-9006 described in the textbook, a key step was the formation of **229** from **228** and *para*-aminophenol. Diaryl ethers are not normally made by reaction between an aryl chloride and a phenolate anion. **Problem 32.13**: Why does it work so well here?

Answer 32.13: The pyridine nitrogen atom makes the coupling possible. Clearly the acid chloride could not be used so the right amide was made first after the Bakke-style chlorination put chlorine in the right position for the addition-elimination mechanism.[10] The anion of the phenol adds to the pyridine **W61** and elimination of chloride **W62** follows. Note that the carbonyl group does not help and that it is not necessary to use fluoride as leaving group in S_NAr reactions on pyridines.

Preparation of Non-Aromatic Heterocycles via Lithiation of Pyridines

Problem 32.14: This question concerns the lithiation of two pyridines **W63** and **W65** and their reactions with electrophiles. The seven-membered rings can be formed from either **W63** or **W65** but the chlorine atom is necessary in **W65** for a good yield. The six-membered ring product **W68** can be formed only from **W65**: attempted synthesis from **W63** gives a different product. Deduce in each case what kind of lithiation occurs and how the products are formed.

Answer 32.14: In both cases the first lithiation occurs[11] on nitrogen, hence the need for 2.2 equiv of BuLi. Alkylation and cyclisation gives the pyrido[2,3-*b*]azepines **W64** and **W66**. Next

W63 undergoes lateral lithiation to give **W69** while **W65** undergoes *ortho* lithiation to give **W71**. The Me group in **W63** carries more acidic protons than the *ortho* position in **W65** and BuLi tends to add to C-6 of **W65** if the chlorine is not there.

In the second pair of reactions, **W63** again undergoes lateral lithiation but the electrophile is too short and can donate Br to the organo lithium by elimination **W73**. The doubly lithiated starting material **W69** reacts with the bromide **W74** to give the dimeric product **W67**. No problems arise with **W65** as Cl and I are well separated.

Preparation of Bridged Analogues of Epibatidine

Epibatidine is the famous but very toxic pyridine-containing analgaesic for which non-toxic analogues are being sought. We shall look at the synthesis[12] of one of these where an extra bond has been made between the pyridine ring and the bicyclic aliphatic amine. **Problem 32.15**: Suggest reagents for steps (a) and (b) and a mechanism for step (c).

Answer 32.15: Some form of electrophilic iodine is needed for step (a): the chemists used H_5IO_6 and I_2 in aqueous acetic acid to give 71% yield of **W77**. Step (b) was oxidation with *m*CPBA (you might well have suggested some other oxidising agent) followed by salt formation with HCl in ether. Step (c) starts with acetylation of the oxyanion then deprotonation by acetate ion **W81** to form a reactive exocyclic alkene that can add chloride **W81** with cleavage of the weak N–O bond to give the amine **W82** that can be acetylated to give **W79**. If you preferred to do a [3,3]-sigmatropic rearrangement on the exocyclic alkene and substitute chloride for acetate later, who is to say you are wrong? The chloride comes from the hydrochloride salt **W78**.

The alkyl chloride **W82** is now simply coupled with the bicyclic amine **W83** to give **W84** which is cyclised to **W85**. **Problem 32.16**: Suggest a mechanism for the coupling reaction.

Answer 32.16: This looks like a simple Heck reaction at first with Pd(0) inserting in the Ar-I bond and then joining onto the nearer end of the alkene. The twist is that no double bond reappears. Elimination from the Pd(II) intermediate is not possible as one hydrogen atom is on the wrong side and the other would give a bridgehead alkene. Instead it is reduced by transfer hydrogenation from the formate anion giving **W88** and CO_2. Finally, reductive elimination gives **W85** and recycles Pd(0).

The remaining steps convert **W85** into the epibatidine analogue **W89**. Epibatidine itself is **W90** and the analogy should be immediately apparent. Another analogue was also made but sadly both lacked activity. This gives useful information about the active conformation of epibatidine.

References

1. S. A. Harris and K. Folkers, *J. Am. Chem. Soc.*, 1939, **61**, 1245.
2. S. M. McElvain and M. A. Goese, *J. Am. Chem. Soc.*, 1943, **65**, 2233.

3. H. J. den Hertog, L. van der Does and C. A. Landheer, *Rec. Trav. Chim.*, 1962, **81**, 864.

4. G. A. Olah, S. C. Narang and A. P. Fung, *J. Org. Chem.*, 1981, **46**, 2706.

5. D. G. Wishka, D. R. Graber, E. P. Seest, L. A. Dolak, F. Han, W. Watt and J. Morris, *J. Org. Chem.*, 1998, **63**, 7851.

6. C. Bolm, M. Ewald, M. Felder and G. Schillinghoff, *Chem. Ber.*, 1992, **125**, 1169; F. Romero-Salguero and J.-M. Lehn, *Tetrahedron Lett.*, 1999, **40**, 859.

7. X. Wang, P. Rabbat, P. O'Shea, R. Tillyer, E. J. J. Grabowski and P. J. Reider, *Tetrahedron Lett.*, 2000, **41**, 4335.

8. D. L. Comins, M. F. Baevsky and H. Hong, *J. Am. Chem. Soc.*, 1992, **114**, 10971.

9. D. L. Comins and J. M. Nolan, *Org. Lett.*, 2001, **3**, 4255.

10. D. Bankston, J. Dumas, R. Natero, B. Reidl, M.-K. Monahan and R. Sibley, *Org. Process Res. Dev.*, 2002, **6**, 777.

11. A. J. Davies, K. M. J. Brands, C. J. Cowden, U.-H. Dolling and D. R. Lieberman, *Tetrahedron Lett.*, 2004, **45**, 1721.

12. L. E. Brieaddy, S. W. Mascarella, H. A. Navarro, R. N. Atkinson, M. I. Damaj, B. R. Martin and F. I. Carroll, *Tetrahedron Lett.*, 2001, **42**, 3795.

33

Oxidation of Aromatic Rings and of Enol(ate)s

Problems and Further Examples Relating Directly to the Text

You will find it helpful to have chapter 33 from the textbook open as you look at this first section. **Problem 33.1**: Attempt to explain the stereoselectivity in the epoxidation and the hydroxylation of the quassinoid precursor **207**.

Answer 33.1: Conformational drawings are required to make much sense of this. There is of course no reason why reactions in different regions of the same molecule should not occur with different facial selectivities as they will be affected by the local stereochemistry. In the epoxidation step **W1**, which occurs first, there are two nearby axial methyl groups on the top face and only an equatorial methyl (which is practically in the plane of the alkene) on the bottom face. The CH_2CO_2t-Bu group is too far away to have a blocking effect. It is just possible that it and/or the axial oxygen of the acetal might deliver the mCPBA to the bottom face but they are both quite remote.[1,2]

In the hydroxylation step **W2** the dominant group near to the enol is the large and axial CH_2CO_2t-Bu group which is on the next atom along the ring. There is an axial methyl group at the ring junction but it is smaller and farther away. Do not forget that this is a surprising result both in its efficiency and uniquity.

$208 \longrightarrow$ [structure] $\longrightarrow 209$

W2; hydroxylation from the top

Problem 33.2: Outline the remaining operations needed to convert **209** into bruceantin **210**. Suggest the sort of reagents needed (without necessarily specifying the exact reagent) and point out the likely problems.

$207 \xrightarrow[\substack{CH_2Cl_2 \\ \text{room} \\ \text{temperature}}]{\substack{\text{excess} \\ m\text{CPBA}}}$ [structure] $\xrightarrow{?}$ [structure]

209; only product; 100% yield   **210; bruceantin**

Answer 33.2: Incredibly, some further hydroxylation, or at any rate oxidation of positions next to carbonyl groups which could be accomplished by hydroxylation, has still to be done **W3**. Assuming all this can be done stereoselectively by some of the methods we have described, there are still further challenging tasks to be accomplished. We might hope to reach an intermediate like **W4** by hydroxylation.

W3   **W4**

Reduction of the epoxide might be controlled to give an alcohol and then a ketone next to one of the new sites for hydroxylation and hydroxylation on the methyl group could be accomplished by a Barton-style reaction. The rest is not too difficult! It is obviously very difficult to accomplish all these tasks with full control and you may find it interesting to analyse some of the possibilities in detail. The original paper[2] does not go very far along the road, but you might be interested to see the idea for closing the lactone ring (diagram 40 in the paper).

Paquette's Remarkable Epoxide

The full story of the reaction of epoxide **219** with acid is that there is a by-product, the ketone **W5**. Paquette suggests that the formation of this product helps to confirm the mechanism of the rearrangement of **219** to **220**. **Problem 33.3**: Do you agree?

Answer 33.3: The same intermediate from the opening of the epoxide **219** can lead to the major product **220** by silyl transfer or to the rearranged ketone **W5** by C–C bond migration **W6**. The longer lifetime of the silyloxy epoxide **219** is probably the result of the sulfone destabilising the cation. The formation of **W5** does not prove that **W6** is an intermediate in the formation of **220**, but it helps.[3]

The conversion of the ketone **228**; R = *t*-Bu into the lactone **227**; R = *t*-Bu is described in the text. **Problem 33.4**: How might you convert the same ketone **227**; R = *t*-Bu into the related lactone **W7** which is *trans* fused onto the other side of the ketone? The remarkable regioselectivity of enolisation of **227**; R = *t*-Bu is described in the text – it is unlikely that you can reverse this selectivity.

Answer 33.4: One idea is to block the side which favours enolisation and then hydroxylate the other side. The way Lansbury and Vacca[4] did this was to introduce a conjugated alkene by sulfenylation and *syn* elimination **W8** to give the enone **W9**.

They then hydroxylated by an old method using $Pb(OAc)_4$ but you might prefer kinetic enolisation of the enone in the α' position (chapter 11) and one of the methods from this chapter. Olefination and reduction must now include reduction of the introduced alkene too. You are referred to the paper for full details of their method. There are of course many other ways.

The text explains how it is possible for the two α-hydroxy carbonyl compounds **255** and **252** to equilibrate but it does not explain why both stereoisomers of **255** give a single diastereoisomer of **252** with KOH in MeOH. **Problem 33.5**: So, why does **255** isomerise to **252** and not the other way round, and why is that stereoisomer formed?

Answer 33.5: The starting material **255**, however formed, has two conformations **255a** and **255b** neither of which is very favourable as one group must go axial. The product can put both groups equatorial **252a**. There is also a basic ambiguity in conformation when substituents are next to carbonyl oxygen. Equatorial avoids 1,3-interactions but clashes with carbonyl oxygen. Thus **255** has both substituents flanking the ketone while **252** has only one in this awkward position.

Oxidation with MoOPH

Problem 33.6: Fill in the details of the control in the hydroxylation of the aspartic acid derivative **259** with MoOPH. Can you draw the Houk conformation for *syn*-selective hydroxylation? What is the role of the BuLi in the *anti*-selective reaction? What is chelating what?

Answer 33.6: The chemists suggest[5] an open structure for the enolate without BuLi which seems to us to suggest the Houk conformation **W10** with hydrogen eclipsing the enolate double bond. But this conformation could give either *syn* or *anti* product **260** depending on the relative size of CO_2Me and NHR as perceived by MoOPH. They imply that CO_2Me is larger and this is possible if the large R group rotates out of the way. We suggest that NHR might well be larger (R is very large) and this would give *anti*-**260**.

A better alternative seems to us to be chelation by the other ester in the absence of BuLi. This would give an enolate structure **W11** that would definitely react on the face opposite NHR to give **W12**. Rotation of **W12** reveals that it is *syn*-**260**.

They also suggest[5] that the preliminary treatment with BuLi removes the NH proton from the protected amine so that the nitrogen atom now coordinates to the lithium in Zimmerman-Traxler fashion **W13** and this does indeed give *anti*-**261**. Regardless of the explanation, reversal of selectivity by preliminary treatment with BuLi is a remarkable achievement.

Problem 33.7: Account for the 'opposite' stereoselectivity in the hydroxylation of the tetra-cycline precursors **294** and **296**. In the hydroxylation of **294**, what aspect of regioselectivity is also controlled and how?

294

293

273

296

295

297; (+)-dimethoxy-sultam

Answer 33.7: The hydroxylation of **294** with the racemic sulfonyl oxaziridine **273** is under substrate control with racemic reagent attacking the opposite face of the enolate to the large CH_2OBn group. The hydroxylation of **296** is under reagent control where the enantioselectivity of the reagent dominates the prochiral substrate. The other enantiomer of **295** would be produced with the other enantiomer of the chiral sulfonyl oxaziridine.

Regioselectivity arises in the formation of enolate **W14** rather than the less substituted and unconjugated **W15**. It also occurs in the reaction of enolate **W14** which has an extra double bond and is therefore an extended enolate (chapter 11). The reaction is evidently, and predictably, under kinetic control – evidently because it occurs at the α-position and predictably because it is difficult to imagine that hydroxylation would be reversible.

Revision Question

In chapter 7 we referred to Büchi's synthesis[6] of aspersitin. This is given in full here. The first stages involve a Friedel-Crafts acylation and then some selective protection. Now the free phenol **W19** is hydroxylated using an old reagent, $Pb(OAc)_4$, which happens to work very well here.

So much for the background. Now for the questions which are intended to be relevant to this chapter but also to revise material from earlier in the book. **Problem 33.8**: The symmetrical starting material is readily available. Suggest a mechanism for the formation of **W16**.

Answer 33.8: This is a Friedel-Crafts acylation with a Lewis acid but, unusually, using the free acid rather than the acid chloride. No doubt the BF$_3$ forms an acyl borane and maybe also a phenoxyborane to bring the reagents together. You might have proposed something like **W21** to **W24**. However, you might not have bothered to link either or both OH groups to the carboxylic acid and you might have proposed an acylium ion (RCO$^+$) as an intermediate as in a normal Friedel-Crafts acylation. All these variations are reasonable but it is necessary to use the boron to remove the OH from the carboxylic acid.

Problem 33.9: Comment on the selective protection required to make **W19** from **W16**.

Answer 33.9: The three OH groups are rather similar in reactivity but the most reactive is the top one, mostly on steric but also on electronic grounds. The two OH groups *ortho* to the carbonyl group are able to form hydrogen bonds to the carbonyl oxygen. It is necessary to use a large silyl group to block the top OH, then to methylate the other two, and finally to remove the silyl group.

Problem 33.10: Suggest a mechanism for the formation of **W20**. Is there any selectivity in this step?

Answer 33.10: This is the hydroxylation step and is very similar to the hydroxylation of enols discussed in the chapter. We did briefly mention Pb(OAc)$_4$ as an old method, but it really works well here. The reagent must provide an electrophilic acetate group and does so by dropping down from Pb(IV) to Pb(II). One oxygen atom on one acetate become electrophilic as the another leaves **W25**.

Now, is there any selectivity? There certainly is *ortho* selectivity as hydroxylation could have occurred *para* to the OH group. Indeed the aromatic system between the OH and the carbonyl group is rather like a very extended enol. Then what about the two positions *ortho* to the OH group: are they the same, perchance? It is all too easy to forget the sleeping stereogenic centre next to the carbonyl group in the side chain. The two faces of the benzene ring are in fact diastereotopic, as are the two *ortho* positions. However, the stereogenic centre is too far away to exert any influence (it would be 1,5 selectivity) and a 1:1 mixture results. More of that later.

The final step in the synthesis involves replacement of one of the MeO groups by NH_2 to give aspersitin **W25**. **Problem 33.11**: What is happening in the formation of **W25**? Discuss the selectivity.

Answer 33.11: The formation of **W25** involves ammonolysis of the acetate by attack on the carbonyl group (trivial) and a conjugate substitution of one of the OMe groups (which are different in **W20**) by ammonia. Conjugate addition occurs to the external enone **W26** and the methoxide driven out **W27**. The selectivity arises because the oxyanion in the intermediate **W27** is stabilised by both carbonyl groups – outside and inside the ring – while the oxyanion in the intermediate from attack at the other position would be stabilised only by the oxygen atom in the ring. Methoxide is a bad leaving group and the reaction works only because the intermediate is well stabilised.

Problem 33.12: Aspersitin **W25** is formed in low yield as a 1:1 mixture of isomers. What isomers are these? How might the yield of this step and the selectivity of the synthesis be improved?

Answer 33.12: The isomers are of course diastereoisomers **W28** and **W29** which arise during the hydroxylation. The only hope for such a remote pair of stereogenic centres would be to use an optically active acid in the first step (chiral pool strategy, chapter 29) and then a reagent-controlled asymmetric hydroxylation (chapters 30 and 40) with a chiral camphor-derived oxaziridine. There is no guarantee that this would work but both reactions can be carried out in either enantiomeric sense so there is hope. The yield in the reaction with ammonia might be better if the starting material **W20** were a pure compound instead of a mixture of diastereoisomers but other reagents should also be tried. Büchi was able to show that **W29** has the same relative stereochemistry as natural aspersitin but the absolute stereochemistry was not determined.[6]

W28

W29

Enones by Oxidation of Ketones

Tuberiferin **W30** has been synthesised from the ketolactone **W31**. **Problem 33.13**: Suggest a racemic synthesis of **W31**.

W30; tuberiferene   ?   **W31**

Answer 33.13: There are of course many different routes to **W31** but a good starting material, used by Grieco[7] in our featured synthesis, is the Robinson annelation product **W32**. Protection as the acetal also drives the alkene into the other ring as there is no longer any conjugation. Hydroboration now gives the alcohol **W34** with the OH group in the right place but all the relative stereochemistry wrong.

W32   **W33; 56% yield**   **W34; 90% yield**

Oxidation (Collins) to the ketone **W35** allows epimerisation of the ring junction to the more stable *trans*- decalin **W36** and alkylation of the kinetic enolate gives the ketoester **W37**. However, with a horrid inevitability this new centre is also wrong.

W35   **W36; 90% from W34**   **W37; 62% yield**

Fortunately, epimerisation with NaOMe in MeOH again comes to the rescue and ester hydrolysis and lactonisation gives **W39**. Standard methylation of the lithium enolate gives **W40**, a protected form of **W31** with all the relative stereochemistry correct. Make sure you can understand all the stereochemistry. This is all controlled by the six-membered rings and the principles are *exo*-attack, axial alkylation of enolates and thermodynamic control.

W38; 95% yield   **W39**   **W40; 88% yield**

Problem 33.14: How might **W40** be converted into tuberiferin?

Answer 33.14: There are various methods but Grieco[7] chose the selenium route. The first selenium was added to the lithium enolate of the lactone **W40**: addition occurs on the *exo*-face. Then the acetal was removed, the kinetic ketone enolate prepared at −78 °C and selenium added again with the more reactive PhSeCl. Oxidation leads to spontaneous elimination at room temperature.

W41; 85% yield   **W42; 76% yield**

In the textbook we discussed a number of oxidations that appeared side by side in the 1986 volume of *Organic Syntheses*. One we did not discuss used *o*-iodosobenzoic acid.[8] **Problem 33.15**: Suggest a mechanism for the reaction of *o*-iodosobenzoic acid **W44** with acetophenone **W43** to give **W45** and hence the product of hydroxylation **W46**.

W43   **W44**   **W45; 65% yield**   **W46; 83% yield**

Answer 33.15: The essential thing is to use ArIO as an electrophile on the enol(ate) and then use the cleavage of the weak I–O bond to make iodide a leaving group. Formation and opening of the epoxide **W50** does the rest. Conversion of **W45** to **W46** is just acetal hydrolysis.

W43   **W47**   **W44 = ArIO**   **W48**

W49   **W50**   **W51**

The Diazotisation Route to Phenols

In the textbook we gave a sequence from α-pyridone to the 'phenol' **14** using diazotisation. **Problem 33.16**: What is the yield of **14** from **9** by this sequence? Explain any selectivity.

Answer 33.16: Simply multiply all the individual yields together to get the overall yield. We make it 10.8% – not a wonderful result from all that effort! Regioselectivity in the iodination of **9** is simply that OH activates and directs while nitro and the ring N deactivates that position least. You might have also mentioned the last step where ester hydrolysis is preferred to nucleophilic aromatic substitution.[9]

The Friedel-Crafts/Baeyer-Villiger Route to Phenols

The key step in the synthesis of dechlorolecideoidin **27** we described in the textbook was an oxidation of **36** to the spiro compound **37**. **Problem 33.17**: Explain the specific formation of **37** by oxidation of **36**.

Answer 33.17: This is clearly a radical coupling reaction as Fe(III) is a one electron acceptor. You could have converted both OH groups to oxygen-centred radicals and coupled them to give **37**. The problem with this otherwise reasonable approach is: why should the oxygen atom on the left-hand ring cyclise on the other ring rather than viceversa? The authors solution, based on their experience, is that an electron is removed only from the 'ring of lower oxidation potential', i.e. the ring with more electron-rich substituents, to give **W52** which cyclises to **W53** and hence to **37** by loss of an electron from the OH group.[10]

The spiro compound **37** was heated at 170 °C in refluxing PhOEt to give *O*-methyl dechloro-lecideoidin. **Problem 33.18**: How does **37** give **W54** merely on heating?

Answer 33.18: Again we consult the authors[11] who say 'rearrangement via a ketene'. This means **W56**, formed by fragmentation **W55** of the anion of **37**. Cyclisation of the phenoxide onto the ketene **W56a** then gives the anion **W57** of **W54**.

The Borane Route to Heterocyclic 'Phenols'

The most benzene-like of the heterocycles, thiophenes such as **W58** give 'phenols' **W60** that prefer to exist as ketones **W60a** by the boronic acid route.[12]

Pd(II) Oxidation of Silyl Enol Ethers

This reaction was discovered by Tsuji[13] during attempted allylation of silyl enol ethers with allylic carbonates as reagents and Pd(OAc)$_2$ as catalyst. In THF the expected allylation of the enolate occurred but in MeCN the enone **123** was formed with various other products.

Diallyl carbonate forms a π-complex with Pd(0) that decomposes **W62** to give the η^3 Pd-allyl action complex **W63** and the alkoxide **W64**. The alkoxide releases the enolate **124** from **122** and reaction with **W63** gives the enone and Pd(0) ready for the next cycle.

Hydroxylation Via Epoxides

The simple cyclopentenone oxide **W66** reacts with acetic acid to give an enone hydroxylated on the other side of the ketone **W67**. **Problem 33.19**: What is the mechanism?

Answer 33.19: Epoxidation by the anion of H_2O_2 follows the normal path of conjugate addition **W68** and cyclisation with cleavage of the weak O–O bond **W69**. This step is the genuine oxidation.[14]

Now the enol of the keto-epoxide reacts with acetate ion in an S_N2' reaction **W70** to give another enol that can eliminate water **W71** to give the final product.

Woodward's Dictum on Model Compounds (see page 796 in the textbook)

What Woodward actually said[15] was: 'Perhaps also this is the point at which I should emphasise explicitly the importance of the availability of the "unnatural" enantiomer. Much as had been our progress at this point (in the synthesis of vitamin B_{12}), we were not unaware that we still had far to go, and that it might be either necessary or desirable – as indeed it turned out to be – to investigate a considerable number of alternatives for further advance. In these explorations we were able to utilise (the unnatural enantiomer) confident that whatever route we might establish through its study would be applicable to its counterpart of the natural series; our experience has been such that this is about the only kind of model study which we regard as wholly reliable! And in fact, although the reactions I shall describe in the sequel will be presented for compounds in the natural series, almost all of them were first discovered using the enantiomeric substances.'

References

1. R. D. Clark and C. H. Heathcock, *Tetrahedron Lett.*, 1974, 2027.
2. C. H. Heathcock, C. Mahaim, M. F. Schlecht and T. Utawanit, *J. Org. Chem.*, 1984, **49**, 3264.
3. L. A. Paquette, H. Lin and J. C. Gallucci, *Tetrahedron Lett.*, 1987, **28**, 1363.
4. P. T. Lansbury and J. P. Vacca, *Tetrahedron Lett.*, 1982, **23**, 2623.
5. F. J. Sardina, M. M. Paz, E. Fernández-Megía, R. F. de Boer and M. P. Alvarez, *Tetrahedron Lett.*, 1992, **33**, 4637.
6. G. Büchi, M. A. Francisco and P. Michel, *Tetrahedron Lett.*, 1983, **24**, 2531.
7. P. A. Grieco and M. Nishizawa, *J. Chem. Soc., Chem. Commun.*, 1976, 582.
8. R. M. Moriarty, K.-C. Hou, I. Prakash and S. K. Arora, *Org. Synth.*, 1986, **64**, 138.
9. La V. L. Brown, S. Kulkarni, O. A. Pavlova, A. O. Koren, A. G. Mukhin, A. H. Newman and A. G. Hortl, *J. Med. Chem.*, 2002, **45**, 2841.
10. P. M. McEwen and M. V. Sargent, *J. Chem. Soc., Perkin Trans. 1*, 1981, 883.
11. T. Sala and M. V. Sargent, *J. Chem. Soc., Perkin Trans. 1*, 1981, 855.
12. A. B. Hörnfeldt, *Acta Chem. Scand.*, 1965, **19**, 1249.
13. J. Tsuji, I. Minami and I. Shimizu, *Tetrahedron Lett.*, 1983, **24**, 5635, 5639; J. Tsuji, K. Takahashi, I. Minami and I. Shimizu, *Tetrahedron Lett.*, 1984, **25**, 4783.
14. M. P. L. Caton, G. Darnbrough and T. Parker, *Synth. Commun.*, 1978, **8**, 155.
15. R. B. Woodward, *Pure Appl. Chem.*, 1968, **17**, 519, 529.

34

Functionality and Pericyclic Reactions: Nitrogen Heterocycles by Cycloadditions and Sigmatropic Rearrangements

This chapter considers the effect of nitrogen atoms on Diels-Alder reactions, [3,3]-sigmatropic rearrangements such as the Aza-Cope, group transfer reactions such as the Alder Ene reaction and other pericyclic processes.

In the textbook we say that 'the tautomerism between **24** and **27** may be especially easy[1] as it can be drawn as a [1,5]H-sigmatropic shift.' **Problem 34.1**: Can it? There seems to be a problem at first glance. Why is this not a serious problem?

Answer 34.1: The problem is that the required H for the [1,5]-sigmatropic shift is held away from the nitrogen atom by the *E*-alkene (between C-2 and C-3) **24a**. This is not in fact a problem as conjugated imines have low rotation barriers as the π-electrons are polarised towards the N atom. Hence, **24** can isomerise–not much but a bit–into **W1**, the [1,5]H is then very favourable and gives **W2** which is a conformer of **27**.

The Synthesis of Pumiliotoxin

In the textbook we described a synthesis of pumiliotoxin C **102** using the aza-Diels-Alder reaction. The required aldehyde **105** was made from the known[1] enantiomerically pure **106**. **Problem 34.2**: Suggest how **105** might be made and how it might be transformed into **102**.

Answer 34.2: We simply give the published synthesis. The protecting group R in **106** was THP and the other alcohol was oxidised by PCC to the unstable aldehyde **W3** and then combined immediately with the ylid from **W4** to give the diene **W5** after deprotection. The isomers of **W5** could be separated by chromatography.[2]

The chain extension was done in two stages. First the anion of a sulfone was added to the iodide **W6** and the phenyl sulfonyl group removed by reduction. Then a cyanide extension gave **W8** ideal for DIBAL reduction to **105**. This sequence is old fashioned but it works efficiently.

Boeckmann's Lycorine Synthesis

We stated in the textbook that the starting material **118** for lycorine was prepared by elimination on the spirocyclic ammonium salt **117**. **Problem 34.3**: Suggest a mechanism for the formation of **118** from **117**.

Answer 34.3: We hope you enjoyed this simple problem! Base (DBU) forms the enolate of **117** that undergoes E1cB cleavage of the spirocyclic ammonium salt[3] **W9**.

Problem 34.4: The starting material **117** for the last example was made from a cyclopropane-containing ammonium salt **W10**. Suggest a mechanism.

Answer 34.4: Nearly as easy as the last one! This time the ammonium salt is cleaved **W11** first and then recycles to give the 5/5 spiro system by a very favourable 5-*exo-tet* reaction **W12**.

Aza-Cope Rearrangements

The racemic ketone **177**, a starting material for aza-Cope chemistry, was made by the sequence below. **Problem 34.5**: Suggest mechanisms for the reactions.

W13 → W14 → 177; 70% yield

W15 → W16

Answer 34.5: The formation of **W14** is standard silyl-acyloin chemistry.[4] The formation of **W16** is a Mannich (or Strecker) style of reaction much like the ones used in the textbook.[5] The interesting step is the coupling of **W14** and **W16**. Methanol releases **W17** which is the enol of the acyloin **W18** itself. Addition of the amine **W16** gives the imine salt that tautomerises **W19** to the enamine **W20**, also the enol of the product **177**. The asymmetric version uses a single enantiomer of 1-phenylethylamine.[6]

W14 → W17 → W18 → W19 → W20

One product **W21** from this sequence was used in a synthesis of crinine. **Problem 34.6**: Suggest a possible route from **W21** to crinine **W22**. You may find it easier if you draw **W21** as **W21a**. The original used a sequence of six steps.

W21 = W21a → W22; crinine

Answer 34.6: Again we shall deal only with the published version.[7] Part of the synthesis is straightforward. A Pictet-Spengler reaction with formaldehyde (another relation of the Mannich reaction) forms the last ring **W23**. Bromination and elimination gives the enone **W24** though with alkene and ketone transposed from that needed for crinine.

W21a → W23; 79% yield → W24; 73% yield

Now to reduce and transpose the enone. Reduction of **W24** gave a mixture of alcohols **W25** converted into their tosylates and allowed to solvolyse in aqueous bicarbonate. Water added to the allylic cation derived from **W26** at the far end of the bridge and from the opposite face (axially) to give crinine **W22**. This was a racemic synthesis but clearly the same chemistry would be successful on enantiomerically pure material.

W24 $\xrightarrow{\text{LiAlH}_4}$ **W25** $\xrightarrow[\text{2. TsCl}]{\text{1. BuLi}}$ **W26** $\xrightarrow[\text{H}_2\text{O}]{\text{NaHCO}_3}$ **W22** 42% yield

Asymmetric Aza-Cope Rearrangements: The Chiral Pool Strategy

In the textbook we revealed but did not discuss a simpler route to the [3,3] precursor **191**. **Problem 34.7**: Suggest reagents for the conversion of **189** to **190**.

189; alanine $\longrightarrow$ **190** $\xrightarrow[\text{excess}]{\text{Li}}$ **191; 80% yield** 20:1 diastereoisomers $\longrightarrow$ **192**

Answer 34.7: We need to protect the amino group and convert the acid into a phenyl ketone. You might have suggested a Weinreb amide. The published route comes from earlier work by Rapoport and uses reaction of the protected amino acid **W27** with PhLi or a Friedel-Crafts reaction on the acid chloride. Treatment of **W27** with three molecules of PhLi (one to remove the carbamate proton) gave 85–90% yield of **190**. The Friedel-Crafts route, without isolating the acid chloride **W28**, gave 90% of **190**. Samples of the ketone **190** prepared by either method[8] had >99% ee.

189; alanine $\xrightarrow[\substack{\text{NaOH} \\ \text{pH 9}}]{\text{ClCO}_2\text{Et}}$ **W27** $\xrightarrow[\text{CH}_2\text{Cl}_2]{(\text{COCl})_2}$ [**W28**] $\xrightarrow[\text{AlCl}_3]{\text{PhH}}$ **190**

Problem 34.8: Explain the stereoselectivity in the formation of **191**. Why is excess vinyllithium needed?

Answer 34.8: Your first thought should have been Felkin, but the best conformation would be **W29** and this gives **W30**, the other diastereoisomer **W31** of **191**.

190 $\longrightarrow$ **W29** $\xrightarrow{\text{RLi}}$ **W30** $\longrightarrow$ **W31**

The second point (why is excess vinyl-lithium needed?) gives a clue that chelation control is in action here. The first molecule of the vinyl-lithium removes the NH proton to give **W32**. Chelation control then reverses the direction of addition **W33** and gives **191**.

Problem 34.9: Suggest reagents for the conversion of **191** to **192**.

Answer 34.9: Removal of the protecting group, closure of the oxazolidine ring, and *N*-methylation are all that is needed.[5,9] Hydrolysis of the carbamate gave **W34** with spontaneous decarboxylation. Redrawing **W34** as **W34a** makes it clear that reaction with formaldehyde gives **W35** with the right stereochemistry and reductive methylation gives **192** in 84% yield.

In the textbook we gave a tandem aza-Cope **193** and intramolecular Mannich **194** sequence for the conversion of **192** into **195**. **Problem 34.10**: Propose an alternative tandem sequence for these reactions involving only ionic reactions (a cyclisation and a rearrangement).

Answer 34.10: A simple ionic cyclisation **W36** (a kind of aza-Prins reaction if you really want to know) gives a relatively stable tertiary benzylic cation that can rearrange in a (semi-)pinacol style **W37** to give **195**.

Problem 34.11: How do the experiments outlined in the chapter prove that this mechanism is not correct?

Answer 34.11: The intermediate **194** is achiral and explains why the product **195** is racemic. The intermediate **W37** in the alternative mechanism is chiral and both steps are stereospecific so this mechanism does not explain racemic product.

Electrocyclic Reactions and the Synthesis of Lysergic Acid

In the textbook we showed how an electrocyclic reaction can give **214** which is protonated and reduced to give a stable heterocyclic system **217**.

If the product **214** is not trapped by reduction under the reaction conditions a suprafacial [1,5]-H sigmatropic shift **W38** occurs to give a delocalised version **W39** of the stable amide **W40**. This might well be a useful reaction but not when you are trying to make lysergic acid.

The electrocyclic reaction leads to a mixture of diastereoisomers (7:1) in favour of the one that is wanted whose stereochemistry can be seen from the reduction product **213**.

The minor diastereoisomer from the reaction is actually **W42** which must come from the stereoselective reduction of the minor diastereoisomer **W41**. The stereoselectivity of the reduction is evidently dominated by the remote chiral centre (ringed in **W41**) and not by the stereochemistry of the new ring junction as that is *trans* in **213** and *cis* in **W42**. **Problem 34.12**: Suggest an explanation for this stereoselectivity.

W41   → NaBH₄ →   W42

Answer 34.12: You might like to be reminded of the way we presented the electrocyclic reaction in the textbook. It looks best if drawn on a delocalised form of the starting material **210** and we can now add the extra chiral centre (ringed in **214**) as this step is totally stereoselective. The two flat rings (**A** and **B** in **210b**) must already be tilted so the vital H in ring **B** is above ring **A** and that of ring **A** below ring **B**. This tilting ensures that the furan (ring **B**) is *cis* fused to the amide ring. That the reduction is 7:1 stereoselective is not surprising, we should expect the *trans* ring junction to be preferred in **213**.

210a   210b   *hv* six electron conrotatory electrocyclic   214

Problem 34.13: Outline the chemistry needed to convert **217** into lysergic acid **211**.

217   ?   211; Lysergic acid

Answer 34.13: It is simplest just to give the published version,[10] though it is old work and you may have better ideas. First the dihydrofuran ring must be oxidised. Dihydroxylation and oxidation of the mixture of diols **W43** gave an epimeric mixture of aldehydes **W44**.

217   1. OsO₄   2. H₂S → W43; 53% yield   NaIO₄ / H₂O → W44; 53% yield

Oxidation of the aldehyde gave a mixture of esters **W45** but elimination epimerised them to the equatorial isomer **W46**, a compound that had already been converted into lysergic acid by Ramage: ester hydrolysis and deprotection at nitrogen are all that is needed.[11]

Tandem Aza-Diels-Alder and Ring-Closing Metathesis Reactions

Barluenga has been a major player in the area of aza Diels-Alder reactions and in recent work[12] his group has revealed new ways of making bicyclic alkaloids with a bridgehead nitrogen atom. Both diene and dienophile were prepared with simple alkene side chains ready for subsequent ring-closing metathesis. The dienophiles were simple imines **W48** and **W49** prepared from unsaturated amines and an aldehyde.

The dienes required harder work. Palladium-catalysed coupling of an electron-deficient alkyne **W50** with electron-rich trimethylsilyl ethyne gave, after oxidation state adjustment, the unsaturated aldehyde **W51**. The rest is simple except for the Hg(II)-catalysed addition of an amine to the alkyne.

The cycloaddition between **W53** and **W48** is catalysed by the Lewis acid ytterbium triflate and the product **W54** is easily hydrolysed with acid to give the ketone **W55** with four chiral centres. **Problem 34.14**: Account for the stereochemistry of **W55**.

W53 → W54; 55% yield → W55; 97% yield

Answer 34.14: Partly this is easy to explain by the *endo* arrangement for the Diels-Alder reaction: the *N*-aryl group overlapping with the back of the diene **W56**. It should be clear that the *C*-Ar group is *anti* to the marked hydrogen atom. The stereochemistry of the methoxy group is more tricky and that of the methyl group appears only during the hydrolysis of the enamine and is under thermodynamic control as **W55a** should make clear. Every substituent on the piperidine ring is equatorial.

W56 → W54 → W55a

The ring-closing metathesis requires only the Grubbs I catalyst (chapter 15) and gives very high yields of **W57** from **W55**. The other imine **W49** can be used in the same sequence with the slightly modified diene **W58** to give a new eight-membered ring **W59**.

W55 → W57; 97% yield   W58 → W59

References

1. H. R. Snyder, R. B. Hasbrouck and J. F. Richardson, *J. Am. Chem. Soc.*, 1939, **61**, 3560; H. R. Snyder and J. C. Robinson, *J. Am. Chem. Soc.*, 1941, **63**, 3279.
2. P. A. Grieco and D. T. Parker, *J. Org. Chem.*, 1988, **53**, 3658.
3. R. K. Boeckman, J. P. Sabatucci, S. W. Goldstein, D. M. Springer and P. F. Jackson, *J. Org. Chem.*, 1986, **51**, 3740.
4. Clayden, *Organic Chemistry*, pp. 1032–3.
5. L. E. Overman and S. Sugai, *Helv. Chim. Acta*, 1985, **68**, 745.
6. L. E. Overman, L. T. Mendelson and E. J. Jacobsen, *J. Am. Chem. Soc.*, 1983, **105**, 6629.
7. H. W. Whitlock and G. L. Smith, *J. Am. Chem. Soc.*, 1967, **89**, 3600.
8. T. F. Buckley and H. Rapoport, *J. Am. Chem. Soc.*, 1981, **103**, 6157.

9. L. E. Overman and S. Sugai, *Helv. Chim. Acta*, 1985, **68**, 745; E. J. Jacobsen, J. Levin, and L. E. Overman, *J. Am. Chem. Soc.*, 1988, **110**, 4329.
10. T. Kiguchi, C. Hashimoto, T. Naito and I. Ninomiya, *Heterocycles*, 1982, **19**, 2279.
11. R. Ramage, V. W. Armstrong and S. Coulton, *Tetrahedron*, 1981, **37**, (Suppl. 1), 157.
12. J. Barluenga, C. Mateos, F. Aznar and C. Valdés, *J. Org. Chem.*, 2004, **69**, 7114.

35

Synthesis and Chemistry of Azoles and other Heterocycles with Two or more Heteroatoms

This chapter considers how aromatic heterocyclic compounds with two or more nitrogen (or other heteroatoms, chiefly O and S) atoms can be made. In particular it deals with the addition of new C–C and C–X bonds to previously prepared heterocycles of this kind.

Problem 35.1: Suggest a synthesis for compound **W1** containing both thiazole and furan rings.

Answer 35.1: Disconnecting the thiazole first **W1a** reveals thiourea as a good starting material. The bromoketone **W2** can be made by bromination of **W3** that comes in turn from the furyl ketone **W4** by nitration. Furan is very α-selective in electrophilic substitution: the first (acetyl) group goes in either α-position and the second (nitro) goes in the other α-position in spite of the deactivation by the acetyl group. The most interesting step, the formation of **W1** from **W2** and thiourea actually gives the HBr salt in ethanol in 84% yield. The crystalline free base **W1** was formed just by slurrying the HBr salt in water.[1]

Problem 35.2: Suggest a synthesis of the starting material **118** for the synthesis of aconazole **119**.

W5   W6; aconazole

Answer 35.2: The most obvious disconnections are of the acetal **W6a** followed by a Friedel-Crafts **W7**.

Analysis

W5a   W7   W8   W9

This is the synthesis used in a patent.[2]

Synthesis:

W8   W7   W5

The most remarkable thing about this route is not that they used the triazole anion in the formation of **W5** from **W6** but that the S_N2 reaction works. Neighbouring oxygen atoms slow S_N2 reactions and a tertiary centre next to the reacting centre is also very bad. So is it possible that the acetal (a five-membered heterocycle with two heteroatoms) participates in the reaction? The alternative synthesis of **W7 = 131** given in the textbook avoids this problem and ensures that the asymmetric diol is successfully incorporated.[3]

130   131   132; antifungal drugs

The Synthesis of Isoxazoles

Problem 35.3: Suggest a synthesis for 'sulfamethoxazole' **W10** one of the sulfa drugs that still finds uses today. Notice that it is in fact an isoxazole, not an oxazole reminding us not to place too much reliance of 'chemical' names given by drug companies.

W10
sulfamethoxazole

Answer 35.3: The obvious first disconnection is at the sulfonamide to split the molecule into two similar-sized parts at an easily made bond. The intermediate **W11** is some derivative of available 4-amino-benzenesulfonic acid and the isoxazole **W12** is the interesting bit. There were several syntheses of isoxazoles in the textbook chapter and you might have tried any of them. Two that look promising are the 1,3-dipolar cycloaddition of a nitrile oxide to an alkyne **W12a** and addition of hydroxylamine to a ketoacid derivative **W12b**. Both have potential regioselectivity problems.

One old method that does work is to use the keto-nitrile **W13** and hydroxylamine. A very remarkable method is the treatment[4] of the propargyl bromide **W15** with sodium nitrite and the reduction of the nitro-isoxazole **W14** with Sn (II). **Problem 35.4**: Suggest a mechanisms for the formation of **W14**.

Answer 35.4: Clearly two nitrite anions are added to **W15** to get the right number of atoms and one must be added as a nucleophile to make the nitro compound **W16**. But now some electrophilic source of 'NO[+]' must add the 'enolisable' position of **W16** to give **W17** and hence **W18**. Now we have all the necessary atoms for **W14**.

The authors suggestion[5] is that nitrite ion sometimes adds to **W15** to give the nitrite ester **W19** and that this is the nitrosating agent. You probably didn't see this but any reasonable nitrosating agent will do. Cyclisation **W20** in a favourable 5-*endo-dig* process gives **W14** after protonation. The final sulfonation uses the amide of **W11**; X = Cl and hydrolysis with NaOH removes the amide.

Problem 35.5: Suggest a synthesis of Lundbeck's LU25-109 **W23**, an arecoline analogue proposed as a muscarinic M_1 agonist and M_2 antagonist for the treatment of Alzheimer's disease. Reduction of the pyridinium salt **W22** with $NaBH_4$ in ethanol at room temperature gives **W22**. Suggest a mechanism for this reduction.

Answer 35.5: The standard way to make tetrazoles is the 1,3-dipolar cycloaddition between azide ion and a nitrile, here available 3-cyanopyridine. Alkyl groups are added by alkylation of the tetrazole anion: in this case ethylation gives **W25** regioselectively. Now methylation on the pyridine nitrogen and reduction gives **W23**.

The reduction occurs first at the iminium salt and the dienamine is protonated by solvent **W26** to form a new iminium salt **W27** that is again reduced at the most electrophilic centre. The products of this widely used reaction usually put the remaining alkene where it is stabilised by the substituent.

Aromaticity

Problem 35.6: The imidazole ring in cimetidine **3** is aromatic – why? The seven-membered ring in valium **4** could become conjugated by forming an enol but it does not do so. Why not?

3
**Cimetidine
(Tagamet)
(1970s)**

imidazole

4
**Diazepam
(Valium)**

Answer 35.6: Counting two electrons each for the π-bonds and two for the lone pair in a p-orbital on the pyrrole-like N atom gives six. We do not count the lone pair on the pyridine-like N as that cannot be delocalised. The enol of valium **W28** would be conjugated but also anti-aromatic having six π-electrons plus two in the enamine N. Again we do not count the lone pair on the imine N.

sp² not
counted

2p

2p

2p

3a

4

W28

Nitrile Oxides for 1,3-Dipolar Cycloadditions

In the textbook we revealed that there are two good ways to make nitrile oxides. Dehydration of alkyl nitro compounds **68**, either with PhNCO or with Ph_3P and DEAD ($EtO_2C–N=N–CO_2Et$) in a Mitsunobu elimination gives nitrile oxides **69**, as does the 1,3-elimination of HCl from chloro-oximes **70**. **Problem 35.7**: Draw mechanisms for these reactions.

68

PhNCO
or Ph_3P
DEAD

69

Et_3N

70

Cl_2

71

NH_2OH

Answer 35.7: The dehydration of the nitro-compound obviously needs removal of one oxygen atom and the two Hs marked in **68**. Using the isocyanate as an example, addition to one of the Os **W29** gives an intermediate that can transfer a proton intramolecularly **W30** to give a compound that decomposes with loss of CO_2 and aniline **W31** to give the nitrile oxide. The Mitsunobu version is very similar.

W29

W30

W31

$69 + CO_2 + PhNH_2$

The 1,3-elimination route may have foxed you because you can not draw a continuous chain of connected arrows. The weak base gives some of the anion of **70** that eliminates chloride faster than the neutral compound as the product **69** is neutral.

Regioselectivity of Isoxazole Synthesis by 1,3-Dipolar Cycloaddition

In the textbook we described a successful way to control the regioselectivity of 1,3-dipolar cycloadditions to alkynes with nitrile oxides. **Problem 35.8**: Draw mechanisms for the formation of **83** and **84** from **82** and explain the regioselectivity of each reaction.

Answer 35.8: There was some discussion of this in the textbook. In essence the reaction can occur the 'sensible' way round **W33** with the negative end of the dipole attacking the electron-deficient end of the ynone and the other end of the alkyne attacking the electrophilic end of the nitrile oxide. The other way round **W34** is much less appealing.[6]

We know that this is false as the major product is **84**, not **83**. We know from Baldwin's rules that we do not understand LUMOs of alkynes as well as we might suppose. However, we certainly have no excuse if we suppose we understand HOMOs and LUMOs of nitrile oxides. We only have to redistribute the electrons in **69** in a different way **69a** to change our view. The solution expounded in the textbook is to make a definite change in the dipolarophile. Thus **85** is electron deficient and will use its LUMO while **86** is electron rich and will use its HOMO.

Reactions of Tetrazoles

We described the synthesis of Merrell Dow's anti-allergic agent MSD 427 **99** in the textbook.
Problem 35.9: Draw a mechanism for the cyclisation of **106** to **99** and comment on the
regioselectivity. This might be tricky because the anion of the tetrazole might be expected to
give a different product.

Answer 35.9: Any base will remove the proton from the tetrazole to give **W35** but this cannot
cyclise to **99** and would be expected to react through one of the tetrazole nitrogens. However,
the dianion produced by removal of the other NH proton can cyclise **W36** to **99** if a strong
enough base is used.[7]

How to get Reaction at Nitrogen using Azole Anions

The anti-inflammatory tetrazole broperamole **124** was made from **126** as described in the
textbook. The last stage was a base-catalysed conjugate addition between **125** and acrylic acid.
Problem 35.10: What sort of base would you recommend and what species would actually be
present in the reaction mixture?

Answer 35.10: A weak base should be recommended so that the anion of the tetrazole would
be in equilibrium with that of the carboxylic acid. Then the required reactive species – tetrazole
anion and neutral carboxylic acid – would both be present. The two pK_as should be about the
same.[8]

An Indole with a Tetrazole

We described the synthesis of the indole-based drug **128** by acylation of the dianion **127** of **126**. **Problem 35.11**: Suggest a synthesis of the starting material **126**.

Answer 35.11: The tetrazole is made from the nitrile **W37** and hence from some electrophile **W38** that could be made from indole itself **W39**. You might also have considered a Fischer indole synthesis of **W37** or **W38**.

There is an excellent way to make compounds like **W38**. The Mannich reaction works very well on indole and is entirely regioselective for 'gramine' **W40**. Alkylation with MeI gives the ammonium salt that reacts well with basic nucleophiles to give,[9] in this case, **W37**. This reaction probably occurs *via* elimination of NMe_3 from the anion of **W41**.

The formation of the tetrazole **126** is the usual 1,3-dipolar cycloaddition of azide ion and the acylation completed with a twofold excess of NaH and the right acid chloride.

Reactions at Nitrogen Under Neutral Conditions

In the textbook we discussed using silyl derivatives with cation-like electrophiles to achieve reaction at nitrogen in neutral conditions, as in the synthesis of **155**. **Problem 35.12**: Draw a mechanism for and explain the selectivity in the silylation of **151**.

Cytidine Synthesis:

151 → 152 (with (Me₃Si)₂NAc)

153; ribose → 154 (BzO, BzO, OBz, OAc)

155 protected cytidine

Answer 35.12: Silylation could occur at any N or O atom by simple nucleophilic attack at silicon. However, reaction occurs outside the ring to preserve the aromaticity of the pyrimidine.[10]

W42 → W43 + W44 → 152

Problem 35.13: Suggest how ribose **153** might be converted into the protected form **154** and how **155** might be deprotected to give cytidine **146**.

Answer 35.13: This is an exercise in protection and deprotection. There is no difficulty in benzoylating or acetylating alcohols: the problem is how do we acetylate just one alcohol and benzoylate all the others. The key is the unique properties of the hemiacetal at C 1'. Any alcohol can be added to give an acetal such as **W45** and the remaining hydroxyl groups benzoylated as usual.[11] Hydrolysis of the acetal **W46** and acetylation gives **154**.

153; ribose → W45 (MeOH, H⁺) → W46 (PhCOCl) → 154

The deprotection is straightforward: ammonia in methanol is a good reagent for cleaving esters when the alcohol is wanted – the by-product being benzamide in this case.[12] The silyl groups often drop off during the Vorbrüggen coupling but can be removed with a source of fluoride such as TBAF.

155 protected cytidine → W47 (NH₃, MeOH) → 146 cytidine (TBAF, MeOH)

A Purine Analogue as an Anti-Cancer Drug: Clofarabrine

Antineoplastic agents include nucleoside analogues such as clofarabrine **W48** that could clearly be made by coupling some protected version of the sugar derivative **W49** with the purine **W50**. However, there is a problem. Unlike the ribose compounds discussed here and in the textbook, there is no participating group to guide the stereochemistry. This was the problem faced by chemists at ILEX and Ash Stevens.[13]

W48; clofarabine　　　W49　　　W50

Previous work[14] had established that silylated purine bases gave mixtures of inverted **W52** and retained **W53** products from the protected arabinose derivative **W51**. The S_N2 product **W52** was favoured by less polar solvents but was never the exclusive product.

W51　　silylated purine bases　　W52　　+　　W53

The chemists at ILEX and Ash Stevens decided[13] to try direct S_N2 reactions between **W51** and **W50** under basic conditions. Eventually they found that potassium *t*-butoxide in a mixture of MeCN and *t*-amyl alcohol with a calcium hydride additive (it acts as a drying agent) gave a high yield of a >20:1 ratio of the required stereochemistry **W52** to the unwanted **W53**: a mixture that could be purified without chromatography. Deprotection gave clofarabrine.

W51　　W50 CaH₂　*t*-BuOK MeCN *t*-AmOH　　W54　　NaOMe MeOH　　W48 64% yield

The Synthesis of Cimetidine

Returning to cimetidine **3** the synthesis of one starting material starts with formation of the imidazole **206** by a version of the Bredereck reaction. More generally, an α-hydroxy carbonyl compound **W52** (or derivative) reacts with an excess of formamide to give an imidazole **W53**. **Problem 35.14**: Suggest a general mechanism for the Bredereck reaction giving **W53**. This is tricky as there is no general agreement on the mechanism but it is a good challenge for you at this stage.

208; X = OAc → 206 ; W52 → W53

Answer 35.14: It is generally agreed[15] that formation of an imine **W54** between the ketone and formamide allows tautomerism to the enamine **W55** which is also the enol of the transposed ketone **W56**.

W55 → W57 ⇌ W58 ⇌ W59

In the same way, addition of a second molecule of formamide to **W59** gives a new enamine that can cyclise **W60** to form the five-membered ring **W61** and hence by dehydration, readdition of water to the formyl group and loss of formic acid **W63**, to the product. There are other possible mechanisms including an attractive formation of an oxazole intermediate.

W60 ⇌ W61 ⇌ W62 → W53 ; W63

The Synthesis of Pentostatin

The synthesis of pentostatin described in the textbook starts with the routine nitration of available **228**, the base-catalysed reaction of **227** with benzaldehyde to give **226**; R=Ph and benzylation also in base, to give predominantly the isomer **227** with the longest conjugated system. The bases are different: piperidine at 95 °C as reagent and solvent was used for the formation of **226** while K_2CO_3 in DMF at 75 °C was used for the alkylation. **Problem 35.15**: Draw mechanisms for the formation of **226** and **227** and comment on the choice of bases.

228 (available) → 227 → 226, R = Ph>80% yield

227 >95% yield, 3:1 isomers 228 → 229, 75% yield crystalline acid

Answer 35.15: Base removes a proton **W64** from the methyl group of **227** to create an anion stabilised by the nitro group.[16] Aldol-style reaction with benzaldehyde **W65** and E1cB elimination gives **226**.

The alkylation step looks simpler: the weak base removes the proton from the imidazole ring as the resulting anion is stabilised by the nitro group **W67**. The alkylation can occur through either nitrogen but **W68** is favoured by the longer conjugated system in **230**. This raises the question of what really happens in the first deprotonation **227**. The nitro group is already there so presumably the first proton is removed from nitrogen and the second from the methyl group to give **W69** as the true intermediate. Hence the need for the stronger base: piperidine has a pK_a of about 12.5, K_2CO_3 has a pK_a of about 10. Alkylation occurs at carbon on the 'first in, last out' principle (chapter 2).

References

1. W. R. Sherman and D. E. Dickson (Abbott), *J. Org. Chem.*, 1962, **27**, 1351; O. Dann, H. Ulrich and E. F. Moller, *Z. Naturforsch.*, 1952, **7b**, 344.
2. G. Van Reet, J. Heeres and L. Wals, Janssen Pharmaceutical N.V., *Ger. Pat.*, 2,551,560, *Chem. Abstr.*, 1976, **85**, 94,386.
3. G. J. Tanoury, C. H. Senanayake, R. Hett, Y. Hong and S. A. Wald, *Tetrahedron Lett.*, 1997, **38**, 7839; J. Heeres, L. J. J. Backx and J. Van Cutsem, *J. Med. Chem.*, 1984, **27**, 894.
4. S. Rossi and E. Duranti, *Tetrahedron Lett.*, 1973, 485.
5. H. Kano, H. Nishimura, K. Nakajima and K. Ogata, *U.S. Pat.* 2,888,455 to Shionogi and Co. Ltd, *Chem. Abstr.*, 1959, **53**, 22018i.
6. C. Kashima, S.-I. Shirai, N. Yoshiwara and Y. Omote, *J. Chem. Soc., Chem. Commun.*, 1980, 826.
7. N. P. Peet, L. E. Baugh, S. Sunder, J. E. Lewis, E. H. Matthews, E. L. Olberding and D. N. Shah, *J. Med. Chem.*, 1986, **29**, 2403; A. P. Vinogradoff and N. P. Peet, *J. Heterocycl. Chem.*, 1989, **26**, 97.
8. Miles Laboratories Inc., British Patent 1,319,357, *Chem. Abstr.*, 1973, **79**, 92231.
9. P. F. Juby and T. W. Hudyma, *J. Med. Chem.*, 1969, **12**, 396.
10. H. Vorbrüggen and C. Ruh-Polenz, *Org. React.*, 1999, **55**, 1; R. T. Walker in *Comprehensive Organic Chemistry*, **5**, 53, 64.
11. E. F. Ricordo and H. Rinderknecht, *Helv. Chim. Acta*, 1959, **42**, 1171.
12. U. Niedballa and H. Vorbrüggen, *J. Org. Chem.*, 1974, **39**, 3655; *J. Org. Chem.*, 1976, **41**, 2084.

13. W. E. Bauta, B. E. Schulmeier, B. Burke, J. F. Puente, W. R. Cantrell, D. Lovett, J. Goebel, B. Anderson, D. Ionescu and R. Guo, *Org. Process Res. Dev.*, 2004, **8**, 889.

14. C. H. Tann, P. R. Brodfuehrer, S. P. Brundidge, S. Sapino and H. G. Howell, *J. Org. Chem.*, 1985, **50**, 3644.

15. M. R. Grimmett, *Adv. Het. Chem.*, **12**, 103.

16. D. C. Baker and S. R. Putt, *J. Am. Chem. Soc.*, 1979, **101**, 6127; E. Chan, S. R. Putt, H. D. S. Showalter and D. C. Baker, *J. Org. Chem.*, 1982, **47**, 3457.

36

Tandem Organic Reactions

Problems and Further Examples Relating Directly to the Text

You will find it helpful to have chapter 36 from the textbook open as you look at this first section.

A Problem from the Textbook

We promised that the synthesis of the cyclic enamine **73** would appear as a problem in the workbook and here it is. **Problem 36.1**: Suggest a synthesis of the cyclic enamine **73**.

Answer 36.1: As **73** is an enamine, the obvious disconnection **73a** reveals an amino aldehyde with a 1,4-relationship between the functionalised carbons. Some disconnection such as **W1** might follow but aldehydes like **W2** are tricky to use as they enolise and self-condense very easily. There must be a better way.

The method used was indeed very different and depends on a rearrangement of a cyclopropane. The stable nitrile **W3** was alkylated to give the cyclopropane **W4** and reduced to the aldehyde **W5**. This aldehyde, unlike **W2**, cannot enolise and is stable. The imine **W6** rearranges with catalytic ammonium iodide to give the enamine[1] **73**. **Problem 36.2**: Suggest a mechanism for this last reaction.

Answer 36.2: Iodide can open the cyclopropane ring **W7** of the protonated imine and the resulting amine could cyclise **W8** either to **W6** or to **73**. The five-membered ring is much more stable than the three-membered and this is an example of the general rearrangement of vinyl cyclopropanes (with or without heteroatoms) into cyclopentenes.[2] The original workers[3] are rather coy about the exact mechanism but they are agreed that it is an acid-catalysed rather than a thermal reaction. Stevens says: 'The implication that the thermally induced rearrangement of cyclopropyl imines is analogous to the well documented vinylcyclopropane rearrangement appears on the basis of our experience to be subject to considerable doubt.'

Other Examples of Tandem Reactions

This last chapter is a good place for some general revision and the next section introduces a selection of tandem reactions not specifically treated in the textbook.

A Simple Example

Apparently unrelated reactions can be much more successful as tandem processes than they were when done separately. An asymmetric synthesis of isoquinuclidines starts with a cyclo-addition of a pyrrole **W9** to a tetrabromoketone **W10**. **Problem 36.3**: Given that the intermediate formed from **W10** is the oxyallyl cation **W12**, suggest a mechanism.

Answer 36.3: The oxyallyl cation **W12** does a six-electron $4+2$ cycloaddition **W13** on the deactivated pyrrole as the allyl cation supplies two electrons. The dibromoketone **W14** is the first product and this is debrominated by the Zn/Cu couple.[4] So why bother with four bromines in **W10** when only two are needed? This is a purely practical consideration. Bromoketones are difficult to purify but **W10** can be made directly from acetone in one step and crystallised from the reaction mixture in 98% isolated yield.[5]

The ketone **W11** is of course achiral but asymmetric silyl enol ether **W15** formation with a chiral base followed by bromination gave one diastereoisomer of the bromoketone **W16**. The proposed radical rearrangement (see below) was unsuccessful on the ketone **W16** so it was first reduced to the alcohol **W17** with sodium borohydride. These reactions give one diastereoisomer of **W16** and **W17** in reasonable yield. **Problem 36.4**: Suggest explanations. You are *not* expected to explain the asymmetric induction in the formation of **W16** but you might like to consult chapter 24 of the textbook.

Answer 36.4: There is a chair six-membered ring in **W11** that is more obvious in **W11a**. Bromination occurs axially to retain the chair **W16a**. Reduction of a six-membered cyclic ketone by the small reagent is expected[6] to give the more stable equatorial alcohol **W17a**.

Now finally the promised radical rearrangement can be carried out successfully using Bu_3SnH to generate the radical **W18**. Cyclisation to radical **W19** (3-*exo-tet* is preferred to 4-*endo-tet*) which could simply reopen to give **W18** but prefers the nitrogen-stabilised radical **W20** that gives the product **W21** on hydrogen atom abstraction from Bu_3SnH thus completing the radical chain.

This is all right but the yield of **W17** from **W16** is not very satisfactory (56%). Since Bu_3SnH can be regenerated from Bu_3SnBr (the by-product from the formation of **8**) a much better sequence is to use catalytic Bu_3SnH and $NaBH_4$ to do all the reactions in tandem fashion. The ketone **W16** can be converted into the product **W21** in one pot and in 81% yield.

The Rauhut-Currier Reaction

Why is it that tandem reactions attract such a variety of names? it is probably because of the combination of two reactions in one sequence. So the Rauhut-Currier reaction, which will probably be new to you, has also been called 'Catalytic crossed Michael cycloisomerisation' and even the 'Intramolecular vinylogous Morita-Baylis-Hillman' reaction.' But what is it? The reaction was invented at American Cyanamid[7] as a reaction to dimerise acrylates to e.g. **W22** and later developed at ICI[8] as a route to 1,6-diaminohexane **W24**, a starting material for nylon manufacture. **Problem 36.5**: Suggest mechanisms for these reactions. (You may wish to consult the section on the Baylis-Hillman reaction in chapter 11 of the textbook.)

Answer 36.5: You should have discovered an interesting dilemma—these two products **W22** and **W24** are not formed by the same reaction. Well, not entirely! The formation of **W22** is a Baylis-Hillman reaction in which the intermediate adds in conjugate rather than direct fashion (hence 'vinylogous B-H reaction'). Conjugate addition of Bu_3P to the acrylate **W25** forms an enolate that can add to a second acrylate **W26** to give an enolate **W27** in equilibrium with another that can eliminate Bu_3P **W28** to give the dimer **W22**.

The reaction with acrylonitrile starts the same way but after conjugate addition of the catalyst **W29**, anion exchange gives a phosphorus ylid before the second addition **W31** occurs. Again a proton transfer is needed before the catalyst can be eliminated **W33**. These two reactions often compete and both linear **W23** and branched **W22** products may be formed in the same mixture. The conjugate addition of an enolate, itself formed by reversible conjugate addition of a catalyst, to an unsaturated carbonyl compound is what is usually meant by the Rauhut-Currier reaction whether linear or branched products result.

Two adjacent papers in 2004 described an intramolecular version of the reaction to give cyclic enones. The first,[9] calling the reaction the 'intramolecular Rauhut-Currier reaction,' gave interestingly different results for two homologous *bis*-enone **W34** and **W37**. When a five-membered ring was formed, the two isomeric adducts **W35** and **W36** were formed in a 1:1 mixture. The two six-membered ring products **W38** and **W39** were formed in a 7:1 ratio. **Problem 36.6**: Suggest a mechanism for the reaction(s) and comment on the difference in selectivity.

Answer 36.6: If the catalyst adds to one ketone to give enolate **W40**, cyclisation leads to **W36** but if it adds to the other enone **W41** then **W35** results. As the authors say: 'These data suggest the initial formation of tributylphosphine adducts is indiscriminate.' So why the preference in the cyclisation to give the six-membered rings? They suggest that, because cyclisation is slower in the formation of six- rather than five-membered rings, there is time for the enolates to equilibrate and the more stable conjugated enone **W38** is favoured.

The second[10] describes similar chemistry and includes heterocyclic examples **W42** and those in which the regioselectivity is dominated by a difference in the two carbonyl groups **W44**. **Problem 36.7**: Which unsaturated carbonyl group is attacked by the catalyst?

W42   W43   W44   W45

Answer 36.7: The more electrophilic in each case (i.e. enal in **W42** and enone in **W44**) to give **W46** and **W47**, respectively. We expect cyclisation to a five-membered ring to be faster than enolate exchange.

W42 $\xrightarrow[\text{MeCN}]{\text{Me}_3\text{P}}$ W46   W44 $\xrightarrow[\text{t-AmOH}]{\text{Me}_3\text{P}}$ W47

The Synthesis of the Spinosin A Nucleus

Roush has extended this reaction in combination with other reactions to form many rings in one operation with a high degree of stereoselectivity. Tricyclic **W49** was isolated in 88% yield as a single diastereoisomer and a 96:4 mixture of regioisomers. **Problem 36.8**: One reaction in the sequence is the Rauhut-Currier (or intramolecular vinylogous Morita-Baylis-Hillman as Roush has every right to prefer). What is the other one, and what is the intermediate?

W48 $\xrightarrow{\begin{array}{c}\text{1. 40 °C, 67 hours}\\ \text{t-AmOH}\\ \text{2. Me}_3\text{P}\\ \text{23 °C, 9 hours}\end{array}}$ W49

Answer 36.8: The first reaction is, of course, a Diels-Alder as can easily be proved by isolating the product **W51** when the second step is omitted.[11] Four new centres are created in this step with excellent control. They are all around the new six-membered ring and are arbitrarily numbered in **W51**. The *relative* stereochemistry of C-1 and C-6 comes from the *E*-alkene from which they were formed. The *absolute* stereochemistry of all the new centres is controlled by the two centres already present in **W50**. The dienophile slides underneath the diene to keep away from the two large ethers. This pushes the H at C-2 up. The *trans* arrangement at C-1 and C-2 and the *cis* arrangement at C-5 and C-6 should be obvious from those Hs in **W50**.

W50   →   W51; 93% yield, >30:1 diasts

The second step is the vinylogous Baylis-Hillman reaction and the phosphine evidently adds to the enone as in the previous example. The extra centre created in this reaction puts the CH_2CO_2Me group on the *exo* face of the *cis* fused rings.

Tandem Grubbs Metathesis and Hydrogenation

A clear indication for a tandem reaction comes when two reactions can have the same catalyst. Ruthenium-based catalysts are used both for metathesis and hydrogenation and Grubbs has found[12] that both Grubbs 1 and Grubbs II (see chapter 15) can be used for hydrogenation. The amide **W52** gives the cyclic amide **W53** and then the saturated equivalent **W54** by tandem metathesis and hydrogenation with the same catalyst.

Erythrina Alkaloids by Tandem Allyl Cation and Michael Addition

Palladium catalyses many reactions under conditions that are easily compatible with many more. A striking example[13] is the palladium-catalysed allylation and Michael addition strategy for the synthesis of *Erythrina* alkaloids such as erythramine **W55**. It is simple to add and then disconnect an amide to give the tricyclic intermediate **W57**. What next?

The published solution[13] is surprising. Removing the NE corner of the molecule completely reveals that a nitro-compound **W59** could be an excellent starting material. As a double nucleophile it could perhaps attack the double electrophile **W60**, once by Pd-catalysed allylation and once by Michael addition.

Catalytic Pd(0) is needed to combine the anion of **W59** with the allylic acetate **W60** and the Michael cyclisation occurs spontaneously to give **W58** as a single diastereoisomer with the large groups equatorial **W58a**.

Reduction of the nitro group required another tandem sequence as Zn/NH$_4$Cl gave the hydroxamic acid and TiCl$_3$ was needed to get the amine which spontaneously cyclised to the amide **W61**. The final ring was closed by addition of vinyl sulfoxide and Pummerer cyclisation of **W62**. This was not straightforward and you can read about their adventures in the paper.[13]

The Synthesis of a Naphthyridone MAP Inhibitor

The kinase inhibitor **W63** is produced by Merck as a potential treatment for rheumatoid arthritis, Crohn's disease and psoriasis. There are some obvious disconnections **W64** but it is not obvious how these bonds could easily be made.

Indeed an early synthesis required 18 steps in the longest linear sequence and gave only 2% yield of **W63**. In this synthesis, bond **a** was made by a Suzuki coupling, with **W65**, bond **b** by a Stille coupling with **W66**, and bond **c** by amide **W67** synthesis before the naphthyridine ring was assembled. Both the synthesis of **W66** and the removal of the alkene from the piperidine ring were troublesome while the closure of the naphthyridone ring was difficult.

W65   W66   W67; Ar = 2,4-diflurophenyl

The new synthesis clearly required for production of quantities of **W63** contains very interesting chemistry as well as a tandem reaction. The idea was to carry out a Heck reaction between the iodopyridine **W69** and the acrylamide **W70**. This did form the expected product but, under the right conditions with Pd(OAc)₂ and NaOAc in ethylene carbonate, an 80% yield of the naphthyridone **W71** was the result. This is the tandem reaction. **Problem 36.9**: What is the structure of the intermediate and why is palladium necessary for the second step?

W68   W69   W70   W71

Answer 36.9: The Heck reaction between **W69** and **W70** starts with oxidative insertion of Pd(0) into the pyridine–iodine bond followed by carbo-palladation of the alkene and β-elimination of palladium. This gives the Heck product **W72**. The problem is that **W72** cannot cyclise to **W71** as the alkene has the wrong (*E*) geometry. The Heck reaction is notably *E*-selective but only the *Z* compound **W73** can cyclise to **W71**. Further palladium catalysis is needed to isomerise the double bond and to catalyse the coupling between the amide nitrogen and the pyridyl bromide.

W72   W73   W71

The rest of the synthesis is less exciting. Having displaced in turn iodine and bromine from the pyridine ring, we now need to displace chlorine and the Suzuki coupling was used for this. Finally, the piperidine ring needs to be added at an unactivated position on the pyridine ring of **74**. After various trials they used the *N*-oxide **W75**.

W71   W74; 96% yield   W75; 96.6% yield

Addition of the Grignard reagent of the saturated piperidine (contrast **W66**) and trapping with *iso*-butyl chloroformate gave a 1:1 mixture of **W76** and the drug itself **W63**. A disaster? Not at all. Heating the mixture in pyridine at 110 °C gave a 92% yield of **W63**.

W76; Ar = 2,4-difluorophenyl   **W63; Ar = 2,4-difluorophenyl**

The chemists suggest[14] that the Grignard edition is not stereoselective: the *syn*-compound undergoes a base-catalysed *anti*-elimination at the temperature of the coupling but the *anti*-compound requites heating for a concerted *syn*-elimination of CO_2 and *i*-BuOH.

anti-**W76; Ar = 2,4-difluorophenyl**   *syn*-**W76; Ar = 2,4-difluorophenyl**

References

1. R. V.Stevens, P. M.Lesko, and R.Lapalme, *J. Org. Chem.*, 1975, **40**, 3495; C. P. Forbes, J. D. Michau, T. van Ree, A. Wiechers and N. Woudenberg, *Tetrahedron Lett.*, 1976, 935.
2. *The Disconnection Approach*, p. 286.
3. R. V.Stevens and M. C.Ellis, *Tetrahedron Lett.*, 1967, 5185; R. V. Stevens, M. C. Ellis and M. P. Wentland, *J. Am. Chem. Soc.*, 1968, **90**, 5576; S. L. Keely and F. C. Tahk, *J. Am. Chem. Soc.*, 1968, **90**, 5584.
4. D. M. Hodgson and J. M. Galano, *Org. Lett.*, 2005, **7**, 2221; J. Mann and L.-C. de Almeida Barbosa, *J. Chem. Soc., Perkin Trans. 1*, 1992, 787.
5. H. Kim and H. M. R. Hoffmann, *Eur. J. Org. Chem.*, 2000, 2195.
6. Textbook chapter 21; Clayden, *Organic Chemistry*, chapter 33.
7. M. Rauhut and H. Currier, *U.S. Pat.*, 3,074,999, 1963, *Chem. Abstr.*, 1963; **58**, 11224b.
8. C. D. Hall, N. Lowther, B. R. Tweedy, A. C. Hall and G. Shaw, *J. Chem. Soc., Perkin Trans. 2*, 1998, 2047.
9. L.-C. Wang, A. L. Luis, K. Agapiou, H.-Y. Jang and M. J. Krische, *J. Am. Chem. Soc.*, 2002, **124**, 2402.
10. S. A. Frank, D. J. Mergott and W. R. Roush, *J. Am. Chem. Soc.*, 2002, **124**, 2404.
11. D. J. Mergott, S. A. Frank and W. R. Roush, *Org. Lett.*, 2002, **4**, 3157.
12. J. Louie, C. W. Bielawski and R. H. Grubbs *J. Am. Chem. Soc.*, 2001, **123**, 11312.
13. C. Jousse-Karinthi, C. Riche, A. Chiaroni and D. Desmaële, *Eur. J. Org. Chem.*, 2001, 3631.
14. J. Y. L. Chung, R. J. Cvetovich, M. McLaughlin, J. Amato, F.-R. Tsay, M. Jensen, S. Weissman and D. Zewge, *J. Org. Chem.*, 2006, **71**, 8602.

Index

Workbook for Organic Synthesis: Strategy and Control Paul Wyatt and Stuart Warren
© 2008 John Wiley & Sons, Ltd

9 780471 929642